An Introduction to the Physics of
INTENSE CHARGED
PARTICLE BEAMS

An Introduction to the Physics of
INTENSE CHARGED PARTICLE BEAMS

R. B. Miller
Sandia Laboratories
Albuquerque, New Mexico

PLENUM PRESS • NEW YORK AND LONDON

Library of Congress Cataloging in Publication Data

Miller, R. B.
  An introduction to the physics of intense charged particle beams.

  Bibliography: p.
  Includes index.
  1. Particle beams. I. Title.
QC793.3B4M54   539.7′3   82-557
ISBN 0-306-40931-3   AACR2

First Printing—April 1982
Second Printing—March 1985

© 1982 Plenum Press, New York
A Division of Plenum Publishing Corporation
233 Spring Street, New York, N.Y. 10013

Printed in the United States of America

Preface

An intense charged particle beam can be characterized as an organized charged particle flow for which the effects of beam self-fields are of major importance in describing the evolution of the flow. Research employing such beams is now a rapidly growing field with important applications ranging from the development of high power sources of coherent radiation to inertial confinement fusion. Major programs have now been established at several laboratories in the United States and Great Britain, as well as in the USSR, Japan, and several Eastern and Western European nations. In addition, related research activities are being pursued at the graduate level at several universities in the US and abroad.

When the author first entered this field in 1973 there was no single reference text that provided a broad survey of the important topics, yet contained sufficient detail to be of interest to the active researcher. That situation has persisted, and this book is an attempt to fill the void. As such, the text is aimed at the graduate student, or beginning researcher; however, it contains ample information to be a convenient reference source for the advanced worker.

Most of the phenomena involving the transport of charged particle beams can be understood within the framework of a macroscopic fluid model based on moments of the Vlasov and Maxwell equations; this description is used throughout the book. In certain important situations, however, which depend on the detailed momentum space structure, a plasma kinetics approach based on the Vlasov equation is introduced. I have also adopted the Gaussian system of units, except as specifically noted in a few isolated instances.

The book is divided into two main sections. In the basic chapters 1–5, much of the essential material pertaining to the generation and transport of intense beams has been assembled into a logical format and presented in a consistent fashion. In general, this material is relatively well understood, and in its description I have strived to maintain a balance between physical clarity and mathematical development; the basic physics of each important situation or concept is described in detail before proceeding with the analysis.

In contrast, the material covered in the special topics chapters 6–8, is generally less well understood, and the treatment more descriptive, reflecting the fact that these areas are in a state of rapid development. In some cases there is considerable debate concerning the important physical mechanisms, and it is more difficult to formulate a methodical presentation that is both complete and well-balanced. The selection of special topics is based on the author's interests, as well as practical time constraints. It is anticipated that individual monographs will be forthcoming on these, as well as several other special topics requiring a knowledge of intense beam phenomena.

The problems given at the end of each chapter are arranged in the order of presentation of the chapter material, and not according to the degree of difficulty. Some of the problems are relatively trivial, but are included because they demonstrate important physical concepts. On the other hand, a few problems are quite difficult, and will require some tedious algebra.

I have been aided by several colleagues in the preparation of this work. I especially wish to thank B. Godfrey, R. Adler, J. Poukey, T. Genoni, J. Freeman, C. Olson, K. Prestwich, S. Humphries, M. Widner, D. Straw, C. Clark, T. Martin, G. Kuswa, G. Yonas, A. Mondelli, P. Sprangle, W. Barletta, B. Newberger, S. Putnam, L. Sloan, K. Brueckner, and N. Rostoker for comments on specific questions, and/or reviews of portions of the manuscript. I also wish to thank D. Woodall (for the opportunity to teach a graduate level course based on this work at the University of New Mexico), and my hardworking students. Finally, I am pleased to acknowledge Karen Adler, Carla Barela, and Nancy Lee for typing the manuscript, and Cindy Miller and Margaret Clark for assistance with the figures and other last minute details.

Albuquerque, New Mexico   R. B. Miller

Contents

Introduction

1.1. Background

The rapid development since the early 1960s of a high-voltage pulsed power technology has resulted in methods for generating very-high-current pulses ($\gtrsim 10$ kA) of electrons and ions with particle kinetic energies in the range from ~ 100 keV to $\gtrsim 10$ MeV. Although such techniques were originally developed for materials testing, X-radiography, and nuclear weapon effects simulation applications, they have since found widespread use in such diverse fields as thermonuclear fusion, microwave generation, collective ion acceleration, and laser excitation.

While the goals and objectives of such varied research areas may necessarily differ, there is nonetheless a common need for understanding the physical principles which govern the motion of intense charged particle beams. The central aim of this book is to provide a basic description of intense beam transport in a variety of situations of practical interest. In this introductory chapter we first present a brief summary of the pulse power technology that has made possible the remaining topics. In Section 1.3 we present an elementary description of intense beam behavior with the use of single-particle beam envelope equations, while the basic equations relevant for a macroscopic fluid treatment are developed in Section 1.4.

Chapters 2–5 form the essential core of the book, with Chapter 2 presenting the various mechanisms for generating intense electron and ion beams in high-voltage diodes. The important electron emission processes are described in Section 2.2, while electron flow in high-power diodes is studied in Section 2.3. As discussed in Section 2.4, the same high-voltage diodes are also capable of generating very intense ion beams provided that the electron distribution in the diode is suitably controlled or modified.

Chapter 3 deals with the transport of intense beams in vacuum. For electron beams particular emphasis is placed on the equilibrium and stability requirements, and the phenomenon of virtual cathode formation. For ion beams attention is focused on the mechanisms of charge neutralization by an electron space charge. The general equations for laminar flow equilibria are developed in Section 3.2, and the important concepts of space-charge-limiting currents and virtual cathode formation are examined in Sections 3.3 and 3.4. Several relativistic electron beam equilibria are described in Section 3.5, while electron neutralized transport of intense ion beams is studied in Section 3.6. Various stability topics are examined in Section 3.7.

Chapter 4 is concerned with the propagation of intense beams in a charge-neutralizing plasma background. The important phenomenon of current neutralization is described in Section 4.2, while macroscopic beam–plasma equilibria are considered in Section 4.3. Several macroscopic and microscopic beam–plasma instabilities are studied in Sections 4.4 and 4.5. Finally, plasma heating by linear relativistic electron beams is examined in some detail in Section 4.6.

Chapter 5 concludes the basic study of beam transport processes with a discussion of beam propagation through a neutral gas background. In Section 5.2 the important ionization processes are summarized, while in Sections 5.3–5.5 neutral gas transport is characterized according to various beam current and background pressure regimes.

The remaining chapters of the book contain analyses of selected special topics. Chapter 6 deals with high-power sources of coherent radiation, including relativistic magnetrons (Section 6.2), electron cyclotron masers (Section 6.3), and free electron lasers (Section 6.4). Chapter 7 presents a survey of collective methods of ion acceleration, including an historical review (Section 7.1), followed by techniques for controlling the potential well associated with an unneutralized electron beam front (Section 7.2), and methods for growing and controlling large-amplitude beam waves (Section 7.3). The final special topic chapter presents an overview of several approaches for achieving inertial confinement fusion with intense particle beams. Pellet implosion requirements are summarized in Section 8.1, and possible methods of meeting these requirements are examined in Sections 8.2–8.4.

1.2. Pulsed Power Technology

The development of high-voltage pulsed power systems was initiated in the early 1960s by J. C. Martin at the Atomic Weapons Research Establishment in England. The essential feature of this research was the successful

Table 1.1. Output Characteristics of Several Representative Intense Beam Generators

Machine	Laboratory	Voltage (MV)	Current (kA)	Pulse duration (nsec)
Proto II	SNLA	1.5	6000	24
Hermes II	SNLA	10	100	80
OWL	PI	1.5	750	110
Aurora	PI/HDL	14	1600	120
Blackjack	Maxwell	1.3	1100	50
Gamble II	NRL	1.0	1000	50
FX-100	IPC	8.0	100	120
PBFA-I	SNLA	2.0	15,000	40

development of techniques for using an existing Marx generator technology to pulse charge a high-speed transmission line section in order to produce short-duration (10–100 nsec), high-power pulses. Since that time progress in pulse power technology has been very rapid, with development work being performed at a number of major laboratories throughout the world. In the United States these include Ion Physics Corp., Physics International Co., Maxwell Laboratories, Cornell University, the Naval Research Laboratory, and Sandia National Laboratories. Excellent reviews of pulse power technology development are now available.[1, 2]

Several representative devices are listed in Table 1.1. In general such systems can be either dc charged or pulse charged. A block diagram of the dc charged systems, such as the FX-25 and FX-100 machines, is presented in Fig. 1.1; they consist of a high-voltage terminal that is charged by a Van de Graaff column. In this case the pulse-forming line is the high-voltage terminal itself. Energy is coupled to the high-voltage diode by means of a single-output switch. The dielectric medium is high-pressure gas that usually contains an electronegative molecule such as sulfur hexafluoride. While the simplicity of these systems is obvious, and they can provide very stable, reproducible pulses, they are generally less flexible than pulse-charged systems and have an energy output limitation. They will not be discussed further.

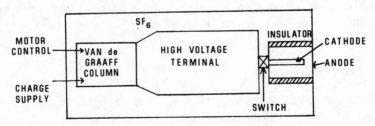

Figure 1.1. Block diagram of a Van de Graaff-charged high-voltage pulsed power system.

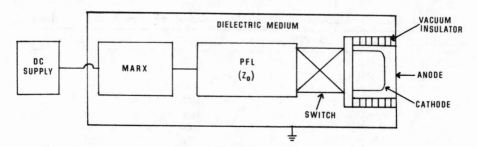

Figure 1.2. Block diagram of a typical pulse-charged high-voltage system.

A block diagram of the more common pulse-charged systems is shown in Fig. 1.2. Successful designs prevent unwanted arcs, yet utilize controlled high-voltage breakdowns for reliable switching. While the basic charging unit can be a pulse transformer, low-inductance Marx generators predominate by far. In well-designed Marx systems the time required to pulse charge a pulse-forming line is typically 1 μsec. In some cases an intermediate storage capacitor is introduced between the Marx generator and the pulse-forming line (PFL) in order to decrease the charging time to a few hundred nanoseconds or less. The PFL can be a simple transmission line, although Blumlein transmission lines are somewhat more common because they are able to deliver the full charge voltage to a matched load. The dielectric energy storage medium is usually a liquid, with transformer oil and deionized water being the most common. A fast output switch is used to rapidly transfer the energy from the PFL into the diode load. In some applications another transmission line (a pulse transformer) is inserted between the PFL output switch and the vacuum diode in order to increase or decrease the characteristic pulse impedance. A solid dielectric insulator is used to separate the liquid-insulated transmission line sections from the evacuated beam generation region of the diode. In high-voltage applications ($\gtrsim 2$ MV) this interface usually consists of a stacked, voltage-graded ring assembly, while low-voltage, low-impedance applications generally use a low-inductance radial insulator. Each of these separate components will now be discussed in more detail.

1.2.1. Marx Generators

The term Marx generator has now come to be associated with all systems which have the general feature of capacitors being charged in

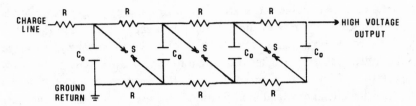

Figure 1.3. Simplified schematic of a four-stage Marx generator.

parallel, then discharged in series by a number of switches. To illustrate the action of a Marx generator, consider the four-stage unit in Fig. 1.3. All capacitors are initially charged by a dc power supply to V_0. Triggering of the spark gap switches causes the Marx to "erect" by connecting the capacitors in series, thereby producing an output voltage of $4V_0$.

In the simple circuit of Fig. 1.3 it is necessary to trigger all the switches; however, a relatively minor modification can permit self-erection following the triggering of at least the first switch gap. Consider the schematic given in Fig. 1.4. It is assumed that there is a stray capacitance C_g across each spark gap and a coupling capacitance $C_c \gg C_g$ across every two spark gaps. If the voltage at point A is V_A, then closure of SW1 implies that the voltage at point B is $V_B = V_A + 2V_0$, where V_0 is the capacitor charge voltage. The gap capacitance and the coupling capacitance act as a voltage divider network so that the potential difference across gap BD is $V_{BD} = 2V_0(C_c/C_c + C_g)$. The discharge of the coupling capacitance through the resistor in a characteristic time RC_c gives $V_{BD} \approx 2V_0$. Hence, if SW2 had been adjusted to prevent breakdown at voltages $V_0 < V < 2V_0$, then SW2 is overvolted at $V_{BD} = 2V_0$ and it self-fires.

The configuration of Fig. 1.4 is termed an $n = 2$, RC-coupled, self-erecting Marx, since V_{BD} tends to $2V_0$. For an $n = 3$ Marx, the coupling

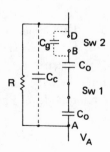

Figure 1.4. Simplified schematic of a self-erecting $n = 2$ Marx generator.

capacitance is made large across every third gap, and the gap voltage tends to $3V_0$ following the firing of the preceding two gaps. In this case the Marx can self-erect with voltages of about one-third of the self-breakdown voltage of the individual gaps. Higher n configurations are even more stable with respect to self-breakdown problems, but the erection times tend to be longer. A beneficial practical modification is the use of plus–minus charging of alternate capacitors to reduce the required number of switches by a factor of 2.

The switches in the Marx generators are usually gas-filled spark gaps of fixed dimensions with the operating range being covered by varying the pressure of the gas (usually SF_6). The charge and trigger resistors are made of flexible plastic hose filled with a $CuSO_4$ solution of the appropriate resistivity. For output voltages in excess of a megavolt, the Marx is usually immersed in transformer oil to prevent flashover.

When properly constructed the Marx generator is a very reliable and consistent system. In addition to the large single Marx devices, such as Hermes II,[3] multiple Marx generators can be operated in parallel to achieve a further reduction in output inductance. For example, in the Proto II device[4] eight 112-kJ Marx generators are used to parallel charge the intermediate stores and the pulse-forming lines.

1.2.2. Pulse-Forming Lines

For some long pulse applications it is desirable to couple the Marx generator directly to the vacuum diode; however, the pulse rise time is then limited by the Marx inductance and capacitance, and the generator impedance is greater than typically several tens of ohms. In order to produce short, fast-rising, low-impedance beam outputs it is customary to use the Marx to charge a pulse-forming line (PFL). Although the PFLs may be constructed in a variety of shapes (strip, coaxial, radial, etc.), they are typically used in only two types of circuits—the simple transmission line, and the double, or Blumlein,[5] line.

A simple coaxial transmission line PFL schematic is shown in Fig. 1.5. If the electrical transit time of the line is short compared to the Marx charge time then the line acts as a lumped capacitance according to the simple circuit model of Fig. 1.6. The line reaches a peak charge voltage V_0 when the

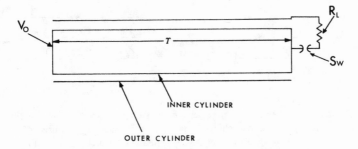

Figure 1.5. Schematic diagram of a simple coaxial transmission line.

current in the circuit is zero. It is easy to show that V_0 is related to the erected Marx voltage V_m according to

$$V_0 = \frac{2V_m C_m}{C_m + C_l} \tag{1.1}$$

where C_m is the series capacitance of the Marx generator, and C_l is the capacitance of the PFL. The factor $2C_m(C_m + C_L)^{-1}$ is termed the lossless circuit ringing gain.

When the PFL is discharged it must be treated as a transmission line with a characteristic impedance given by $Z_0 = (L_l/C_l)^{1/2}$. The essential features of PFL operation can be understood by considering the diagrams of Fig. 1.7. The open circuited, charged transmission line of electrical length $\tau = (LC)^{1/2}$ can be considered to consist of forward and backward waves as illustrated in Fig. 1.7a. In order to satisfy the open circuit boundary conditions the voltage polarity of each wave is the same, but the current polarity inverts on reflection from the open terminations. Hence, the transmission line has a dc voltage of V_0 and there is no net current flow. When the switch S is closed (Fig. 1.7b) a load impedance Z_L is connected across the output of the line and energy is extracted out of the positive-going

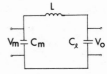

Figure 1.6. Lumped capacitance circuit model for the simple transmission line PFL.

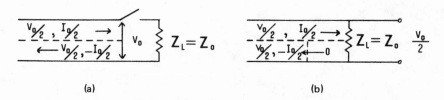

(a)   (b)

Figure 1.7. Analysis of the pulse-forming behavior of a charged transmission line. (a) Prior to switch closure the open-circuited, charged line can be considered to consist of equal and opposite forward- and backward-traveling waves. (b) After switch closure a load impedance is connected across the output of the line and energy is extracted from the positive-going wave.

wave. If $Z_L = Z_0$ (matched load) there are no reflections at the output, and all of the stored energy of the line is extracted in a time 2τ. Since the current flowing through the load resistor is $i_0/2$, the voltage across the load is $V_0/2$, i.e., half the charge voltage.

In contrast to the simple transmission line, an alternate circuit invented by A. D. Blumlein[5] is capable of producing an output pulse into a matched load that equals the charge voltage. A cylindrical version of the Blumlein circuit is schematically represented in Fig. 1.8. It consists of three coaxial cylinders with the intermediate cylinder being charged by the Marx generator. The center cylinder is connected to the outer grounded cylinder by an

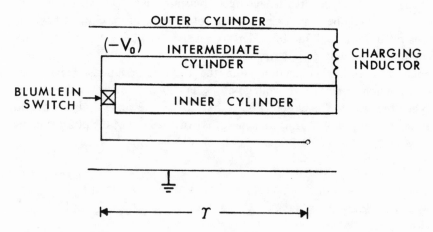

Figure 1.8. Schematic diagram of a coaxial Blumlein pulse-forming line.

inductor. Ideally the inductor acts as a short during the charge cycle, and then as an open for the short duration of the output pulse. Operation of the Blumlein circuit is illustrated in Fig. 1.9. During the slow Marx charge the inductor shorts the inner and outer conductors so that the summation of voltages is zero (Fig. 1.9a). When the inner and intermediate lines are shorted by means of the switch closure, the polarity of the forward-going voltage wave is inverted, and after a time τ the voltage summation yields $2V_0$ at the open circuited output (Fig. 1.9b) for a time 2τ. (Note that the inductance has been eliminated from the schematic since it represents a high-impedance load over the short pulse duration.) If a matched impedance Z_L is switched into the output circuit at time τ after the switch closure, then the output voltage drops to V_0, but all the energy can be extracted in the additional time 2τ.

Both of the PFL circuits examined above generate a diode prepulse voltage as they are being charged. Depending on the detailed diode design the prepulse can produce a diode plasma that may deleteriously affect the main output pulse. In the case of the simple transmission line the equivalent circuit is that of Fig. 1.10a, and the source of the prepulse is seen to be the capacitive coupling across the output switch. For the Blumlein PFL, the situation is a bit more complicated; the equivalent circuit is given in Fig. 1.10b, where C_2 represents the capacitance between the intermediate charged cylinder and the grounded outer cylinder, and C_1 denotes the capacitance between the charged cylinder and the inner cylinder. L represents the inductor that ties the inner cylinder to the grounded outer cylinder. If the inductor were a true short, then there would be no prepulse voltage; however, L must appear as a large impedance during the rapid discharge of the PFL, i.e., $\omega L \gg Z_D$, where Z_D is the diode impedance. If $Z_D \sim 100\Omega$ and the diode rise time is of the order of 10 nsec, then the inductance must be at least a few microhenries to avoid appreciable diode loading.

There are several practical methods available for reducing the prepulse voltage to acceptable levels. In the case of the simple transmission line a resistor or an inductor can be connected from the high-voltage diode electrode to ground in order to discharge the diode capacitance. An alternate technique, especially applicable for the Blumlein circuit, is the use of a prepulse switch prior to the diode. If the switch has a low capacitance compared to the diode then voltage division will occur in the same manner as indicated previously in Fig. 1.10a. Other useful techniques include surface flashover switches in the high-voltage diode electrode to provide

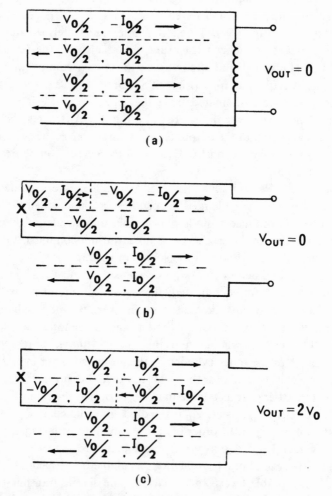

(a)

(b)

(c)

Figure 1.9. Analysis of the pulse-forming behavior of a Blumlein line. (a) During the slow Marx charging cycle the inductor shorts the inner and outer conductors so that the summation of voltages is zero. (b) Following switch closure the Blumlein circuit generates a short pulse—for which the inductor now represents a high impedance load.

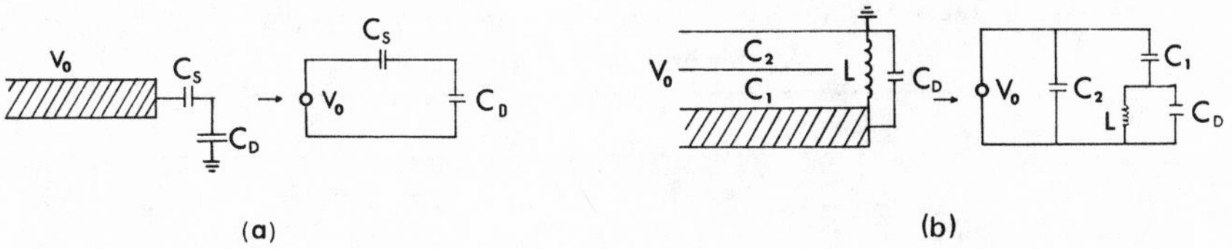

Figure 1.10. (a) In the simple transmission line a prepulse is generated as a result of the capacitive coupling across the output switch. (b) In the Blumlein circuit a prepulse is developed across the inductor during the Marx charging cycle.

further capacitive voltage division, and the use of plasma erosion switches[6] which conduct current during the prepulse phase, but open on arrival of the main voltage pulse.

The electrical parameters of the PFL are determined by its geometrical construction and by the choice of dielectric. Using the simple coaxial transmission line of Fig. 1.5 for illustration, the line impedance is given by

$$Z = \left(60/\sqrt{\varepsilon} \right) \ln(b/a) \tag{1.2}$$

while the 2τ pulse length is determined by

$$\tau_p = 2l\sqrt{\varepsilon}/c \tag{1.3}$$

where ε is the dielectric constant, l is the length of the coaxial line, a and b are the inner and outer radii, and c is the speed of light in vacuum.

The inner and outer radii of the line are determined by the desired line impedance and charge voltage, and by electrical breakdown limitations. The most commonly used dielectrics are water ($\varepsilon \simeq 80$) and transformer oil ($\varepsilon \approx 2.4$). For these liquids J. C. Martin has shown that the breakdown electric field strength (in MV/cm) varies approximately as

$$E_{br} = \frac{k}{t_{eff}^{1/3} A^{1/10}} \tag{1.4}$$

where A is the electrode area in cm^2 over which the electric field does not vary by more than 10% from the maximum value, and t_{eff} is the time for the electric field intensity to change from $0.63\, E_{br}$ to E_{br}, expressed in microseconds. The constant k has the values given in Table 1.2. Note that for water k

Table 1.2. Electrical Breakdown Constants k_+ and k_- for Water and Transformer Oil

Dielectric	k_+	k_-
Oil	0.5	0.5
Water	0.3	0.6

is polarity dependent. For electrode areas of the order of 1000 cm^2 and microsecond charging times, typical breakdown strengths are about 15 MV/m for water (+) and 25 MV/m for transformer oil. For much shorter duration pulses (tens of nanoseconds) the simple empirical formula of Eq. (1.4) can underestimate the breakdown field by almost a factor of 2.

An examination of the characteristics of simple transmission lines versus Blumlein lines, and of water versus transformer oil leads to the summary of applicable voltages and load impedances in Fig. 1.11. For high-voltage applications the Blumlein PFL has an obvious advantage in terms of the lower required charge voltage; however, at lower voltages the simple lines are much more compact. In low-impedance applications, the high dielectric constant of water makes it the more suitable dielectric medium.

1.2.3. Switching

In order to rapidly transfer energy from the PFL to the diode the switch should have low inductance and low resistive losses, and should also be capable of synchronization to nanosecond accuracy. In self-breaking

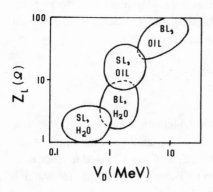

Figure 1.11. Summary of applicable voltages and load impedances for simple lines and Blumlein lines for water and transformer dielectrics.

switches both oil and water, as well as high-pressure gas, perform well yielding fast-rising pulses. For applications which require triggered operation gas gaps (SF_6) are most commonly used.

The breakdown phenomena in such switches are very complicated (Ref. 21, for example), and a detailed theoretical description is beyond the scope of this book. The main characteristic is the formation of positive and negative "streamers," i.e., advancing regions of net space charge, similar to the leader stroke in lightning. In nonuniform field switching configurations (point–plane, blade–plane), the mean streamer velocity in oil for voltages less than 1 MV is described empirically by

$$\langle v_s \rangle = d/t \approx kV^n \tag{1.5}$$

were $\langle v_s \rangle$ is in units of centimeters per microsecond and V is the gap voltage in MV. The constants are given by $k_+ = 90$, $n_+ = 1.75$ for positive polarity pulses, and $k_- = 31$, $n_- = 1.28$ for negative pulses. For voltages in the range of 1 to 5 MV both polarities obey a relationship of the form

$$\langle v_s \rangle = 80V^{1.6}d^{-1/4} \tag{1.6}$$

For water the relationships are best described by the empirical formulas

$$\langle v_s \rangle t^{1/2} = 8.8V^{0.6} \quad \text{(positive)}$$

$$\langle v_s \rangle t^{1/3} = 16V^{1.1} \quad \text{(negative)} \tag{1.7}$$

In self-break gas switches four gases (air, nitrogen, freon, and sulfur hexafluoride) are in common usage. For air or nitrogen in relatively uniform fields, the breakdown field can be expressed in the form

$$E_{br}(kV/cm) = \left[24.6p + 6.7(p/d_{eff})^{1/2} \right] F^{-1} \tag{1.8}$$

where p is the pressure in atmospheres, d_{eff} is an effective electrode gap separation, and F represents the field enhancement factor, i.e., the ratio of the maximum field on the electrodes to the mean field. Table 1.3 defines d_{eff} and F for a gap with a spacing equal to the electrode diameter d. Values for other geometries are given by Alston.[7] The breakdown of SF_6 and freon occurs at fields approximately two and one-half and five times greater than

Table 1.3. Effective Gap Separation and Field Enhancement Factors
for Typical Gas Switch Geometries

Electrode shape	d_{eff}/d	F
Cylinders	0.115	1.3
Spheres	0.057	1.8

given by Eq. (1.8), although SF_6 is preferred because carbon is not formed during the discharge.

For the very divergent fields associated with the point–plane or blade–plane gaps, the breakdown can become both time and polarity dependent. This case may be approximately described by the expression

$$\langle E_{br} \rangle = k_{\pm} \, p^n (dt)^{-1/6} \tag{1.9}$$

where $\langle E_{br} \rangle = V/d$ is the mean field in kV/cm, d is the gap distance in centimeters, t is the time in microseconds, and p is the gap pressure in atmospheres. Values of $k_{\pm}$ and n are given in Table 1.4. Equation (1.9) is generally valid over the range of 1–10 atm and for gap distances on the order of 10 cm. Additional nonlinearities occur for higher-pressure operation, generally requiring experimental testing for each new configuration.

The most common triggered switch configurations are the trigatron and field distortion (or midplane) gaps. The cross section of a field distortion gap is shown in Fig. 1.12. The trigger electrode is centrally located between the spherical or cylindrical electrodes, and takes the form of a disk with a hole. Application of a trigger pulse distorts the fields and causes one half of the gap to fire. As a result the entire gap voltage is developed across the remaining half of the gap, which then self-fires.

The cross section of a trigatron switch is shown in Fig. 1.13. In this case application of the trigger pulse to the trigger pin causes an arc to occur

Table 1.4. Electrical Breakdown Constants for Air and SF_6
in Strongly Divergent Switch Geometries

Gas	k_+	k_-	n
Air	22	22	0.4
SF_6	44	72	0.4

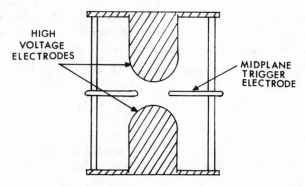

Figure 1.12. Field distortion switch gap geometry.

between the pin and the electrode. The arc distorts the fields near the electrode and also generates photons, in addition to the initiating electrons that promote streamer formation and gap closure.

In addition to these triggered switches, UV, laser, soft x-ray, and electron beam triggered switches have received some attention because of the possibilities for reducing the energy required to trigger the gap, and for reducing the switch jitter. In particular, recent results obtained with a KrF laser in SF_6 are especially promising.

The performance of a switch must be characterized not only by its breakdown characteristics, but also by the pulse rise time. The latter is

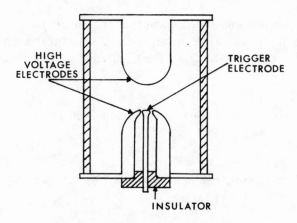

Figure 1.13. Trigatron switch gap geometry.

determined by the gap inductance and the resistance of the heated channel, with the relative importance of each depending on the detailed switch conditions. The e-folding time for the inductive current rise (or, equally, the voltage fall across the gap) through a switch of inductance L being driven by a circuit with impedance Z is

$$\tau_L \simeq L/Z \qquad (1.10)$$

Although an accurate calculation of the channel inductance would require a detailed knowledge of the channel radius a, it is usually sufficient to approximate L by

$$L \approx 2d\ln b/a \approx 14d \quad \text{(nanohenries)} \qquad (1.11)$$

where d is the channel length in centimeters. Since $b \gg a$, L is relatively insensitive to a, and for rise times on the order of a few nanoseconds, the value of the log term is usually taken to be seven.

The second component is termed the resistive phase term and is due to the energy absorbed by the arc channel. As the channel heats and expands, however, the resistive impedance decreases with time. For gases the e-folding time for this resistive term obeys the empirical relationship[8] developed by J. C. Martin

$$\tau_R = \frac{88}{Z^{1/3}E_{\mathrm{br}}^{4/3}}(\rho/\rho_0)^{1/2} \quad \text{(nanoseconds)} \qquad (1.12)$$

where (ρ/ρ_0) denotes the ratio of density of the gas to that of air at STP, Z is again the impedance of the driver, and E_{br} is the breakdown field along the channel in units of $10\ kV/cm$. For liquids and solids a similar relationship exists:

$$\tau_R = \frac{5}{Z^{1/3}E_{\mathrm{br}}^{4/3}} \quad \text{(nanoseconds)} \qquad (1.13)$$

where E_{br} is now in units of MV/cm.

The effective pulse rise time is obtained by summing Eq. (1.10) with either Eq. (1.12) or Eq. (1.13); i.e.,

$$\tau_{\mathrm{tot}} = \tau_L + \tau_R \qquad (1.14)$$

Equation (1.14) gives the e-folding rise time for exponentially rising pulses; for nonexponential pulses τ_{tot} is approximately equal to the maximum voltage divided by the maximum slope of the output voltage waveform, i.e.,

$$\tau_{tot} \approx V/(dV/dt)_{max} \tag{1.15}$$

As is apparent from Eqs. (1.10), (1.12), and (1.13) the rise time can effectively limit the useful pulse duration for low-impedance machines. Hence, it has been found useful to switch the PFL at high impedance to obtain a fast pulse rise time, and then transform the impedance down to the desired level using a tapered transmission line. The Gamble II accelerator is a notable example.[9]

An alternate method for reducing the pulse rise time is to achieve multiple, rather than single, switch channels. If there are n channels through which current can flow, the Z must be replaced by nZ in Eqs. (1.10), (1.12), and (1.13). Hence, the inductive term decreases as n^{-1}, while the resistive phase term drops as $n^{-1/3}$. The number of switch channels that will close can be predicted reasonably well using an expression proposed by Martin[8]:

$$2\sigma_V \left(\frac{V}{dV/dt} \right) \leqslant 0.1(\tau_R + \tau_L) + 0.8T_s \tag{1.16}$$

In Eq. (1.16) σ_V is the normalized standard deviation of the voltage breakdown for a rapidly rising pulse. The $[V/(dV/dt)]$ term converts the left-hand side of the equation to the deviation in time for streamer closures. It is essentially the time interval during which useful current-carrying channels may form. The time T_s is the electromagnetic wave transit time between arcs and is directly proportional to n^{-1}. Thus, the right-hand side of Eq. (1.16) corresponds to the time required for the voltage across an arc to drop to about 90% of its peak value and for this information to be transmitted to adjacent arcs.

From Eq. (1.16) it is apparent that the number of current-carrying channels can be increased in two ways: (1) increasing (dV/dt), and (2) decreasing σ_V. In fact σ_V is also a decreasing function of (dV/dt), and the two effects can combine to markedly improve switch performance. Hence, the use of intermediate storage capacitors inserted between the Marx generator and the PFL to increase (dV/dt) has become widespread for applications which require very fast rising output pulses.

1.2.4. Vacuum Diodes

The conversion of the short-duration, high-power pulses supplied by the PFL into useful particle beams requires the use of a vacuum diode. The two most commonly used configurations are illustrated in Fig. 1.14. For high-voltage (>1 MV) applications the vacuum insulator consists of a stack of solid dielectric insulating rings (usually Lucite) alternated with annular aluminum disks. The metal rings help to "grade" the insulator stack by distributing the diode voltage more uniformly along the insulator length. They also help to prevent a surface flashover cascade. On the diode side the Lucite insulators are usually cut at an angle of 45° to the diode axis. This permits any stray electrons emitted from the surface to propagate freely without encountering additional plastic, thus preventing further secondary emission. In short pulses the average electric field strength along the insulator stack which results in vacuum flashover is given by the empirical formula[8]

$$\langle E_{\mathrm{br}} \rangle t^{1/6} A^{1/10} = 175 \qquad (1.17)$$

where t is the time in nanoseconds during which $E > 0.9 \langle E_{\mathrm{br}} \rangle$ in kV/cm, and A is the area of the insulator in square centimeters. According to Eq. (1.17) $\langle E_{\mathrm{br}} \rangle$ can reach field stresses on the order of 100–150 kV/cm for very short pulses.

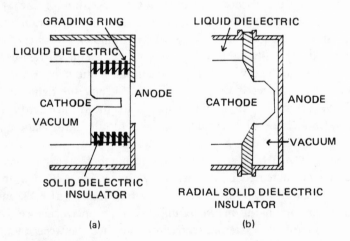

Figure 1.14. Vacuum diode insulators: (a) graded insulator ring stack, and (b) low-inductance radial insulator.

In an equivalent circuit model the insulator stack can be described by an effective inductance which determines the beam current rise time according to $\tau_r \sim L_i/(Z + Z_d)$, where Z_d is the diode impedance and Z is the line impedance. τ_r should be less than the rise time of the PFL pulse to prevent overvoltage of the insulator stack. Hence, for low-impedance applications it is desirable to reduce the inductance as much as possible.

A single radial insulator (Fig. 1.14b) has been used successfully at a number of laboratories for voltages of $\lesssim 2$ MV. In these designs it must be ensured that the electric field lines make a $\sim 45°$ angle with respect to the insulator surface, and that triple points (the junction of the insulator wall with the metal and a vacuum) are electrically shielded along the dielectric.

Regardless of the particular design the vacuum insulator is almost always the weakest accelerator link with respect to electric breakdowns. Consequently, methods for increasing the breakdown strength of the insulator surface, such as magnetic flashover inhibition, hold future promise. Also, for certain inertial confinement fusion applications it is desirable to remove the insulator a substantial distance from the actual beam generation region, yet keep the inductance of the power feed as low as possible. A recent successful approach has been developed by using self-magnetically insulated vacuum transmission lines.[10] Although the electric field exceeds the vacuum breakdown limit, the magnetic field also associated with the electromagnetic wave is of sufficient intensity to cause the electrons to follow cycloidal-like trajectories along the cathode surface without reaching the anode. Detailed theories of self-magnetic insulation are now available.[11-13]

1.3. Qualitative Behavior of Charged Particle Beams

Using the techniques described in Section 1.2 it is possible to apply multi-megavolt pulses to a vacuum diode, which in turn can result in the production of intense ($>1\,\text{kA/cm}^2$) beams of electrons and/or ions. The latter processes are described in detail in Chapter 2. Once the high-current beams have been generated it is often useful to be able to quickly predict the qualitative motion of an intense charged particle beam under various projected operating conditions. In this regard beam envelope equations, which neglect the detailed internal state of the beam, serve a very useful function. In this section we present a simple derivation of an envelope equation, and then apply it to several situations of practical interest. Other derivations can be found in most books on electron optics.

The motion of a single particle of mass γm and charge e is described by the Lorentz force law

$$\frac{d}{dt}\mathbf{p} = e\left[\mathbf{E} + \frac{1}{c}(\mathbf{v} \times \mathbf{B})\right] \tag{1.18}$$

$\mathbf{p} = \gamma m \mathbf{v}$ is the relativistic momentum, with $\gamma = (1 - \beta^2)^{-1/2}$, $\beta = |\mathbf{v}|/c$, where $\mathbf{v}$ is the particle velocity vector and c is the speed of light. $\mathbf{E}$ and $\mathbf{B}$ represent the macroscopic electric and magnetic fields. Taking the dot product of $\mathbf{v}$ with Eq. (1.18) indicates that the relativistic particle energy changes according to

$$\frac{d}{dt}(\gamma mc^2) = e(\mathbf{v} \cdot \mathbf{E}) \tag{1.19}$$

It is now assumed that the beam system is cylindrically symmetric about the z axis as shown in Fig. 1.15. The velocity is decomposed as

$$\mathbf{v} = v_r \hat{e}_r + v_\theta \hat{e}_\theta + v_z \hat{e}_z \tag{1.20}$$

It is further assumed that $|v_r|, |v_\theta| \ll v_z$, in which case

$$\gamma \simeq (1 - v_z^2/c^2)^{-1/2} \tag{1.21}$$

$$\frac{d}{dt}(\gamma mc^2) = ev_z E_z \tag{1.22}$$

The variables v_z and γ are thus regarded as known functions of z and t.

The perpendicular components of the equation of motion are given by

$$\dot{\gamma} m \mathbf{v}_\perp + \gamma m \dot{\mathbf{v}}_\perp = e\left[\mathbf{E}_\perp + \frac{1}{c}(\mathbf{v} \times \mathbf{B})_\perp\right] \tag{1.23}$$

where the dot notation implies total time differentiation. With the assumed cylindrical symmetry, the important field components are E_r, B_θ, and B_z. We further assume that the longitudinal magnetic field is generated by external magnetic field coils only, and is constant and uniform across the beam radial profile; i.e., $B_z = B_0$. In this case the radial component of Eq.

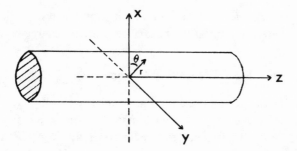

Figure 1.15. Cylindrical charged particle beam propagating parallel to a uniform externally generated magnetic field. Cylindrical polar coordinates (r, θ, z) are indicated, with the z axis corresponding to the axis of symmetry. θ denotes the polar angle in the $x-y$ plane, while $r = (x^2 + y^2)^{1/2}$ is the radial distance from the z axis.

(1.23) yields

$$\ddot{r} + \frac{\dot{\gamma}\dot{r}}{\gamma} - \frac{v_\theta^2}{r} = \frac{e}{\gamma m}\left[E_r + \frac{1}{c}(v_\theta B_0 - v_z B_\theta) \right] \tag{1.24}$$

while the azimuthal component becomes a statement of conservation of canonical angular momentum

$$\gamma r v_\theta + \tfrac{1}{2}\Omega r^2 = P_\theta / m = \text{const} \tag{1.25}$$

where $\Omega = eB_0/(mc)$ is the cyclotron frequency. Solving Eq. (1.25) for v_θ yields

$$v_\theta = \left(P_\theta/m - \Omega r^2/2 \right)(\gamma r)^{-1} \tag{1.26}$$

Restricting attention to a single particle at the edge of the beam, substitution of Eq. (1.26) into Eq. (1.24) gives the equation for the radial envelope of the beam

$$\ddot{r}_b + \frac{\dot{r}_b \dot{\gamma}}{\gamma} + \left(\frac{\Omega}{2\gamma} \right)^2 r_b = \frac{e}{\gamma m}\left(E_r - \frac{v_z B_\theta}{c} \right) + \frac{(P_\theta/m)^2}{\gamma^2 r_b^3} \tag{1.27}$$

In the following sections Eq. (1.27) will be applied to several cases of practical importance. Although these cases will be examined in great detail in subsequent chapters, the envelope equation analysis will give a good qualitative understanding of the behavior of intense charged particle beams.

1.3.1. Solenoidal Field Transport in Vacuum

Consider the case of an intense beam propagating in vacuum with constant axial velocity v_0 along a longitudinal magnetic field. The beam kinetic energy is also assumed to remain constant. Eq. (1.27) then becomes

$$\ddot{r}_b + \left(\frac{\Omega}{2\gamma_0}\right)^2 r_b = \frac{e}{\gamma_0 m}(E_r - \beta_0 B_\theta) + \frac{(P_\theta/m)^2}{\gamma_0^2 r_b^3} \tag{1.28}$$

The field components E_r and B_θ at the edge of the beam depend upon the beam charge density n according to Gauss' and Ampere's laws

$$\int \mathbf{E} \cdot d\mathbf{A} = 4\pi e \int n \, dV \tag{1.29}$$

$$\int \mathbf{B} \cdot d\mathbf{l} = \frac{4\pi e}{c} \int n\mathbf{v} \cdot d\mathbf{A} \tag{1.30}$$

For a beam of constant density out to the radius r_b, Eqs. (1.29) and (1.30) give

$$E_r = 2\pi e n_0 r_b$$
$$B_\theta = 2\pi e n_0 r_b \beta_0 = \beta_0 E_r \tag{1.31}$$

and

$$\frac{e}{\gamma_0 m}(E_r - \beta_0 B_\theta) = \frac{2\pi e^2 n_0 r_b}{\gamma_0^3 m} = \frac{\omega_e^2 r_b}{2\gamma_0^3} \tag{1.32}$$

where $\omega_e = (4\pi n_0 e^2/m)^{1/2}$ is the beam plasma frequency. Substituting Eq. (1.32) into Eq. (1.28) yields

$$\ddot{r}_b + \left(\frac{\Omega}{2\gamma_0}\right)^2 r_b - \frac{\omega_e^2}{2\gamma_0^3} r_b = \frac{(P_\theta/m)^2}{\gamma_0^2 r_b^3} \tag{1.33}$$

Hence, the unneutralized beam self-fields cause the beam to expand, while the B_0 supplies a restoring force. The equilibrium radius $r_{b0}(\ddot{r}_b = 0)$ is given by

$$r_{b0}^2 = \frac{2P_\theta/m}{\left(\Omega^2 - 2\omega_e^2/\gamma_0\right)^{1/2}} \tag{1.34}$$

and there can be no equilibrium unless

$$\Omega^2 \geqq 2\omega_e^2/\gamma_0 \tag{1.35}$$

For a 1-MeV electron beam of density $10^{12}/\text{cm}^3$, the magnetic field strength required for equilibrium is only 2.6 kG; however, for a 1-MeV proton beam of the same density B_0 must exceed 188 kG. Hence, solenoidal magnetic fields can easily satisfy the equilibrium condition for vacuum transport of intense electron beams, but charge neutralization is usually required for intense ion beam transport.

The general solution of Eq. (1.33) is oscillatory. Eliminating ω_e^2 in favor of the total beam current $I = \pi r_b^2 n_0 e \beta_0 c$, which does not vary as the beam expands or contracts, Eq. (1.33) becomes

$$\ddot{r}_b + \left(\frac{\Omega}{2\gamma_0}\right)^2 r_b - \frac{2eI}{\gamma_0^3 m\beta_0 cr_b} = \frac{(P_\theta/m)^2}{\gamma_0^2 r_b^3} \tag{1.36}$$

Assuming small perturbations about the equilibrium radius, $r_b = r_{b0} + \delta r$, the linearization of Eq. (1.36) yields

$$\delta\ddot{r} + \left[\left(\frac{\Omega}{\gamma_0}\right)^2 - \frac{4eI}{\gamma_0^3 m\beta_0 cr_{b0}^2}\right]\delta r = 0 \tag{1.37}$$

which has sinusoidal solutions with frequency

$$\omega = \left[\left(\frac{\Omega}{\gamma_0}\right)^2 - \frac{4eI}{\gamma_0^3 m\beta_0 cr_{b0}^2}\right]^{1/2}$$

and wavelength $\lambda = 2\pi\beta_0 c/\omega$.

1.3.2. Charge[14] and Current[15] Neutralization by a Background Plasma

When an intense charged particle beam is injected into a plasma, the plasma charges tend to move in such a fashion as to neutralize the beam self-fields. These phenomena can be described qualitatively by introducing the charge and current neutralization fractions f_e and f_m into Eq. (1.31) as

$$E_r = 2\pi e n_0 r_b (1 - f_e) \tag{1.38}$$

$$B_\theta = 2\pi e n_0 r_b \beta_0 (1 - f_m) \tag{1.39}$$

In the absence of an external magnetic field and assuming $P_\theta = 0$, Eq. (1.28) becomes

$$\ddot{r}_b - \frac{2eI}{\gamma_0 m \beta_0 c r_b}\left[1 - f_e - \beta_0^2 (1 - f_m)\right] = 0 \tag{1.40}$$

and the equilibrium condition is identified as

$$f_e = \gamma_0^{-2} + \beta_0^2 f_m \tag{1.41}$$

If the charge neutralization fraction exceeds the equilibrium criterion, the beam will constrict, or "self-pinch"; in the opposite limit the beam will expand. The various possibilities are described in Table 1.5.

All of the interesting phenomena described in Table 1.5 can be observed when an electron beam is injected into a neutral gas at various background pressures (Fig. 1.16).[16, 17] At very low pressures the beam-generated ionization is small ($f_e = f_m = 0$), and the beam expands radially. At somewhat higher pressures, $f_e > \gamma^{-2}$, but the conductivity remains too low to provide significant current neutralization ($f_m = 0$), and pinched beam propagation is observed. As the pressure is further increased both the charge and current neutralization fractions can increase to unity, and the beam free streams in a force neutral fashion with beam expansion depending on the beam

Table 1.5. Beam Behavior for Various Values of the Charge and Current Neutralization Fractions

f_m/f_e	0	$1/\gamma^2$	1
0	Expands	Equilibrium	Contracts
1	—	—	Equilibrium

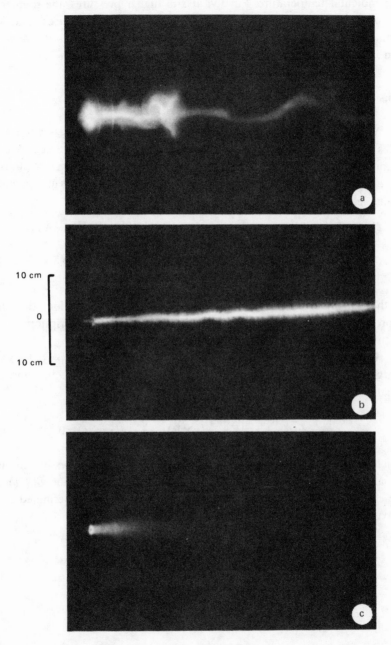

Figure 1.16. Intense relativistic electron beams injected into neutral gas as a function of background pressure. (a) $f_e = f_m = 0$ (at 0.03 Torr); (b) $f_e > 1/\gamma^2$, $f_m \simeq 0$ (at 0.7 Torr); (c) $f_e = f_m = 1$ (at 1.6 Torr); ($V = 2$ MV, $I = 20$ kA).

perpendicular temperature. Finally, at still higher pressures, the frequency of collisions between plasma electrons and neutrals increases, and the conductivity decreases ($f_m < 1$). Since the beam space charge is still neutralized ($f_e = 1$) pinched beam propagation is again possible.

1.4. The Macroscopic Fluid Description

While the envelope equation developed in the preceding section provides a useful qualitative guide to the general behavior of a charged particle beam, a more detailed understanding of intense beam physics requires a more sophisticated theoretical treatment. Two such approaches are generally available[18]:

(1) a plasma kinetics (microscopic) approach based on the Vlasov and Maxwell equations; and
(2) a macroscopic fluid approach based on moments of the Vlasov and Maxwell equations.

In the remainder of this book we will almost exclusively adopt the fluid approach because it provides a relatively simple, lucid description of the essential physics.

The evolution of the distribution function $f(\mathbf{x}, \mathbf{p}, t)$ for particles with charge e and mass m is described by the collisionless Boltzmann transport equation, or Vlasov equation[19]

$$\left\{ \frac{\partial}{\partial t} + \mathbf{v} \cdot \frac{\partial}{\partial \mathbf{x}} + e\left[\mathbf{E} + \frac{1}{c}(\mathbf{v} \times \mathbf{B}) \right] \cdot \frac{\partial}{\partial \mathbf{p}} \right\} f(\mathbf{x}, \mathbf{p}, t) = 0 \qquad (1.42)$$

which is an expression of Liouville's theorem for the incompressible motion of particles in the six-dimensional phase space $(\mathbf{x}, \mathbf{p})$. $E(\mathbf{x}, t)$ and $B(\mathbf{x}, t)$ are the electric and magnetic fields that are self-consistently determined from Maxwell's equations

$$\nabla \cdot \mathbf{E} = 4\pi e \int d^3 p f(\mathbf{x}, \mathbf{p}, t) \qquad (1.43)$$

$$\nabla \cdot \mathbf{B} = 0 \qquad (1.44)$$

$$\nabla \cdot \mathbf{E} = -\frac{1}{c} \frac{\partial \mathbf{B}}{\partial t} \qquad (1.45)$$

$$\nabla \times \mathbf{B} = \frac{4\pi e}{c} \int d^3 p \mathbf{v} f(\mathbf{x}, \mathbf{p}, t) + \frac{1}{c} \frac{\partial \mathbf{E}}{\partial t} \qquad (1.46)$$

In the fluid description[20] the particle density $n(\mathbf{x}, t)$, mean velocity, $\bar{\mathbf{v}}(\mathbf{x}, t)$, momentum $\bar{\mathbf{p}}(x, t)$, and pressure $\mathbf{P}(\mathbf{x}, t)$ are defined by

$$n(\mathbf{x}, t) \equiv \int d^3p\, f(\mathbf{x}, \mathbf{p}, t) \tag{1.47}$$

$$n(\mathbf{x}, t)\bar{\mathbf{v}}(\mathbf{x}, t) \equiv \int d^3p\, \mathbf{v} f(\mathbf{x}, \mathbf{p}, t) \tag{1.48}$$

$$n(\mathbf{x}, t)\bar{\mathbf{p}}(\mathbf{x}, t) \equiv \int d^3p\, \mathbf{p} f(\mathbf{x}, \mathbf{p}, t) \tag{1.49}$$

$$\mathbf{P}(\mathbf{x}, t) \equiv \int d^3p\, [\bar{\mathbf{p}} - \mathbf{p}(\mathbf{x}, t)][\bar{\mathbf{v}} - \mathbf{v}(\mathbf{x}, t)] f(\mathbf{x}, \mathbf{p}, t) \tag{1.50}$$

where the velocity $\mathbf{v}$ and momentum $\mathbf{p}$ are related by

$$\mathbf{v} = \frac{\mathbf{p}/m}{\left(1 + p^2/m^2c^2\right)^{1/2}} \tag{1.51}$$

With these definitions the Maxwell equations become

$$\nabla \cdot \mathbf{E} = 4\pi e n \tag{1.52}$$

$$\nabla \cdot \mathbf{B} = 0 \tag{1.53}$$

$$\nabla \times \mathbf{E} = -\frac{1}{c}\frac{\partial \mathbf{B}}{\partial t} \tag{1.54}$$

$$\nabla \times \mathbf{B} = \frac{4\pi e n \mathbf{v}}{c} + \frac{1}{c}\frac{\partial \mathbf{E}}{\partial t} \tag{1.55}$$

Taking the moments $\int d^3p$ and $\int d^3p\, \mathbf{p}$ of the Vlasov equations yields

$$\frac{\partial}{\partial t} n + \nabla \cdot (n\bar{\mathbf{v}}) = 0 \tag{1.56}$$

$$\frac{\partial}{\partial t}\bar{\mathbf{p}} + \bar{\mathbf{v}} \cdot \nabla \bar{\mathbf{p}} + \nabla \cdot \mathbf{P}/n = e[\mathbf{E} + (\mathbf{v} \times \mathbf{B})/c] \tag{1.57}$$

Equation (1.56) is the continuity equation, while Eq. (1.57) is the equation of motion. For a cold fluid, $\nabla \cdot \mathbf{P} = 0$. If the effects of finite temperature were to be included, this term would be retained, and it would be necessary to make an assumption about an equation of state in order to calculate the stress tensor $\mathbf{P}(\mathbf{x}, t)$ and close the system of equations. Alternatively, further moments of the Vlasov equation would be required, and it would then be necessary to make appropriate assumptions about the heat flow tensor. In addition to the difficulty of including finite temperature effects, all momentum space information is lost in taking the moments. As a result, certain waves and instabilities which depend on the detailed momentum space distribution of the particles cannot be treated with the macroscopic fluid approach.

An equilibrium analysis based on the macroscopic fluid equations results from setting $\partial/\partial t$ terms in Eqs. (1.52)–(1.57) to zero. Assuming cold fluid equilibria this gives

$$\nabla \cdot \mathbf{E} = 4\pi en \tag{1.58}$$

$$\nabla \cdot \mathbf{B} = 0 \tag{1.59}$$

$$\nabla \times \mathbf{E} = 0 \tag{1.60}$$

$$\nabla \times \mathbf{B} = 4\pi en\mathbf{v}/c \tag{1.61}$$

$$\nabla \cdot (n\mathbf{v}) = 0 \tag{1.62}$$

$$\mathbf{v} \cdot \nabla \mathbf{p} = e[\mathbf{E} + (\mathbf{v} \times \mathbf{B})/c] \tag{1.63}$$

Note that the bars over the mean quantities have been dropped for simplicity.

To analyze the system stability, the fluid and field quantities are expressed as the sum of their equilibrium values plus a perturbation

$$\xi(\mathbf{x}, t) = \xi_0(\mathbf{x}) + \delta\xi(\mathbf{x}, t) \tag{1.64}$$

Substituting these expressions into Eqs. (1.52)–(1.57) and eliminating the zero-order equilibrium solution yields the linearized set of cold fluid equations given by

$$\frac{\partial}{\partial t}\delta n + \nabla \cdot (n_0\delta\mathbf{v} + \delta n\mathbf{v}_0) = 0 \tag{1.65}$$

$$\frac{\partial}{\partial t}\delta\mathbf{p} + \mathbf{v}_0 \cdot \nabla \delta\mathbf{p} + \delta\mathbf{v} \cdot \nabla \mathbf{p} = e[\delta\mathbf{E} + (\mathbf{v}_0 \times \delta\mathbf{B} + \delta\mathbf{v} \times \mathbf{B}_0)/c] \tag{1.66}$$

$$\nabla \cdot \delta \mathbf{E} = 4\pi e \delta n \tag{1.67}$$

$$\nabla \cdot \delta \mathbf{B} = 0 \tag{1.68}$$

$$\nabla \times \delta \mathbf{E} = -\frac{1}{c}\frac{\partial}{\partial t}\delta \mathbf{B} \tag{1.69}$$

$$\nabla \times \delta \mathbf{B} = \frac{4\pi e}{c}(\delta n \mathbf{v}_0 + n_0 \delta \mathbf{v}) + \frac{1}{c}\frac{\partial}{\partial t}\delta \mathbf{E} \tag{1.70}$$

If the perturbed quantities grow in time or space, the equilibrium is unstable; if the perturbed quantities damp, the system returns to equilibrium and is stable. In subsequent chapters the fluid equations are used to analyze a wide variety of intense beam topics. Our general procedure will be to first develop important equilibrium configurations, and then to analyze the stability of the equilibrium to small-scale perturbations.

Problems

1.1. Derive Eq. (1.1).

1.2. Calculate the voltage waveform produced by the simple transmission line for the case of (a) $Z_L = 2Z_0$, and (b) $Z_L = \frac{1}{2}Z_0$.

1.3. Calculate the voltage waveform produced by the Blumlein PFL assuming that a load impedance Z_L appears at the output at time τ after switch closure, where τ is the one-way wave transit time in the Blumlein.

1.4. Derive Eqs. (1.2) and (1.3).

1.5. For the case of planar strip lines with water dielectric, what is the equation that describes the electrical breakdown strength?

1.6. Consider the case of a 20-Ω coaxial transmission line that is linearly charged to 3 MV in 200 nsec. The line is switched into the diode via a multichannel, point–plane, self-breaking, oil dielectric output switch. Assume that the switch gap separation is 5 cm, and that the individual switch points are separated from each other by 10 cm. (a) What is the switch breakdown voltage? (b) How many switch channels will form if $\sigma_v \approx 2\%$? (c) What will be the pulse rise time?

1.7. Consider an intense relativistic electron beam propagating in vacuum along a longitudinal magnetic field of strength B with equilibrium radius r_{b_0}. Suppose that at a particular axial position Z_0 the field strength increases nonadiabatically to $B_0 + \Delta B$. Describe the beam behavior for $Z > Z_0$.

1.8. Derive Eqs. (1.56) and (1.57).

2

Intense Electron and Ion
Beam Generation

2.1. Introduction

The development of the high-voltage pulsed power technology described in Chapter 1 has provided the capability for generating intense
particle beams with power levels up to 10^{13} W for time durations of the
order of 100 nsec. In order to take advantage of this capability it is
necessary to have a good understanding of the important intense beam
generation processes. The material in this chapter is devoted to a basic
description of high-power diode phenomenology.

While electron emission can result from any of several processes
(thermionic emission, photoemission, secondary emission, or field emission),
the dominant mechanism of concern in this chapter is termed explosive
emission. When a strong electric field is applied across a diode (Fig. 2.1)
intense electron flow (field emission) through microscopic whiskerlike projections on the cold cathode surface leads to resistive heating and explosive
vaporization of the micropoints resulting in the formation of localized
regions of metallic plasma. The expansion and merger of these cathode
flares dramatically increase the effective electron emission area; the current
emitted from the resulting cathode plasma is not limited by surface area
(supply limited), but by the space charge cloud in the anode–cathode gap
(space charge limited).

There are two types of flow in electron beam diodes which have great
practical importance: (1) electron flow essentially parallel to the electrostatic lines of force; and (2) focused or pinched flow characterized by strong
beam contraction due to the large self-magnetic field. In approximate terms,
if the gyroradius of an electron emitted from the outer edge of the cathode
surface is less than or equal to the anode–cathode gap separation, beam

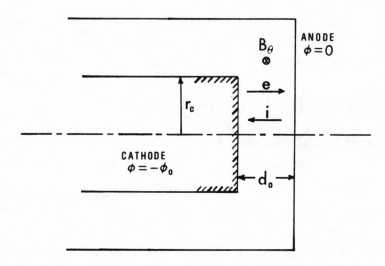

Figure 2.1. Schematic diagram of a high-voltage cold cathode diode.

self-focusing is expected. If the electron current density at the anode is very intense, an anode plasma is formed and the flow of electron current can be dramatically affected by the intense ion current emitted from the plasma.

For applications which require electron beam extraction into vacuum regions, a valuable diode configuration is the magnetically insulated foilless diode. In this device the beam is formed and transported from the diode region by the action of an external magnetic field. Since the beam is not extracted through a grounded anode, problems associated with anode vaporization and anode foil scatter are eliminated.

Although intense ion beams can also be directly produced in high-power diodes, provided that a plasma covers the anode surface, the efficient generation of such beams usually requires the suppression or control of electron flow across the diode. Two solutions for this problem are provided by (1) electron reflexing, and (2) magnetic insulation. In the first case an electron reflexes through an anode placed between two cathodes at the same potential. The effect of the reflexing action is to increase the electron space charge in the diode region allowing the extraction of higher ion currents. In the second method the electrons are prevented from crossing the diode by a strong magnetic field applied transverse to the electric field lines. Under certain conditions it is also appropriate to consider strongly pinched electron diodes as self-magnetically insulating, and such diodes can also be optimized to produce intense ion beams. In the following sections of this chapter various aspects of the production of intense particle beams in cold cathode, plasma emission diodes are discussed in detail.

2.2. Electron Emission Processes

Under normal conditions the potential barrier at the surface of a cold metal electrode prevents the escape of electrons from the conduction band. In thermionic emission and photoemission electrons are given sufficient energy to overcome the potential barrier. On the other hand, in field emission the barrier is sufficiently deformed by an applied electric field that unexcited electrons can tunnel through it. While each of the above processes has its important applications, they will be described only briefly in this section. The topics of primary interest are plasma formation on the cathode surface as the result of a very strong applied electric field, and the subsequent extraction of very intense electron current densities from the zero work function plasma.

2.2.1. Thermionic Emission[1] and Photoemission[2]

Detailed considerations of the atomic energy level structure in solids show that the energy levels coalesce into bands separated by forbidden regions corresponding to the gaps between atomic levels. Each band also has a fine structure, corresponding to and equal in number to the original atomic levels. The energy ε_F of the highest filled level, the Fermi level, is typically several electron volts above the bottom of the band in all metals.

The energy difference between the Fermi level and a field-free region near the surface is called the work function ϕ_w, and has values of 2–5 eV for metals (Fig. 2.2). In photoemission, surfaces are irradiated with light of energy $h\nu \geq \phi_w$. The threshold frequency for clean metals is in the visible or near ultraviolet.

In the case of thermionic emission the metal is heated until a sufficient number of electrons acquire kinetic energies in excess of $\phi_w + \varepsilon_F$. The electron current is related to the cathode temperature by the Richardson–

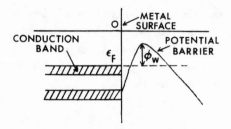

Figure 2.2. Potential energy diagram for electrons in a metal. ε_F is the Fermi energy and ϕ_w is the surface work function.

Dushman equation (for metals)[1]:

$$j \approx 120 T^2 e^{-e\phi_w/kT} \quad (\text{A}/\text{cm}^2) \tag{2.1}$$

where j is the emission current density and T is the absolute cathode temperature. Pure metal cathodes, oxide-coated cathodes, and composite cathodes have all been employed as thermionic emitters, with the best capable of producing current densities $\lesssim 100$ A/cm^2.

2.2.2. Field Emission[3]

The phenomenon of field emission from a cold metal can be described as a quantum mechanical tunneling of conduction electrons through the potential barrier at the surface of the metal.[4-6] Because of its importance in the initial turn-on phase of cold cathode explosive emission, it will be described in more detail. The main characteristic of the field emission process, the emitted current density, j, can be computed by integrating the product of the electron flux, $n(\varepsilon_z)$, incident on the surface [the potential barrier, $V(z)$], and the barrier penetration probability $P(\varepsilon_z)$, over all incident electron energies, i.e.,

$$j = \int_{-\infty}^{\infty} en(\varepsilon_z) P(\varepsilon_z)\, d\varepsilon_z \tag{2.2}$$

where $\varepsilon_z = p_z^2/2m + V(z)$ is the energy associated with the z-directed electron motion and p_z is the z-directed momentum. Assuming that the conduction electrons in the metal form a gas of free particles which obey Fermi–Dirac statistics

$$n(p_z)\,dp_z = \frac{2\,dp_z}{mh^3} \int_{-\infty}^{\infty} dp_x\, dp_y \left[\exp\left(\frac{\varepsilon - \varepsilon_F}{kT} \right) + 1 \right]^{-1}$$

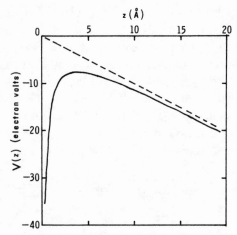

Figure 2.3. Effective potential energy $V(z)$ of an electron near a metallic surface, as given by Eq. (2.5), assuming an electric field strength of $E = 10^8$ V/cm.

or

$$n(\varepsilon_z) = \frac{4\pi mkT}{h^3} \ln\left\{ 1 + \exp\left[-\frac{(\varepsilon_z - \varepsilon_F)}{kT} \right] \right\} \qquad (2.3)$$

where ε is the electron energy, ε_F is the energy of the Fermi level, k is the Boltzmann constant, $\hbar = h/2\pi$ where h is Planck's constant, and T is the absolute temperature.

The barrier penetration probability can be determined by considering the time-independent Schrödinger equation for the electron motion in the z direction:

$$\frac{d^2\psi}{dz^2} + \frac{2m}{\hbar^2}\left[\varepsilon_z - V(z) \right]\psi = 0 \qquad (2.4)$$

An approximate form for the effective one-dimensional potential is (see Fig. 2.3)[7]

$$V(z) = \begin{cases} -W_0, & z < 0 \\ -eEz - \dfrac{e^2}{4z\varepsilon_p\pi}, & z > 0 \end{cases} \qquad (2.5)$$

where $\varepsilon_p = 8.85 \times 10^{-12}$ f/m is the permittivity of free space (MKS units). The three contributions to the effective potential are (i) some constant value $(-W_0)$ within the metal which is lower than the Fermi level, (ii) a contribution $(-eEz)$ arising from an applied electric field, and (iii) a contribution $-e^2/4z\varepsilon_p\pi$ due to the image charge induced on the surface of the metal by an electron outside the metal. The potential is not expected to have significance at $z = 0$, and the main features of the emission process are to be understood by the shape of the potential outside the metal.

Employing the WKB method, $P(\varepsilon_z)$ is given approximately by

$$P(\varepsilon_z) = \exp\left\{ -\int_{z_1}^{z_2} dz \left[\frac{8m}{\hbar} \left[V(z) - \varepsilon_F \right] \right]^{1/2} \right\} \tag{2.6}$$

where z_1 and z_2 are the turning points of Eq. (2.4),

$$z_{1,2} = \frac{|\varepsilon_z|}{2eE} \left[1 \mp \left(1 - \frac{e^3 E}{\varepsilon_z^2 \varepsilon_p \pi} \right)^{1/2} \right] \tag{2.7}$$

Evaluation of the integral of Eq. (2.6) yields

$$P(\varepsilon_z) = \exp\left[-\frac{4v(y)}{3\hbar eE} \left(2m|\varepsilon_z|^3 \right)^{1/2} \right] \tag{2.8}$$

where $y = (e^3 E)^{1/2}/|\varepsilon_z|$. The function v is given by

$$v(y) = 2^{-1/2} \left[1 + (1 - y^2)^{1/2} \right]^{1/2} \left\{ E(k) - \left[1 - (1 - y^2)^{1/2} \right] K(k) \right\} \tag{2.9}$$

where

$$k^2 = \frac{2(1 - y^2)^{1/2}}{1 + (1 - y^2)^{1/2}},$$

and K and E are the complete elliptic integrals of the first and second kinds.

Substitution of Eqs. (2.3) and (2.8) into Eq. (2.2) gives the emission current as

$$j = \int_{-\infty}^{\infty} e\, d\varepsilon_z \left\{ \frac{4\pi mkT}{h^3} \ln\left[1 + \exp\left(-\frac{\varepsilon_z - \varepsilon_F}{kT} \right) \right] \right\}$$

$$\times \exp\left[-\frac{4v(y)}{2\hbar eE} \left(2m|\varepsilon_z|^3 \right)^{1/2} \right] \tag{2.10}$$

In the low-temperature limit, $T \to 0$, the field emission electrons generally have energies approximately equal to the Fermi level and the integration yields

$$j = \frac{e^3 E^2}{8\pi h |\varepsilon_F| t^2(y_F)} \exp\left[-\frac{4(2m)^{1/2}|\varepsilon_F|^{3/2}}{3\hbar eE} v(y_F) \right] \tag{2.11}$$

where $y_F = (e^3 E)^{1/2}/|\varepsilon_F|$, and the function t is defined by

$$t = v - \frac{2}{3} y \frac{dv}{dy} \tag{2.12}$$

The functions v and t are tabulated in Ref. 3.

For low temperatures, the emission differs only slightly from the case of emission at zero temperature, but for higher temperatures the main contribution to the current comes from electrons above the Fermi level and the character of the emission changes. Relatively simple analytical formulas have been obtained in the two limiting cases of high fields and relatively low temperatures ($E-T$ emission), and for weak fields and high temperatures ($T-E$ emission).[8]

2.2.3. Explosive Electron Emission

On almost any surface there exist microscopic surface protrusions (whiskers) which are typically on the order of 10^{-4} cm in height with a base radius of less than 10^{-5} cm and tip radius usually much smaller than the base radius.[9, 10] Reported estimates of whisker concentrations have ranged from approximately 1 to 10^4 whiskers/cm².[11, 12] When high voltage is applied to the diode, the electric field at the whisker tips on the cold cathode

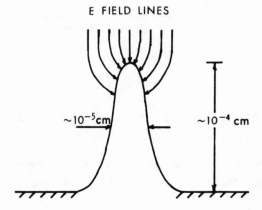

E FIELD LINES

$\sim 10^{-5}$cm

$\sim 10^{-4}$ cm

Figure 2.4. Microscopic view of the local electric field enhancement at the tip of a whiskerlike protrusion on the cathode surface.

surface may be enhanced by a factor of several hundred over the macroscopic electric field (Fig. 2.4). Such field enhancement leads to significant electron field emissions from the micropoints, and can cause whisker evaporation due to excessive Joule heating. The four basic energy exchange phenomena which control the temperature at the whisker tip include the following[13]:

(1) volumetric resistive heating due to the current flowing toward the emitter tip;
(2) emissive heating or cooling because the average energy of the emitted electrons is generally different from the average energy of the conduction electrons in the cathode (the Nottingham effect)[14];
(3) radiative cooling; and
(4) cooling due to thermal conduction toward the support structure.

For the temperatures and thermal conductivities of interest, radiative cooling is usually negligible compared to heat dissipation by conduction. Also, although it is not necessarily negligible, the Nottingham effect is always dominated by Joule heating for cases of practical interest.

A basic understanding of the heat flow problem in a field-emitting micropoint may be obtained by considering the mathematically simple case of a cylindrical emitter (Fig. 2.5). The change in temperature of the cylindrical volume element $dV = \pi r_0^2 \, dl$ is calculated from

$$\frac{\partial T}{\partial t} = \left(\eta c \pi r_0^2 \, dl \right)^{-1} \frac{dQ}{dt} \qquad (2.13)$$

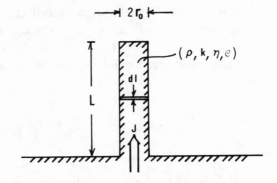

Figure 2.5. Heat flow in a cylindrical field-emitting micropoint assuming only resistive heat generation and thermal conduction. ρ, k, η, and e denote the electrical resistivity, the thermal conductivity, the mass density, and the specific heat, respectively, of the whisker material.

where η is the mass density, e is the specific heat of the emitter material, and dQ/dt is the net heat flow in the volume element. Considering only the resistive generation of heat and thermal conduction (neglecting radiative heat loss and the Nottingham effect) the net heat flow is given by[15]

$$\frac{dQ}{dt} = \frac{dQ}{dt}\bigg|_{\text{joule}} + \frac{dQ}{dt}\bigg|_{\text{conduction}} \tag{2.14}$$

where

$$\frac{dQ}{dt}\bigg|_{\text{joule}} = j^2 \rho \pi r_0^2 dl \tag{2.15}$$

$$\frac{dQ}{dt}\bigg|_{\text{conduction}} = k\pi r_0^2 \left(\frac{\partial T}{\partial l}\bigg|_+ - \frac{\partial T}{\partial l}\bigg|_- \right) \tag{2.16}$$

j is the electron current density while ρ and k are the electrical resistivity and thermal conductivity of the material. Substituting Eqs. (2.15) and (2.16) into Eq. (2.13) yields the differential equation

$$\frac{\partial^2 T}{\partial l^2} - \frac{\eta e}{k} \frac{\partial T}{\partial t} = -\frac{j^2 \rho}{k} \tag{2.17}$$

For the steady state case the solution is

$$T = -\tfrac{1}{2} a l^2 + c_1 l + c_2 \tag{2.18}$$

where $a = j^2\rho/k$. With the origin chosen such that $l=0$ at the base of the whisker (the cool end) and $l=L$ at the tip of the whisker, the boundary conditions are (1) $T=0$ when $l=0$; and (2) $\partial T/\partial l=0$ at $l=L$. With these conditions, the equation for the maximum value of T (at the tip of the emitter) is

$$T_{max} = \frac{1}{2}\left(\frac{j^2\rho}{k}\right)L^2 \tag{2.19}$$

In deriving Eq. (2.19), the variation of the physical constants has been neglected. For the case of tungsten, choosing intermediate values of $\rho = 50 \times 10^{-6}$ Ω cm and $k = 0.25$ cal/sec cm °C yields

$$T_{max} = (2.5 \times 10^{-5})j^2L^2 \quad (°C)$$

where j is in units of A/cm^2 and L is in centimeters.

For typical whisker lengths of 10^{-4}–10^{-3} cm, the current density required to bring the tip of the whisker to the melting point (3400°C) is of the order of 10^7–10^8 A/cm^2.

Electron micrographs of whisker geometry indicate that a wide variety of shapes are present on the surface of any material. A consideration of conical emitters indicates that current densities as high as 10^8 A/cm^2 should be attainable before destruction of the emitter surface occurs.[15, 16] Experimentally observed values of the critical current density for arc initiation lie in the range of 10^7–10^8 A/cm^2. Although these current densities are rather large, the effective emission area is quite small (whisker radii of order 10^{-5} cm), and the total current density emitted by a cathode surface undergoing stable field emission from microscopic projections is typically limited to 1–10 A/cm^2.

Current densities substantially in excess of the critical density for field emission cause explosive vaporization of the whiskers and the formation of local plasma bursts (cathode flares).[8] The rapid hydrodynamic expansion (a few cm/μsec) and merger of the cathode flares quickly form a plasma sheath which covers the entire surface of the cathode; the effective electron emission area increases dramatically, and the expanding cathode plasma emits an electron current which is limited by the associated space charge cloud in the anode–cathode gap.

2.3. Electron Flow in High-Power Diodes

Since the cathode plasma can be considered as a metal surface whose work function is effectively zero, the cathode electron current supply is essentially unlimited. In this case the current that flows in the diode is determined by the modification of the equipotential contours in the diode due to the space charge of the electron current. To illustrate this process assume that a plasma completely covers the cathode surface at the instant that high voltage is impressed on the diode. The evolution of the potential variation between the electrodes is described in Fig. 2.6. Initially, there is no electron space charge in the diode gap and the potential variation is simply a linear function of the distance from the cathode to the anode (curve I). As electrons are drawn into the gap the potential at all positions becomes lower (curve II). As more and more electrons enter the gap a potential minimum can form outside the cathode surface (curve III), and only those electrons which possess an initial energy greater than the maximum height of the barrier (due to electron thermal spread in the cathode plasma) can escape from the cathode and reach the anode. Finally, the condition is reached when most of the electrons which leave the cathode surface are reflected by the potential barrier (curve IV). Since the applied field causes those electrons which pass the barrier to leave the interelectrode space, the barrier breaks down (curve V) and the process repeats. Such time-dependent behavior can be considered as a self-regulating switch that allows a certain average current (the space-charge-limited current) to flow. (For a more detailed discussion of this phenomenon see Section 3.3.)

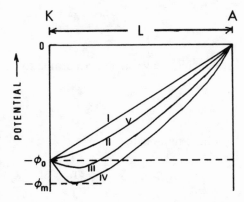

Figure 2.6. Electrostatic potential in a one-dimensional diode including the effects of the electron space charge. A time-dependent potential minimum forms near the cathode surface.

For high-voltage diodes the magnitude of the potential minimum is negligible compared with the high applied potential. In this case the location of the potential minimum is practically coincident with the cathode surface. Hence, it may be assumed that the electric field *vanishes* at the cathode surface and that the emitted electrons have zero initial velocity, in which case the velocity of an electron at any position between the electrodes is determined from the conservation of energy. This is the essential characteristic of space-charge-limited flow.

2.3.1. The Relativistic Planar Diode

As a first example consider the case of a relativistic planar diode.[17] The anode potential is assumed to be zero, while the cathode is at potential-ϕ_0. In this case conservation of energy implies that

$$\gamma = \gamma_0 + e\phi/mc^2 \tag{2.20}$$

For the planar geometry, Poisson's equation is

$$\frac{d^2\phi}{dz^2} = 4\pi e n_e \tag{2.21}$$

where z is the distance from the cathode plasma and n_e is the electron number density. Conservation of electronic charge requires $\nabla \cdot j = 0$, or

$$j = en_e v = \text{const} \tag{2.22}$$

Combining Eqs. (2.20)–(2.22) yields

$$\frac{d^2\phi}{dz^2} = \frac{4\pi j}{v} = \frac{4\pi j}{c}\left(1 - \gamma^{-2}\right)^{-1/2} \tag{2.23}$$

Introducing the variable y according to

$$y = \frac{d\gamma}{dz} \tag{2.24}$$

Eq. (2.23) reduces to

$$y\frac{dy}{d\gamma} = K\left[\frac{\gamma}{\left(\gamma^2 - 1\right)^{1/2}}\right] \tag{2.25}$$

where

$$K = 4\pi j \left(\frac{e}{m_0 c^3} \right) \tag{2.26}$$

Integration of Eq. (2.25) yields

$$y^2 = 2K(\gamma^2 - 1)^{1/2} + C_1 \tag{2.27}$$

The constant of integration C_1 is zero because $y = d\gamma/dz = d\phi/dz = 0$ at the cathode. Taking the square root and integrating, Eq. (2.27) becomes

$$\int_1^\gamma (\gamma^2 - 1)^{-1/4} d\gamma = (2K)^{1/2} \int_0^z dz = (2K)^{1/2} z \tag{2.28}$$

From standard integral tables, the integral over γ may be expressed as

$$\int_1^\gamma (\gamma^2 - 1)^{-1/4} d\gamma = F\left(\delta, \frac{\sqrt{2}}{2} \right) - 2E\left(\delta, \frac{\sqrt{2}}{2} \right) + \frac{2\gamma(\gamma^2 - 1)^{1/4}}{1 + (\gamma^2 - 1)^{1/2}}$$

$$\tag{2.29}$$

where

$$\delta = \cos^{-1} \left[\frac{1 - (\gamma^2 - 1)^{1/2}}{1 + (\gamma^2 - 1)^{1/2}} \right]$$

$F(\delta, \sqrt{2}/2)$ and $E(\delta, \sqrt{2}/2)$ are elliptic integrals of the first and second kind, defined by

$$F(\xi, k) = \int_0^{\sin \xi} [(1 - x^2)(1 - k^2 x^2)]^{-1/2} dx \tag{2.30}$$

$$E(\xi, k) = \int_0^{\sin \xi} (1 - k^2 x^2)^{1/2} (1 - x^2)^{-1/2} dx \tag{2.31}$$

Substituting for K and solving Eq. (2.28) for j yields the relationship between current density and the applied voltage for the relativistic planar diode

$$j = \frac{\left(m_0 c^3/e\right)}{8\pi d^2}\left[F\left(\delta_0, \sqrt{2}\,/2\right) - 2E\left(\delta_0, \sqrt{2}\,/2\right) + \frac{2\gamma_0\left(\gamma_0^2 - 1\right)^{1/4}}{1 + \left(\gamma_0^2 - 1\right)^{1/2}}\right]^2$$

(2.32)

where d is the electrode spacing and γ_0 corresponds to the applied potential $|\phi_0|$. In the nonrelativistic limit, $e\phi_0/m_0 c^2 \ll 1$, Eq. (2.32) reduces to the familiar Child–Langmuir expression[18]

$$j = \frac{\sqrt{2}}{9\pi}\left(\frac{e}{m_0}\right)^{1/2}\frac{\phi_0^{3/2}}{d^2}$$

(2.33)

A comparison of the normalized diode impedance $\{Z_n = U/\xi^2$, where $U = e\phi/mc^2$ and $\xi^2 = [je/(mc^3\varepsilon_p)]d^2(MKS)\}$ for the relativistic and non-relativistic solutions is presented in Fig. 2.7. Note that the diode impedance decreases as the applied potential is increased.

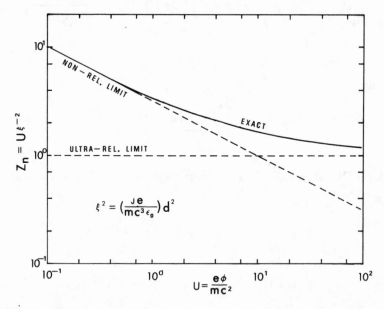

Figure 2.7. A comparison of the relativistic and nonrelativistic solutions for the impedance of a one-dimensional planar diode.

The electron velocity as a function of position may be determined from Eq. (2.20) with the aid of Eqs. (2.23) and (2.32), and the charge density is then determined from Eq. (2.22). The physically impossible result that the electron density is infinite at the cathode is a consequence of the zero initial velocity assumption. In reality, the initial velocities are small but nonzero, and the charge density is large though finite.

Although the above expressions have been derived for the case of the infinite planar diode, they should remain approximately correct for large aspect ratio ($r_c/d \gg 1$) cylindrically symmetric diodes, provided the electron current flow is not too large (see Section 2.3.2). For small aspect ratio diodes undergoing space charge limited flow, the dominant electron motion is still along the electrostatic lines of force (which may have a large radial component), but the geometrical factor in the perveance relationships is altered.

2.3.2. Parapotential Flow

When the effects of the reduction of the diode gap spacing d due to the expanding cathode plasma are included[19] the previous discussions of this chapter provide a self-consistent description of the dominant processes acting within nonpinching diodes. However, in the limit of high-current electron flow the effect of the azimuthal self-magnetic field on the electron trajectories cannot be neglected and may result in self-focusing of the beam in the diode. A rough transition criterion[20] is that diode pinching will occur when the relativistic gyroradius $r_B = \gamma mc^2/eB$ of an electron emitted at the radius of the cathode is equal to the anode–cathode gap separation. Assuming $B \approx 2I/cr_c$, then the diode critical current is approximately given by

$$I_c = \frac{\gamma mc^3}{e} \frac{r_c}{2d} \qquad (2.34)$$

When the diode current exceeds the critical current no exact analytical solution for the electron flow is known. The initial treatments[20-22] of $I/I_c > 1$ diodes assumed that the dominant feature of such diodes was self-consistent electron flow ($E \times B$ drift) along equipotential surfaces within a region extending from somewhat in front of the cathode surface to somewhat in front of the anode surface. (As such, they represent the relativistic generalization of Brillouin flow[23] including the self-magnetic field.) The solutions were termed parapotential[21] to emphasize the departure from normal diode flow orthogonal to the equipotentials.

To illustrate the parapotential flow model and its application to high-current diodes, consider the case of electron flow in a diode with a conical cathode surface with a hollow well which does not distort the conical equipotential surfaces (Fig. 2.8). Following Ref. 22, it is possible to derive the Brillouin solutions assuming azimuthal symmetry and force balance along each equipotential. In this case, the equations which describe the steady state solution are

$$E = -\frac{v}{c} \times B \qquad (2.35)$$

$$\nabla \cdot E = -4\pi n_e e \qquad (2.36)$$

$$\nabla \times B = \frac{4\pi}{c} j = -\frac{4\pi}{c} e n_e v \qquad (2.37)$$

where v is the electron velocity, and E and B are the electric and magnetic fields expressed in terms of the electron number density n_e and the current density j. E may be derived from a potential function ϕ according to

$$E = -\nabla \phi \qquad (2.38)$$

It is assumed that $\phi = -\phi_0$ at the cathode surface and $\phi = 0$ at the anode. If the electrons are presumed to leave the cathode with zero initial velocity

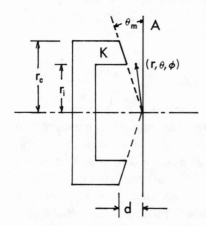

Figure 2.8. Schematic diagram of a low-impedance parapotential diode.

then conservation of energy yields

$$\gamma = \gamma_0 + \frac{e\phi}{mc^2}$$

Taking the divergence of Eq. (2.35) and substituting for $\nabla \times \mathbf{B}$ and n_e yields

$$\nabla \cdot \mathbf{E} = -\frac{(\nabla \times \mathbf{v}) \cdot \mathbf{B}}{c(1 - \beta^2)} \tag{2.39}$$

For the conical diode of Fig. 2.8, the appropriate coordinate system is a spherical polar system whose origin is centered on the anode surface. The polar axis is assumed coincident with the diode axis of symmetry. With the assumed symmetry, $E_r = E_\phi = B_r = B_\theta = v_\phi = v_\theta = 0$, and all nonzero quantities are independent of r and ϕ. In this case Eq. (2.39) reduces to

$$\frac{1}{\sin\theta} \frac{d}{d\theta}(E_\theta \sin\theta) = \frac{B_\phi}{c(1 - \beta^2)} \frac{dv_r}{d\theta} \tag{2.40}$$

and substitution of Eqs. (2.35) and (2.38) into Eq. (2.40) yields

$$-\frac{1}{\sin\theta}\left(\sin\theta\frac{d^2\phi}{d\theta^2} + \cos\theta\frac{d\phi}{d\theta}\right) = \frac{1}{v_r(1 - \beta^2)} \frac{d\phi}{d\theta} \frac{dv_r}{d\theta} \tag{2.41}$$

Carrying out the differentiation of v_r and noting that

$$\frac{d\phi}{d\theta} = \frac{mc^2}{e} \frac{d\gamma}{d\theta}$$

gives

$$\frac{d^2\gamma}{d\theta^2} + \cot\theta\frac{d\gamma}{d\theta} = \frac{\gamma}{\gamma^2 - 1}\left(\frac{d\gamma}{d\theta}\right)^2 \tag{2.42}$$

Making the change of variable

$$\psi = \ln(\tan\theta/2) \tag{2.43}$$

Equation (2.42) may be expressed as

$$\frac{d^2\gamma}{d\psi^2} = \frac{\gamma}{\gamma^2-1}\left(\frac{d\gamma}{d\psi}\right)^2 \tag{2.44}$$

Equation (2.44) may be solved by elementary differential equation methods. Setting $y = d\gamma/d\psi$ and performing the indicated integration yields

$$C_1\frac{d\gamma}{d\psi} = (\gamma^2-1)^{1/2} \tag{2.45}$$

while the integration of Eq. (2.45) gives

$$\psi = C_1\ln\left[\gamma+(\gamma^2-1)^{1/2}\right]+C_2 \tag{2.46}$$

In the parapotential model, it is not necessary that the electron flow patterns extend to the anode; however, since it may be shown that the beam current is maximized in this instance it will be assumed that the flow patterns do extend completely to the anode. In this circumstance the appropriate boundary conditions are

$$\gamma = \gamma_0 = 1 + e|\phi_0|/mc^2, \qquad \theta = \pi/2$$

$$\gamma = 1, \qquad \theta = \theta_m \tag{2.47}$$

where θ_m denotes the angle of the (conical) cathode surface. Substituting these boundary conditions into Eq. (2.46) yields

$$\frac{\ln\left[\gamma+(\gamma^2-1)^{1/2}\right]}{\ln\left[\gamma_0+(\gamma_0^2-1)^{1/2}\right]} = 1 - \frac{\ln(\tan\theta/2)}{\ln(\tan\theta_m/2)} \tag{2.48}$$

In the parapotential model, the current flowing in the diode is calculated using Eq. (2.37) and Stokes' theorem

$$\int_\Gamma \mathbf{B}\cdot d\mathbf{l} = \frac{4\pi}{c}\int_S \mathbf{j}\cdot d\mathbf{A} = \frac{4\pi}{c}I(\theta) \tag{2.49}$$

where S is a surface of constant r, and Γ is the curve of constant r and θ bounding this surface. $I(\theta)$ is, therefore, the total current flowing within the angle θ. Thus

$$\int_{\Gamma} \mathbf{B} \cdot d\mathbf{l} = B_{\phi}(2\pi r \sin\theta) = \frac{4\pi}{c} I(\theta) \tag{2.50}$$

Substituting for B_{ϕ} and E_{θ} from Eqs. (2.35) and (2.38) and making the transformation from ϕ to γ [Eq. (2.20)] yields

$$-\frac{d\gamma}{d\theta}\sin\theta = \beta \frac{2e}{mc^3} I(\theta) \tag{2.51}$$

Performing the indicated differentiation of γ with respect to θ [Eq. (2.48)] gives

$$I(\theta) = -\frac{mc^3}{2e} \frac{\gamma \ln\left[\gamma_0 + \left(\gamma_0^2 - 1\right)^{1/2}\right]}{\ln(\tan\theta_m/2)} \tag{2.52}$$

For large diodes an excellent approximation to the geometrical factor may be derived by noting that $\tan\theta_m/2 = (1 - \cos\theta_m)/\sin\theta_m \approx 1 - \cot\theta_m = 1 - d/r_c$. Expanding the logarithm gives $[-\ln(\tan\theta_m/2)]^{-1} \approx r_c/d$.

Encompassing the total flow pattern $(\theta = \pi/2, \gamma = \gamma_0)$ yields the saturated parapotential current I_p given by

$$I_p = \left(\frac{mc^3}{2e}\right)\left(\frac{r_c}{d}\right)\gamma_0 \ln\left[\gamma_0 + \left(\gamma_0^2 - 1\right)^{1/2}\right] \tag{2.53}$$

It may be seen that the parapotential model requires a bias current flowing interior to $\theta = \theta_m$. Setting $\gamma = 1$ in Eq. (2.52) yields

$$I_b = I(\theta_m) = \left(\frac{mc^3}{2e}\right)\left(\frac{r_c}{d}\right)\ln\left[\gamma_0 + \left(\gamma_0^2 - 1\right)^{1/2}\right] = I_p/\gamma_0 \tag{2.54}$$

The effect of the bias current is to establish a magnetic field which is analogous to the external magnetic field required for nonrelativistic Brillouin flow.

The parapotential flow model has been found to give good agreement with experimental measurements of diode impedance, and allows useful diode design based on its scaling law predictions. Nevertheless, the model suffers from a number of deficiencies. In particular, the parapotential flow model is unable to treat electron flow near the cathode or the anode, i.e., the mechanics of how electrons get on and off equipotential surfaces. In addition, the model does not include the effects of ion flow, which have been found to be very important for large aspect ratio diodes (Section 2.4.4).

2.3.3. Foilless Diodes[24]

For applications which require electron beam extraction into vacuum regions, foilless diodes have proven to be extremely useful diode configurations. In such diodes the beam does not pass through a grounded anode, and beam expansion and transport must be controlled by an externally imposed magnetic field. In addition to eliminating the problems associated with anode vaporization and beam heating due to anode foil scatter, very thin annular beams are easily generated with such diodes. As shown in Chapter 3 the space charge current limit for an annular beam can be substantially larger than that of a solid beam carrying the same number of particles.

Typical foilless diode geometries[24-26] are shown in Fig. 2.9. The designs (a) and (b) somewhat resemble a magnetron gun,[27] since the magnetic field of the external solenoid is approximately perpendicular to the electric field in the diode region. The third configuration, Fig. 2.9c, corresponds to a planar foilless diode which is designed to inject electrons approximately parallel to the uniform applied field.

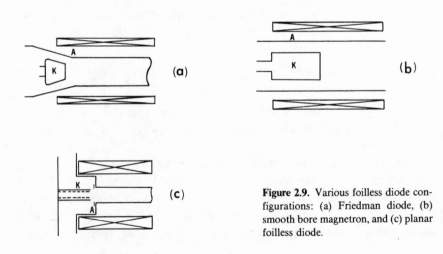

Figure 2.9. Various foilless diode configurations: (a) Friedman diode, (b) smooth bore magnetron, and (c) planar foilless diode.

Theoretical analyses of beam generation in foilless diodes have generally followed two separate approaches. One model assumes that the diode current can be determined by the equilibrium that the beam attains in the hollow drift tube[28, 29]; however, it is more logical to view the allowed beam equilibrium states as depending on the current supplied by the diode geometry in the vicinity of the cathode.[30, 31] The latter approach is presented here.[31]

Consider the idealized geometry of Fig. 2.10; a circular cathode with surface potential $\phi_c(r)$ is joined at a right angle to a grounded cylindrical anode of radius R. Electron emission is assumed to occur over a limited (roughened) region of the cathode surface; the electron motion is constrained to be one-dimensional by a very strong applied axial magnetic field. In addition, the applied voltage is assumed to be ultrarelativistic so that the narrow cathode sheath region can be ignored. Under these assumptions a self-consistent solution for the electron current density can be found which satisfies the criterion of space-charge-limited emission over the roughened portion of the cathode surface.

Since the electrostatic potential satisfies Poisson's equation

$$\nabla^2 \phi(r, z) = -4\pi\rho(r, z) \tag{2.55}$$

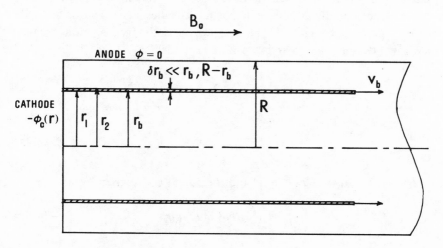

Figure 2.10. Idealized geometry for the planar foilless diode.

$\phi(r, z)$ can be found from

$$\phi(r,z) = \int r'dr'dz'G(r,z;r',z')\rho(r',z') + \frac{1}{2}\int r'dr'\phi_c(r')\left(\frac{\partial G}{\partial z'}\right), \qquad z'=0$$

(2.56)

where $\phi_c(r)$ is the potential variation on the cathode surface, and $G(r, z; r', z')$ is the Green's function given by (assuming axial symmetry)

$$G(r,z;r',z') = \frac{4}{R}\sum_{n=1}^{\infty}\frac{J_0(k_n r)J_0(k_n r')}{k_n[J_1(k_n R)]^2}$$

$$\times \begin{cases} e^{-k_n z}\sinh(k_n z'), & z' < z \\ e^{-k_n z'}\sinh(k_n z), & z' > z \end{cases}$$

(2.57)

In Eq. (2.57) the k_n correspond to the roots of the zeroth-order Bessel function, i.e., $J_0(k_n R)=0$ for $n=1,2,\ldots$. It is easily verified that the Green's function vanishes for all r and z on the boundaries, including the surface at infinity.

Since $\rho(r, z)$ is related to the potential $\phi(r, z)$ through the equation of continuity, Eq. (2.56) is actually a complicated integral equation for the electrostatic potential. In order to make analytical progress, it is assumed, following Chen and Lovelace, that the diode is ultrarelativistic. In this case, the electrons rapidly acquire relativistic velocities and the thin cathode sheath region can be ignored. The equation of continuity then indicates that $\rho(r, z)=\rho(r)$ only, and ρ and ϕ_c can be expanded as

$$\rho(r) = \frac{2}{R^2}\sum_{n=1}^{\infty}\rho_n\psi_n(r)$$

(2.58)

$$\phi_c(r) = \frac{2}{R^2}\sum_{n=1}^{\infty}\phi_n\psi_n(r)$$

(2.59)

where $\psi_n(r) = J_0(k_n r)[J_1(k_n R)]^2$ and the ρ_n and ϕ_n are given by

$$\rho_n = \int_0^R r\,dr\,\rho(r)J_0(k_n r)$$

(2.60)

$$\phi_n = \int_0^R r\,dr\,\phi_c(r)J_0(k_n r)$$

(2.61)

Substituting Eqs. (2.57), (2.58), and (2.59) into Eq. (2.56) yields

$$\phi(r,z) = \frac{8\pi}{R^2} \sum_{n=1}^{\infty} \frac{\rho_n}{k_n^2} (1 - e^{-k_n z}) \psi_n(r) + \frac{2}{R^2} \sum_{n=1}^{\infty} \phi_n e^{-k_n z} \psi_n(r) \quad (2.62)$$

The assumption of space charge limited emission imposes the restriction that $E_z(r,0) = 0$ over the emission surface, and, in principle, yields the ρ_n once the cathode surface potential $\phi_c(r)$ has been specified. Differentiation of Eq. (2.62) with respect to z gives

$$E_z(r,z) = -\frac{8\pi}{R^2} \sum_{n=1}^{\infty} \frac{\rho_n}{k_n} e^{-k_n z} \psi_n(r) + \frac{2}{R^2} \sum_{n=1}^{\infty} k_n \phi_n e^{-k_n z} \psi_n(r) \quad (2.63)$$

For emission over an annulus defined by $r_1 < r < r_2$, $E_z(r,0) = 0$ leads to the integral equation

$$\frac{1}{4\pi} \sum_{n=1}^{\infty} k_n \phi_n \psi_n = \int_{r_1}^{r_2} r' \, dr' \, K(r,r') \rho(r') \quad (2.64)$$

where the kernel is defined by

$$K(r,r') = \sum_{n=1}^{\infty} \frac{1}{k_n} \psi_n(r) J_0(k_n r') \quad (2.65)$$

Equation (2.64) is the basic integral equation for the cylindrical foilless diode.[32] It can be formally solved by expanding the Bessel functions in a set of basis functions which are complete and orthonormal on the interval $[r_1, r_2]$. This task is left as Problem 2.10 for the case $r_1 = 0$.[31]

As an example of the qualitative behavior of foilless diodes we examine the simple case of space-charge-limited emission from the entire cathode surface. In this case Eq. (2.63) indicates that ρ_n and ϕ_n are related according to

$$\rho_n = \frac{k_n^2}{4\pi} \phi_n \quad (2.66)$$

and the beam current is given by

$$I = \frac{c}{R^2} \sum_{n=1}^{\infty} k_n^2 \phi_n \int_0^R r \, dr \, \psi_n(r)$$

$$= \frac{cR}{2} \left| \left(\frac{\partial \phi_c}{\partial r} \right)_R \right| \quad (2.67)$$

For a cathode potential specified by

$$\phi_c(r) = \begin{cases} -\phi_0, & 0 \leqslant r \leqslant r_b \\ -\phi_0(\ln R/r_b)^{-1}(\ln R/r), & r_b < r \leqslant R \end{cases} \tag{2.68}$$

the diode current in the ultrarelativistic limit is calculated to be

$$I = \frac{(\gamma_0 - 1)(mc^3/e)}{2\ln R/r_b} \tag{2.69}$$

where $(\gamma_0 - 1)mc^2 = |e\phi_0|$. An *ad hoc* assumption which extends the range of validity of Eq. (2.69) to nonrelativistic voltages is to replace the quantity $(\gamma_0 - 1)$ by $(\gamma_0^{2/3} - 1)^{3/2}$, i.e.,

$$I \simeq \frac{(\gamma_0^{2/3} - 1)^{3/2}(mc^3/e)}{2\ln R/r_b} \tag{2.70}$$

While not strictly correct Eq. (2.70) does indicate the appropriate scaling, $I \sim \phi_0^{3/2}$, in the nonrelativistic limit.

Although the assumption of space-charge-limited emission over the entire cathode surface is a drastic one, the analysis nevertheless indicates the essential features of foilless diode behavior, including the suppression of electron emission at interior cathode regions due to the beam space charge. Further, as the emitting cathode region nears the anode wall, the beam current rapidly increases.

2.4. Ion Flow in High-Power Diodes

In recent years the state of the art of intense ion beam sources has undergone a dramatic improvement as the result of coupling new techniques for ion beam generation to the high-voltage pulsed power technology.[33] The success of these new methods depends largely on the degree of control or modification of the electron distribution in the high-voltage diode gap.

In conventional ion diodes which do not use electron control the ion current density that can be extracted from an anode plasma is easily obtained using the method detailed in Section 2.3.1 (in the nonrelativistic limit). The space-charge-limited result is given by

$$j_i = (9\pi)^{-1} \left(\frac{2Ze}{m_i} \right)^{1/2} \phi_0^{3/2} d^{-2} \tag{2.71}$$

where Z is the ionic charge and m_i is the ion mass.

Since transverse space charge forces tend to defocus the ion beam, the diode spacing d is kept as small as possible, typically a few millimeters. At the same time the applied voltage must be kept sufficiently low to avoid explosive electron emission from the cathode ($\lesssim 100$ kV/cm). Hence, $\phi_0 \lesssim$ 20–50 kV, and for protons j_i is typically limited to ~ 1 A/cm^2. In the new methods of ion beam generation, the voltage on the diode is substantially increased and various techniques are used to control or suppress electron flow in the diode.

2.4.1. Bipolar Space-Charge-Limited Flow

As an introduction to ion flow in high-power diodes consider the case of a steady state, nonrelativistic diode in which both electrodes behave as high-density plasmas.[18] When a voltage ϕ_0 is impressed on the diode, the electric field causes bipolar flow, i.e., electrons are accelerated toward the anode and ions are accelerated toward the cathode. If both the electrons and ions are emitted with zero velocities, then conservation of energy implies

$$\tfrac{1}{2} m_e v_e^2 = e\phi \tag{2.72}$$

$$\tfrac{1}{2} m_i v_i^2 = e(\phi_0 - \phi) \tag{2.73}$$

where m_e and m_i are the electron and ion masses, v_e and v_i are the electron and ion velocities, and ϕ is the space-dependent potential. (The ions are assumed to be singly charged.)

For the planar geometry Poisson's equation is

$$\frac{d^2\phi}{dz^2} = 4\pi e(n_e - n_i) \tag{2.74}$$

where n_e and n_i are the electron and ion number densities, and z is the distance from the cathode plasma.

Conservation of both electronic and ionic charge yields

$$j_e = e n_e v_e \qquad (2.75)$$

$$j_i = e n_i v_i \qquad (2.76)$$

Substituting Eqs. (2.72), (2.73), (2.75), and (2.76) into Eq. (2.74) and defining the constant α as

$$\alpha = \frac{j_i}{j_e} \left(\frac{m_i}{m_e} \right)^{1/2} \qquad (2.77)$$

yields

$$\frac{d^2\phi}{dz^2} = 4\pi j_e \left(\frac{m_e}{2e} \right)^{1/2} \left[\phi^{-1/2} - \alpha(\phi_0 - \phi)^{-1/2} \right] \qquad (2.78)$$

Defining the dimensionless variables ψ and x according to

$$\psi = \phi/\phi_0 \qquad (2.79)$$

$$x = z/d \qquad (2.80)$$

Equation (2.78) becomes

$$\frac{d^2\psi}{dx^2} = \frac{4}{9} \left(\frac{j_e}{j_0} \right) \left[\psi^{-1/2} - \alpha(1 - \psi)^{-1/2} \right] \qquad (2.81)$$

where j_0 is the electron current in the absence of positive ion flow [Eq. (2.33)].

Setting $y = d\psi/dx$, noting that $d^2\phi/dx^2 = y\, dy/d\psi$, and performing the indicated integration yields

$$\left(\frac{d\psi}{dx} \right)^2 = \left(\frac{16}{9} \right) \left(\frac{j_e}{j_0} \right) \left[\psi^{1/2} + \alpha(1 - \psi)^{1/2} \right] - C_1 \qquad (2.82)$$

where C_1 is the constant of integration. If both electron and ion flows are space charge limited, then the electric field at the anode, as well as the cathode, is zero. Hence,

$$C_1 = \frac{16}{9}\left(\frac{j_e}{j_0}\right) \tag{2.83}$$

$$\alpha = \frac{j_i}{j_e}\left(\frac{m_i}{m_e}\right)^{1/2} = 1 \tag{2.84}$$

Integrating Eq. (2.82) then gives

$$\left(\frac{j_e}{j_0}\right)^{1/2} = \frac{3}{4}\int_0^1 d\psi\left[\psi^{1/2} - 1 + (1-\psi)^{1/2}\right]^{-1/2} \tag{2.85}$$

and numerical evaluation of the integral yields

$$j_e = 1.86 j_0 \tag{2.86}$$

Thus, from Eq. (2.84)

$$j_i = 1.86 j_0 (m_e/m_i)^{1/2} \tag{2.87}$$

Although the presence of ion space charge in the diode increases both the output electron and ion current densities by the factor 1.86 above that obtained in the single-particle flows, since $(m_i/m_e)^{1/2} \geqslant 43$ the bipolar space-charge-limited diode is not an efficient ion beam source. In order to increase the amount of ion current generated substantially above the bipolar flow limit, it is necessary to either modify the electron density in the diode, or else suppress electron flow across the diode. Two general techniques for accomplishing these changes, electron reflexing and magnetic insulation, are discussed in the following sections.

2.4.2. The Reflex Triode[34, 35]

The geometry of the reflex triode is shown in Fig. 2.11. It consists of an anode plane at high positive voltage placed between two cathode planes at ground potential. The anode consists of a metallic foil a fraction of an electron range thick covered on both sides by a surface plasma. The essential feature of a triode is that an electron emitted from one cathode loses energy traversing the anode foil, and is reflected by the opposite cathode. The electron continues to "reflex" through the anode until its kinetic energy becomes too low to penetrate the foil. The enhanced electron density in the vicinity of the anode substantially increases the ion current density that can be extracted from the anode plasma.

To demonstrate the essential features of reflex triode operation consider the one-dimensional geometry of Fig. 2.11. The applied voltage is assumed to be nonrelativistic, and the triode is assumed to reach the steady state. Under these conditions the electrostatic potential is specified by Eq. (2.74). As in the previous section the ion density (assuming mass m_i and charge Ze) is given by

$$n_i = (j_i/Ze)(m_i/2Ze)^{1/2}(\phi_0 - \phi)^{-1/2} \tag{2.88}$$

where j_i is the ion current density extracted from one side of the anode.

To calculate the electron charge density, assume that an electron current density j_e is emitted from each cathode surface. From conservation of energy and the continuity equation the density of the uncollided flux is given by

$$n_{0e} = (j_e/e)(m_e/2e)^{1/2}\phi^{-1/2} \tag{2.89}$$

When the primary flux traverses the anode foil each electron is assumed to lose an amount of energy, $\Delta\phi_1$, so that the electron kinetic energy is given by $\frac{1}{2}m_e v^2 = e(\phi - \Delta\phi_1)$; i.e., the energy constant of the motion has decreased by the amount $\Delta\phi_1$. Hence, these electrons cannot reach the opposite cathode, and are reflected when $\phi = \Delta\phi_1$.

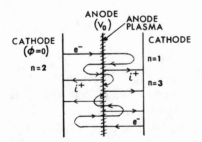

Figure 2.11. Idealized geometry of the reflex triode.

In the steady state the density of electrons traveling in the negative direction is equal to the density traveling in the positive direction. Thus, the electron charge density corresponding to electrons which have made one traversal of the anode foil is given by

$$n_{1e} = (2 j_e/e)(m_e/2e)^{1/2}(\phi - \Delta\phi_1)^{-1/2} \quad (2.90)$$

for all z such that $\phi \geqslant \Delta\phi_1$. Similarly, for electrons which have made n traversals of the anode foil, the electron density is given by

$$n_{ne}(z) = \left(\frac{2 j_e}{e}\right)\left(\frac{m_e}{2e}\right)^{1/2}\left(\phi - \sum_{k=1}^{n} \Delta\phi_k\right)^{-1/2} \quad (2.91)$$

for all z such that $\phi > \sum_{k=1}^{n}\Delta\phi_k$, where $\Delta\phi_k$ is the energy lost in the kth transit through the anode foil. Hence, the electron density can be described by

$$n_e(z) = \left(\frac{j_e}{e}\right)\left(\frac{m_e}{2e}\right)^{1/2}\left[\phi^{-1/2} + 2\sum_{m=1}^{M}\left(\phi - \sum_{k=1}^{m}\Delta\phi_k\right)^{-1/2}\right.$$

$$\left. \times h\left(\phi - \sum_{k=1}^{m}\Delta\phi_k\right)\right] \quad (2.92)$$

where $h(\phi)$ denotes the Heaviside operator, and M is the maximum number of electron transits given by

$$\sum_{m=1}^{M} \Delta\phi_m = \phi_0 \quad (2.93)$$

Since the charge densities and the electrostatic potential are symmetric about the anode, it is sufficient to consider only one side of the triode in the evaluation of the potential distribution. For calculational simplicity it is assumed that $\sum_{k=1}^{m}\Delta\phi_k = m\Delta\phi$; i.e., the energy loss per transit is constant. In this case $M\Delta\phi = \phi_0$, and

$$n_e(z) = (j_e/e)(m_e/2e)^{1/2}\left[\phi^{-1/2} + 2\sum_{m=1}^{M}(\phi - m\Delta\phi)^{-1/2}h(\phi - m\Delta\phi)\right]$$

$$(2.94)$$

Employing Eqs. (2.84), (2.88), and (2.94), and defining the dimensionless
variables $\psi = \phi/\phi_0$ and $x = z/d$, Eq. (2.74) can be expressed as

$$\frac{d^2\psi}{dx^2} = \frac{4}{9}\left(\frac{j_e}{j_0}\right)\left[\psi^{-1/2} + 2\sum_{m=1}^{M}(\psi - m\Delta\psi)^{-1/2}\right.$$

$$\left. \times h(\psi - m\Delta\psi) - \alpha(1-\psi)^{-1/2}\right] \tag{2.95}$$

where j_0 is the Child–Langmuir electron current density in the absence of
ion flow [Eq. (2.33)], and it is assumed that $Z = 1$.

Setting $y = d\psi/dx$ and noting that $d^2\psi/dx^2 = y\,dy/d\psi$, a first integration of Eq. (2.95) yields

$$\left(\frac{d\psi}{dx}\right)^2 = \frac{16}{9}\left(\frac{j_e}{j_0}\right)\left[\psi^{1/2} + 2\sum_{m=1}^{M}(\psi - m\Delta\psi)^{1/2}h(\psi - m\Delta\psi) + \alpha(1-\psi)^{1/2}\right]$$

$$+ C_1 \tag{2.96}$$

where C_1 is a constant of integration. If both the electron and ion emission
processes are space charge limited, then $d\psi/dx = 0$ at $x = 0, 1$. Hence,

$$C_1 = -\frac{16}{9}\left(\frac{j_e}{j_0}\right)\alpha \tag{2.97}$$

$$\alpha = 1 + 2\sum_{m=1}^{M}(1 - m\Delta\psi)^{1/2} \tag{2.98}$$

Performing the second integration, the primary electron current density is
given by

$$\frac{4}{3}\left(\frac{j_e}{j_0}\right)^{1/2} = \int_0^1 d\psi\left\{\psi^{1/2} + 2\sum_{m=1}^{M}h(\psi - m\Delta\psi)(\psi - m\Delta\psi)^{1/2}\right.$$

$$\left. + \alpha\left[(1-\psi)^{1/2} - 1\right]\right\}^{-1/2} \tag{2.99}$$

The results presented thus far have been based on a particularly simple
form of the electron distribution for illustrative purposes. In physical

systems the actual electron distribution must depend on the effects of multiple scattering and energy straggling.[36] In order to account for these processes Eq. (2.94) must be generalized. Define $W_z = \frac{1}{2}m_e v_e^2 - e\phi(z)$ as the total parallel electron energy and let $j = j(W_z)$ be the total electron current density in the negative z direction at the anode surface. Further, define dj/dW_z to be the spectral amplitude of the current density distribution at the anode surface. Since the scattering and energy loss processes occur only in the anode foil, dj/dW_z does not depend on the potential distribution and can be specified independently. With these definitions the summation in Eq. (2.94) can be replaced by an integral as[37, 38]

$$n_e(z) = \left(m_e/2e^3\right)^{1/2}\left[j_e\phi^{-1/2} + 2\int_{-e\phi_0}^{0}\left(\frac{dj}{dW_z}\right)(W_z + e\phi)^{-1/2}\right.$$

$$\left.\times h(W_z + e\phi)dW_z\right] \tag{2.100}$$

and Eq. (2.96) can be written as

$$\frac{d\psi}{dx} = \frac{4}{3}\left(\frac{j_e}{j_0}\right)^{1/2}[G(\psi)]^{1/2} \tag{2.101}$$

where

$$G(\psi) = \psi^{1/2} + \alpha\left[(1-\psi)^{1/2} - 1\right] + F(\psi) \tag{2.102}$$

$$F(\psi) = j_e\left(\frac{e}{\phi_0}\right)^{1/2}\int_0^\phi d\phi'\int_{-e\phi_0}^0\left(\frac{dj}{dW_z}\right)(W_z + e\phi')^{-1/2}h(W_z + e\phi')dW_z$$

$$\tag{2.103}$$

In the calculations performed thus far it has been implicitly assumed that ϕ does not have local maxima or minima. This condition is assured by requiring that $G(\psi) > 0$ for $0 < \psi < 1$.

Employing the boundary condition that $\psi = 0$ at $x = 0$, Eq. (2.101) can be integrated to yield

$$x = \frac{3}{4}\left(\frac{j_e}{j_0}\right)^{-1/2}\int_0^\psi\left\{\psi^{1/2} + \alpha\left[(1-\psi)^{1/2} - 1\right] + F(\psi)\right\}^{-1/2}d\psi$$

$$\tag{2.104}$$

For any particular spectral distribution the value of (j_e/j_0), and j_i, can be found from Eq. (2.104) by applying the boundary condition that $\psi = 1$ at $x = 1$. To illustrate this process let dj/dW_z be a constant function of W_z with the normalization that

$$j = \int_{-e\phi_0}^{0} \left(\frac{dj}{dW_z} \right) dW_z = (\eta - 1) j_e \qquad (2.105)$$

where η is essentially the average number of electron transits through the anode. In this case

$$F(\psi) = (4/3)(\eta - 1)\psi^{3/2} \qquad (2.106)$$

$$G(\psi) = \psi^{1/2} + \alpha\left[(1 - \psi)^{1/2} - 1\right] + (4/3)(\eta - 1)\psi^{3/2} \qquad (2.107)$$

$$\alpha = 1 + F(1) = 1 + (4/3)(\eta - 1) \qquad (2.108)$$

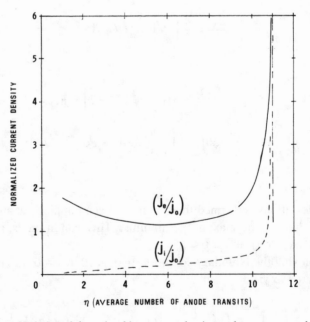

Figure 2.12. (a) Variation of the ratio of ion current density to electron current density with η for the case $dj/dW_z = $const.

The normalized electron and ion currents for this case are presented in Fig. 2.12 as a function of the parameter η. Note that η has an upper limit, η_c, defined by the criterion $G(\psi) \geqslant 0$; as $\eta \rightarrow \eta_c \approx 11$, both the electron and ion current densities increase without bound. This divergence implies that the particle flows become independent of longitudinal space charge effects.

Although the reflex triode can generate extremely high ion current densities, it has a number of practical disadvantages. Since the anode is subject to intense electron bombardment it is usually destroyed in one shot. Also, a strong longitudinal magnetic field is normally required to maintain one-dimensional flow; however, angular momentum considerations make it impossible to extract parallel ion beams into free space. A possible solution is to use virtual cathode extraction with no net current in the output side, but a virtual cathode is not a well-defined concept when the electron distribution is strongly modified by the reflexing process. Finally, the ion production efficiency of the device is given approximately by $[1 + (m_i/m_e)^{1/2}/\eta]^{-1}$. While modification of the electron distribution permits high ion current densities, it also limits the efficiency. For example, for $dj/dW_z = \text{const}$, the maximum efficiency is only 20% when $\eta \rightarrow \eta_c$.

2.4.3. Magnetically Insulated Ion Diodes

A method for increasing the production efficiency of intense ion beams in high-voltage diodes is to prevent electron flow across the diode by applying a strong magnetic field transverse to the gap. A simple one-dimensional magnetically insulated diode configuration is shown in Fig. 2.13. The anode plasma can be preformed by various techniques, e.g., flashboards.[39] While the magnetic field confines the electrons to a sheath region adjacent

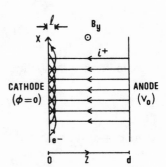

Figure 2.13. Schematic diagram of the magnetically insulated diode.

to the cathode, $0 < z < l$, the more massive ions flow relatively unimpeded. The smallest magnetic field required to insulate the gap, termed the critical field, is given in its relativistically correct form by[40]

$$B_c = \frac{mc^2}{ed}\left[\frac{2e\phi_0}{mc^2} + \left(\frac{e\phi_0}{mc^2}\right)^2\right]^{1/2} \qquad (2.109)$$

Equation (2.109) is easily derived from energy and momentum conservation arguments.

To compute the ion current density in the magnetically insulated diode of Fig. 2.13 we again make use of the Poisson equation, Eq. (2.74). Neglecting curvature of the ion trajectories in the magnetic field, energy conservation and current continuity imply that the ion charge density is again given by Eq. (2.88), i.e.,

$$n_i = (j_i/e)(m_i/2e)^{1/2}(\phi_0 - \phi)^{-1/2}$$

where j_i is the ion current density emitted from the $(Z=1)$ anode plasma.

Similarly, the electron charge density is specified from Eq. (2.75) as

$$n_e = 2j_e/ev_z \qquad (2.110)$$

where j_e is the electron current density emitted from the cathode and the factor 2 accounts for electron reflection at the edge of the sheath. To obtain v_z it is necessary to consider not only conservation of energy, but also conservation of electron canonical momentum in the transverse direction. In the nonrelativistic limit $p_x = mv_x + eB_y z/c = 0$ and $v_z = (2e\phi/m - \Omega^2 z^2)^{1/2}$, where $\Omega = eB_y/mc$. Hence, in the cathode electron sheath region Eq. (2.110) becomes[41]

$$n_e = 2(j_e/e)(2e\phi/m_e - \Omega^2 z^2)^{-1/2} \qquad (2.111)$$

The condition $v_z = 0$ defines the sheath region as $0 \le z \le l$, with

$$l = (2e\phi_l/m_e)^{1/2}\Omega^{-1} \qquad (2.112)$$

where $\phi = \phi_l$ at $z = l$.

Using the expressions for n_i and n_e, Eq. (2.74) becomes

$$\frac{d^2\phi}{dz^2} = 4\pi j_e \left(\frac{m_e}{2e}\right)^{1/2} \left\{ 2[1 - h(z-l)]\left(\phi - \frac{m_e}{2e}\Omega^2 z^2\right)^{-1/2} - \alpha(\phi_0 - \phi)^{-1/2} \right\}$$

(2.113)

where $h(l)$ again denotes the Heaviside operator. In the region not accessible to the electrons, $l < z \leqslant d$, solution of Eq. (2.113) is straightforward and yields

$$\phi(z) = \phi_0 - (9\pi j_i)^{2/3}(m_i/2e)^{1/3}(d-z)^{4/3}$$

(2.114)

In the electron sheath region the solution is somewhat more difficult to obtain. Following Ref. 41 the sheath is assumed to be sufficiently narrow that it is permissible to neglect variations in the ion velocity. Using this assumption and defining the new potential function $\psi(z) = \phi(z) - (m_e/2e)\Omega^2 z^2$, Eq. (2.113) can be reexpressed as

$$\frac{d^2\psi}{dz^2} = 4\pi j_e \left(\frac{m_e}{2e}\right)^{1/2} \left(2\psi^{-1/2} - \alpha\phi_0^{-1/2}\right) - \left(\frac{m_e}{e}\right)\Omega^2$$

(2.115)

A first integration of Eq. (2.115) yields

$$\left(\frac{d\psi}{dz}\right)^2 = 16\pi j_e \left(\frac{2m_e}{e}\right)^{1/2} \psi^{1/2} - 2\left(\frac{m_e}{e}\right)\left(\omega_p^2 + \Omega^2\right)\psi$$

(2.116)

where $\omega_p^2 = 4\pi n_i e^2/m_i = 4\pi\alpha j_e e(2em_e\phi_0)^{-1/2}$. Note that since $\psi = 0$ at $z = l$, $d\psi/dz = 0$ at $z = l$, also. Hence, $(d\phi/dz)_l = m_e\Omega^2 l/e$. Matching these boundary conditions to Eq. (2.114) yields an expression for the sheath edge given by

$$l = (3d/2)\left[1 - \left(1 - \frac{16e\phi_0}{9d^2 m_e\Omega^2}\right)^{1/2}\right]$$

(2.117)

Since the analysis leading to Eq. (2.117) is only valid when $l \ll d$, the appropriate limiting expression is

$$l \approx \tfrac{2}{3}d\xi$$

(2.118)

where ξ is the small dimensionless parameter defined by

$$\xi = \left(\frac{2e\phi_0}{m_e c^2} \right) \Big/ \left(\frac{d^2 \Omega^2}{c^2} \right) \ll 1 \qquad (2.119)$$

In this limit $\phi_l \approx (4/9)\phi_0 \xi$, and the ion current density is given by

$$j_i \approx (9\pi)^{-1} (2e/m_i)^{1/2} \phi_0^{3/2} d^{-2}$$

which is identical to Eq. (2.71). Hence, in this case the use of magnetic insulation permits much higher ion current densities by suppressing electron flow at high diode potential.

From Eq. (2.116) it is straightforward to show that the maximum value of ψ is given by (see Problem 2.14)

$$\psi_m^{1/2} = \frac{16\pi je}{(2m_e/e)^{1/2}\left(\omega_p^2 + \Omega^2 \right)} \qquad (2.120)$$

which occurs at

$$Z_m = \frac{16\pi^2 je}{(2m_e/e)\left(\Omega^2 + \omega_p^2 \right)^{3/2}} = l/2 \qquad (2.121)$$

From Eq. (2.121), the electron current density is given approximately by

$$j_e \approx (12\pi^2)^{-1} (2e/m_e)^{1/2} \xi^{-1/2} \phi_0^{3/2} d^{-2}$$

$$= (8\pi^2)^{-1} (2e/m_e)^{1/2} \phi_l^{3/2} l^{-2} \qquad (2.122)$$

Note that the electron current density scales approximately as that produced by a nonmagnetically insulated diode with voltage ϕ_l and effective gap spacing l.

The treatment given above has emphasized the analytically tractable case of a nonrelativistic diode with a strong transverse magnetic field. When these restrictions are relaxed the self-consistent field equation must be integrated numerically.[42] In the nonrelativistic limit, but for magnetic fields near the critical field, the electron sheath extends nearly to the anode, permitting an ion current density enhancement of approximately three times the Child–Langmuir value. In the electron relativistic limit for $B \lesssim B_c$, the enhancement factor increases to about six because of the greater electron mass.

An alternate approach for examining steady state ion flow in a magnetically insulated diode is the model of Ref. 43, in which the diode voltage is applied adiabatically (compared to electron orbit times). In this case electrons move toward the anode by means of the polarization drift. In the high-applied-field limit, the ion flow is again found to be Child–Langmuir limited. However, for $B \lesssim B_c$ the ion current density becomes unbounded because of the less singular electron distribution (as also found in the case of the reflex triode). In practice, however, it is usually required that $B/B_c \gtrsim 1.5$ to avoid large electron losses, and such large enhancement factors cannot be realized.

2.4.4. Time-Dependent Electron and Ion Flow in Pinched Electron Beam Diodes

Having considered several specific examples of ion flow in controlled configurations we now return to the problem of very intense electron diodes in the absence of external magnetic fields. In this case electron bombardment of the anode rapidly leads to anode plasma formation and the possibility of extracting large ion currents from such diodes.[44]

Although the parapotential flow model (Section 2.3.2) adequately predicts scaling laws of diode impedance it is unable to describe the dynamics of electron flow near the anode or the cathode, and does not include the effects of the ion flow. As a result, a newer, more physically correct theory of large-aspect-ratio diode behavior has been developed which describes the time-dependent evolution of the flow in terms of four distinct phases[45]: (1) pure electron flow at low voltages; (2) weak pinching at large voltages; (3) collapsing pinch due to time-dependent ion emission from the anode plasma; and (4) steady state pinched electron flow and laminar ion flow. These stages are illustrated in Fig. 2.14.

After cathode plasma formation, but early in the voltage pulse, the electron current initially flows orthogonally to the equipotentials, and is approximately described by the Child–Langmuir formula, Eq. (2.33), modified for gap closure due to the expanding cathode plasma

$$j_e = \frac{\sqrt{2}}{9\pi} \left(\frac{e}{m_0} \right)^{1/2} \frac{\phi_0^{3/2}}{\left(d_0 - v_c t \right)^2} \tag{2.123}$$

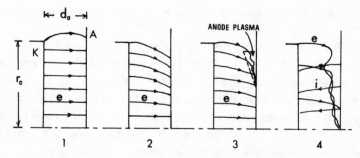

Figure 2.14. Evolution of charge flow in an intense pinched electron beam diode: (1) Child–Langmuir electron flow at low voltage; (2) electron flow at high voltage (weak pinch); (3) anode plasma induced collapsing electron ring; and (4) Quasi-steady-state electron and ion flows.

where d_0 is the initial gap separation, and v_c represents the closure velocity. As the diode voltage continues to rise, the diode current exceeds the critical current and the self-magnetic field bends the outer electron trajectories to form a weak pinch. An estimate for the diode current in the weak pinch phase is given by[46]

$$I(kA) = 8.5(r_1/d)\gamma_0 \ln\left[\gamma_0 + (\gamma_0^2 - 1)^{1/2}\right] \qquad (2.124)$$

where r_1 is estimated as $r_1 \sim r_c \gamma_0^{-1/2}$.

Anode plasma formation due to electrons striking the anode surface may occur at either low voltage (phase 1) or high voltage (phase 2) depending on the generator characteristics (voltage rise time) and the anode material. If the anode-plasma forms during phase 1, the nonrelativistic Child–Langmuir bipolar solution corrected for plasma motion is a good approximate description of the flow (Section 2.4.1); however, if anode plasma formation occurs during phase 2, the bipolar Child–Langmuir solution is inappropriate because the outer electron orbits drift radially to form a pinch; hence, a new model of pinch formation is required which accounts for reflection of electron orbits near the anode plane and provides a time-dependent description of the ion dynamics.

If the plasma formation occurs during phase 2, a dense anode plasma created at least partially from adsorbed gases on the anode surface[47] will form initially at large radii because the grazing incidence of the electron orbits permits more efficient heating of the anode material. The anode plasma provides a space-charge-limited source of ions to be accelerated across the gap. The space charge of the ion flow neutralizes the electron space charge, while the ion current adds to the electron current resulting in strongly-pinched electron flow. When such electrons enter the anode plasma at grazing incidence there is no electric force to counteract the $v_r B_\theta$ magnetic force, and they are reflected back toward the cathode while continuing to drift radially inward until a region is reached where anode plasma has not formed. This process is then repeated until eventually the condition is reached in which the electrons follow complicated orbits drifting inward with multiple reflections thereby resulting in a tight pinch (phase 3). Because of the complexity of this process, the role of numerical simulations in elucidating the important details cannot be overstated.[49]

Assuming that a steady state condition is eventually obtained which can be characterized by laminar ion flow and pinched electron flow (phase 4), it is possible to estimate some of the properties of such a steady state. Since the total charge inside the diode is nearly zero (the electric field is zero at both the cathode and anode plasmas), the ratio of the ion current to the electron current is inversely proportional to the ratio of the average diode crossing times for each species.[48] For the pinched electron flow, the electrons move radially inward and their characteristic crossing length scale is the diode radius r_c, while for the laminar ion flow the length scale remains the diode gap separation d. Hence, the ratio of the crossing time is enhanced by the factor r_c/d yielding

$$\frac{I_i}{I_e} \sim \frac{r_c}{d} \frac{\langle 1/v_e \rangle}{\langle 1/v_i \rangle} \tag{2.125}$$

A more rigorous calculation which includes an estimate of the radial dependence of the ion charge density gives[24]

$$\frac{I_i}{I_e} \gtrsim \frac{1}{2} \frac{r_c}{d} \left(\frac{\beta_i}{\beta_e} \right) \tag{2.126}$$

where $\beta_i = [2(\gamma_0 - 1)m_e/m_i]^{1/2} = 0.033(\gamma_0 - 1)^{1/2} A^{-1/2}$, and A is the atomic weight of the ion of mass m_i. Hence, depending on the aspect ratio (r_c/d), a substantial ion current, in excess of the electron current, may characterize the output diode current.

On the basis of numerical simulations[49] a good approximation for the total current carried by a high-current diode is

$$I_p = I_i + I_e \tag{2.127}$$

where I_i and I_e are the diode ion and electron currents, and I_p is the saturated parapotential current given by Eq. (2.53). Combining Eqs. (2.126) and (2.127) gives

$$I_e = I_p \left[1 + \tfrac{1}{2} (\beta_i / \beta_e) r_c / d \right]^{-1} \tag{2.128}$$

Fixing the diode voltage and maximizing $I_e(r_c/d \to \infty)$ yields

$$I_e^{\max} = 516 A^{1/2} (\gamma_0 + 1)^{1/2} \ln \left(\gamma_0 + (\gamma_0^2 + 1)^{1/2} \right) \quad \text{(kA)} \tag{2.129}$$

Consequently, the injection of high-atomic-weight plasma into the diode may be required to achieve very high diode electron currents. Alternatively, returning to Eqs. (2.126) and (2.127) the solution for I_i is

$$I_i = I_p / \left[1 + 2(d/r_c)(\beta_e / \beta_i) \right] \tag{2.130}$$

In this case there is no upper limit for the diode ion current since $I_i \sim r_c/d$.

In analogy with Section 2.4.3 the pinched electron beam diode can be considered as a self-magnetically insulated diode in which the insulating field is produced by the particle flows themselves. In this view the electron current that strikes the diode in the pinch region can be regarded as a resistive loss mechanism. In fact, the pinched electron beam diode is one example of a general class of self-magnetically insulated diodes.[50] In order for the effective insulating field to exceed the critical field, a bias current is required to flow across the diode. The ratio of ion to electron current density for these diodes generally resembles Eq. (2.126), corrected for specific geometric effects.

2.5. Summary

When a high-voltage pulse is impressed on a cold cathode field emission diode the local electric field is enhanced at the tips of microscopic whiskerlike projections on the cathode surface. The high electric field intensity causes stable field emission of electrons from the whisker tip. As a result of the intense electron flow the whisker is resistively heated and explosively vaporizes to form local plasma bursts on the cathode surface. The rapid hydrodynamic expansion and merger of the cathode flares quickly form a plasma sheath which covers the entire cathode surface. Since the effective electron emission area is dramatically increased, extremely large currents can be extracted from the zero work function cathode plasma.

After explosive electron emission from the cathode surface has been established, the subsequent diode behavior depends largely on the magnitude and duration of the applied voltage pulse and the diode geometry. In the limit of high diode impedance (which limits the current flow) the behavior of large-aspect-ratio diodes is well approximated by the space-charge-limited planar diode model in which the dominant electron motion occurs along the electrostatic lines of force which are essentially perpendicular to the surfaces of the anode and cathode. For small-aspect-ratio diodes operating in a space-charge-limited fashion, the dominant electron motion also occurs along electrostatic field lines which generally have a large radial component, especially near the cathode.

In the limit of high-current electron flow the self-magnetic field cannot be neglected. This is especially true for large-aspect-ratio diodes in which the close proximity of the anode and cathode surfaces shorts out most of the radial electric field associated with the beam space charge. In this instance the beam can self-pinch in the diode. An approximate criterion for such self-focusing is that the relativistic gyroradius of an electron emitted at the outer edge of the cathode be equal to or less than the anode–cathode gap separation. The level of current flow which corresponds to this criterion is termed the diode critical current.

When the diode current exceeds the critical current no exact analytical solution for the electron flow is known. Early theoretical models of the flow in such diodes were termed parapotential because they assumed that the dominant characteristic was self-consistent electron flow along equipotential surfaces (essentially $E \times B$ drifts). Although such a model gives good agreement with experimental measurements of diode impedance it does not consider the mechanics of how electrons get on and off equipotential surfaces. In addition, the parapotential model requires that a bias current be postulated, the effect of which is to create a magnetic field in analogy with the external magnetic field required for Brillouin flow.

For applications which require the extraction of cold electron beams into vacuum regions the foilless diode is a convenient configuration. In such diodes an external magnetic field insulates the gap and forms the electron beam. Since the beam is not extracted through an anode foil problems associated with anode vaporization and scattering are eliminated. In addition, foilless diodes naturally tend to produce annular electron beams which are advantageous from the standpoint of space-charge-limit considerations.

To obtain ion current densities substantially in excess of the 1 A/cm² level available from conventional ion diodes requires higher-voltage operation and techniques for controlling the resultant electron current flow. Two useful methods are electron reflexing and magnetic insulation. In the reflex triode the electron density is enhanced in the vicinity of the anode owing to reflexing electron trajectories, and very high ion current densities can be extracted from the anode plasma. In magnetically insulated ion diodes electron flow across the diode gap is inhibited by the application of a magnetic field transverse to the gap. The suppression of electron flow allows much higher voltage operation with a resultant increase in the Child–Langmuir ion current.

In an intense pinched electron beam diode electron bombardment heating of the anode leads to rapid anode plasma formation, and the possibility for intense ion flow. The time-dependent behavior of such diodes can be described in terms of four stages: (1) pure electron flow at low voltage; (2) weak pinching at high voltage; (3) a collapsing electron flow due to time-dependent ion emission from the anode plasma; and (4) steady state pinched electron flow and laminar ion flow. Assuming that stage (4) is attained, the total charge contained in the diode region is nearly zero, and the ratio of the ion current to the electron current is inversely proportional to the ratio of the average diode crossing time for each species. Since the characteristic electron crossing length scale is the diode radius while that of the laminar ion flow is the diode gap separation, ion currents in excess of the electron current may be possible for sufficiently large r_c/d. The pinched electron beam diode can, in fact, be described as one example of a class of self-magnetically insulated ion diodes in which the insulating field is produced by the intense particle flows within the diode.

Problems

2.1. Show that the maximum of the Schottky potential barrier occurs at

$$Z_{max} = \left(\frac{e}{4\pi\varepsilon_p E}\right)^{1/2}$$

and is given by

$$V_{max} = -\left(\frac{e^3 E}{\pi \varepsilon_p}\right)^{1/2}$$

2.2. The cathode surface of a large planar diode is a plane conducting sheet which has a small hemispherical boss of radius a on its inner surface. Far from the boss the electric field between the electrodes is E_0.

(a) Show that the potential at a point $\mathbf{r}$ between the electrodes is described by

$$V(\mathbf{r}) = E_0\left(r - \frac{a^3}{r^2}\right)\cos\theta$$

where the origin of coordinates is the origin of the boss and the angle θ is measured with respect to the axis of symmetry. (Hint: Use the method of images.)

(b) Show that the total charge on the boss has the magnitude $3E_0 a^2/4$.

(c) Calculate the electric field at the surface of the boss and show that the field at the tip is enhanced by a factor of 3.

2.3. Estimate the total diode current emitted by a 10-cm² cathode surface undergoing stable field emission from only microscopic projections. Assume that the whisker concentration is $10^4/\text{cm}^2$, with each whisker having a tip radius of 10^{-6} cm, and that each whisker emits a current density of 10^8 A/cm².

2.4. For the cathode surface described in Problem 2.3, how rapidly will a plasma sheet form if the cathode flares expand at the rate of 3 cm/μsec? Assume that the whiskers are uniformly distributed in a square array.

2.5. Show that Eq. (2.32) reduces to Eq. (2.33) in the nonrelativistic limit, $e\phi/mc^2 \ll 1$. [Hint: It may be easier to start with Eq. (2.28).]

2.6. Assuming that the cathode plasma expands at a constant velocity v_p, how does the impedance of a relativistic, one-dimensional Child–Langmuir diode vary with time?

2.7. Reexpress the diode critical current relation, Eq. (2.34), in terms of the Lawson parameter, ν/γ.

2.8. Calculate the magnetic field due to the parapotential bias current, Eq. (2.54), flowing interior to θ_m.

2.9. Graph the ratio of the saturated parapotential current, Eq. (2.53), divided by the estimated diode current in the weak pinch phase, Eq. (2.124), as a function of γ_0 over the interval (1, 20).

2.10. Obtain the formal solution to Eq. (2.64) for the case of space-charge-limited emission over the cathode surface $0 \leqslant r \leqslant r_b$, i.e., $r_1 = 0$ and $r_2 = r_b$.

(a) The Bessel functions must be expanded in a basis set that is complete and orthonormal in the interval $[0, r_b]$. Show that this set is defined by

$$u_j(r) \equiv \frac{\sqrt{2} J_0(\kappa_j r)}{r_b J_1(\kappa_j R)}$$

where $j = 1, 2, \ldots$ and $\kappa_j = k_j(R/r_b)$.

(b) Show that the coefficients of the expansion of the Bessel functions, i.e., the a_{m_j} where

$$J_0(k_m r) = \sum_{j=1}^{\infty} a_{m_j} u_j(r)$$

are given by

$$a_{m_j} = \frac{\sqrt{2}\, \kappa_j}{\kappa_j^2 - k_m^2} J_0(k_m r_b)$$

(c) Expand $K(r, r')$ as

$$K(r, r') = \sum_{i=1}^{\infty} \sum_{j=1}^{\infty} b_{i_j} u_i(r) u_j(r')$$

Show that the coefficients b_{i_j} are given by

$$b_{i_j} = \sum_{m=1}^{\infty} \frac{a_{m_i} a_{m_j}}{k_m [J_1(k_m R)]^2}$$

(d) Similarly expand $\tau(r)$ and $\rho(r)$ as

$$\tau(r) = \frac{1}{4\pi} \sum_{j=1}^{\infty} c_j u_j(r)$$

$$\rho(r) = \sum_{j=1}^{\infty} d_j u_j(r)$$

Show that

$$c_j = 4\pi \sum_{i=1}^{\infty} b_{ij} d_i$$

where the b_{ij} are given by

$$b_{ij} = \sum_{m=1}^{\infty} \frac{2\kappa_i\kappa_j[J_0(k_m r_b)]^2}{k_m(\kappa_i^2 - k_m^2)(\kappa_j^2 - k_m^2)[J_1(k_m R)]^2}$$

2.11. (a) Calculate the reflex triode electron and ion current densities when the spectral distribution of the electron flux at the anode is specified by

(1)
$$\frac{dj}{dW_Z} = -\frac{2(\eta - 1)}{(e\phi_0)^2}\left(\frac{j_e}{e}\right)W_Z$$

(2)
$$\frac{dj}{dW_Z} = \frac{2(\eta - 1)}{(e\phi_0)^2}\left(\frac{j_e}{e}\right)(W_Z + e\phi_0)$$

(b) What is the value of η_c for each case? In general, the more randomized the spectrum, the lower the value of η_c.

(c) For the very singular distribution used to obtain Eq. (2.99) what is the value of η_c?

2.12. Show that the ion production efficiency of the reflex triode scales approximately as $[1 + (m_i/m_e)^{1/2}/\eta]^{-1}$.

2.13. (a) Derive the critical magnetic field strength required for magnetic insulation, Eq. (2.109).

(b) Show that the deflection angle of an ion after crossing the magnetically insulated diode is approximately given as

$$\theta_i \approx (Bd/c)(2m_i\phi_0/e)^{-1/2}$$

2.14. Using Eq. (2.116) derive Eqs. (2.120) and (2.121), and show that the variation of $\psi(z)$ in the region $0 < z < l/2$ is given by

$$z = \frac{32\pi j_e}{(2m_e/e)(\Omega^2 + \omega_p^2)^{3/2}}\{x - 2\sin 2x\}$$

where

$$x = \sin^{-1}\left[\frac{(2m_e/e)^{1/2}(\Omega^2 + \omega_p^2)}{16\pi j_e}\right]^{1/2}\psi^{1/4}$$

2.15. The anode temperature rise during electron bombardment heating is given approximately by

$$\Delta T = \frac{j \Delta t}{\eta \mathcal{C} e} \frac{\Delta E}{\min(R, \Delta x)}$$

where j is the current density at the anode, Δt is the bombardment time, η and $\mathcal{C}$ are the mass density and specific heat of the anode material, and ΔE is the energy lost by an electron after traversing a distance equal to the anode foil thickness Δx, or the electron range R, whichever is smaller. For electron energies 1 MeV $< E < 20$ MeV, R (mg/cm^2) ≈ 530 E (MeV) $- 106$. For the case of a 3-MeV electron beam penetrating a 10-μm aluminum anode foil ($\mathcal{C} \approx 9$ J/g °K), calculate the $j \Delta t$ product required to raise the temperature of the foil to its melting point (933°K).

2.16. For a large-aspect-ratio diode undergoing steady state pinched electron flow and laminar ion flow, why is it reasonable to assume that the total charge inside the diode is nearly zero?

2.17. Qualitatively assess the effects of an externally applied longitudinal magnetic field on pinched electron beam diode behavior.
    (a) Are the azimuthal currents which result from the radial electric field near the cathode diamagnetic or paramagnetic?
    (b) Are the azimuthal currents resulting from the self-magnetic pinch force diamagnetic or paramagnetic?
    (c) Estimate the value of longitudinal field strength that would inhibit pinching in large-aspect-ratio diodes.
    (d) For $B_z \gg B_\theta$, why is the diode impedance expected to follow the Child–Langmuir law rather than the parapotential law?

3

Propagation of Intense Beams
in Vacuum

3.1. Introduction

For an intense unneutralized charged particle beam propagating in vacuum without external fields present, the radial electric force due to the beam space charge always dominates the self-magnetic pinching force. This is easily seen by making the transformation to a reference frame moving with the beam, in which case the magnetic field is transformed away. (E_r does not change sign.)

To demonstrate the required degree of space charge neutralization for the existence of a radial force equilibrium, consider a cylindrically symmetric electron beam with velocity $v_e = \beta c$ and density n_e propagating in a stationary ion background of density n_i; the fractional electron space charge neutralization is denoted by $f_e = n_i / n_e$. The radial force F_r on a test electron is computed according to

$$F_r = -e(E_r - \beta B_\theta) \tag{3.1}$$

From Gauss' law and Ampere's law E_r and B_θ can be found as

$$E_r = -\frac{4\pi e}{r} \int n_e(1 - f_e) r \, dr \tag{3.2}$$

$$B_\theta = -\frac{4\pi e}{r} \int \beta n_e r \, dr \tag{3.3}$$

Substitution of Eqs. (3.2) and (3.3) into Eq. (3.1) yields

$$F_r = \frac{4\pi e^2}{r} \int (1 - f_e - \beta^2) n_e r \, dr \tag{3.4}$$

If the beam is unneutralized, $f_e = 0$, and, since $1 - \beta^2 > 0$, the force is radially outward and the beam expands. The amount of fractional space charge neutralization required for $F_r = 0$ is given by

$$f_e = 1 - \beta^2 = \gamma^{-2} \tag{3.5}$$

Thus, for propagation of intense beams in vacuum an external means of containing the beam must be employed. For electron beams this is most easily accomplished by using a strong longitudinal magnetic field. For intense ion beams, however, the required strength of such a field is too high to be practical, and the ion space charge must be neutralized by the injection of electrons into the beam. In analogy with Eq. (3.5) the required degree of neutralization is $f_i = 1 - \beta_i^2 \approx 1$.

While the magnetic field requirements for the existence of stable electron beam equilibrium configurations are of major importance, even if the beam is placed in an infinitely strong magnetic field there still exists a limit to the amount of current that can propagate. If the magnitude of the electrostatic potential associated with the beam space charge exceeds the beam kinetic energy, then an intense beam equilibrium with total transmission of the beam current is clearly not possible; a virtual cathode forms and reflects a fraction of the injected beam current.

In this chapter the propagation of an intense beam through an evacuated drift space is examined. For electron beams particular emphasis is placed on the various equilibrium and stability requirements, and the phenomenon of virtual cathode formation. For ion beams attention is focused on the mechanisms of charge neutralization by either colinear or transverse introduction of electrons into the drift space.

3.2. General Equations for Laminar Flow Equilibria

In this section the general equations are developed which describe the self-consistent equilibria of unneutralized intense beams propagating in a metallic drift tube along a uniform magnetic field. The analysis is performed in cylindrical geometry (r, θ, z) with the z axis aligned with the beam and the external magnetic field (Fig. 3.1). The beam is assumed to have axisymmetric equilibrium radial density and velocity profiles (no θ dependence), and is assumed to be uniform in the z direction. The beam front

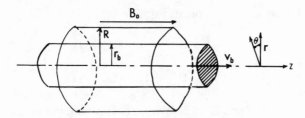

Figure 3.1. Axisymmetric equilibrium configuration for an intense charged particle beam propagating in a metallic drift tube, aligned parallel to a uniform external magnetic field. Cylindrical polar coordinates are employed with the z axis coinciding with the axis of symmetry.

physics are not included, and the beam pulse is assumed to be much shorter than the time required for diffusion of the beam self-fields through the drift tube walls. The assumption of laminar flow implies that all particles (fluid elements) move on nonintersecting helical paths. Hence, the characteristic Larmor radius, r_L, of the beam electrons must be small compared with the beam radius, i.e.,

$$r_L = \frac{mc^2}{eB_0} \gamma \Delta\theta \ll r_b \tag{3.6}$$

where $\Delta\theta$ is the angular spread of the particle velocities which can arise from, e.g., scattering in the anode foil.

In order to gain an appreciation of this assumption, consider the detailed orbits of the single electrons of a uniform electron beam in a uniform, axial magnetic field $\mathbf{B} = B_0 \hat{e}_z$. In Cartesian coordinates, from Eqs. (3.2) and (3.3), the beam self-fields are given by

$$E_x = \frac{2Ix}{\beta_z c r_b^2}, \qquad E_y = \frac{2Iy}{\beta_z c r_b^2}$$

$$B_x = -\frac{2Iy}{c r_b^2}, \qquad B_y = \frac{2Ix}{c r_b^2}$$

where r_b is the beam radius, and the current $I = -\pi r_b^2 (en_0)\beta_z c$. Hence, the equation of motion in component form becomes

$$\ddot{x} = \left(-\Omega^2/\gamma^3\right)x - \left(\Omega_0/\gamma\right)\dot{y}$$

$$\ddot{y} = -\left(\Omega^2/\gamma^3\right)y + \left(\Omega_0/\gamma\right)\dot{x}$$

where $\Omega_0 = eB_0/mc$ and $\Omega^2 = 2eI/m\beta_z cr_b^2$. Defining the parameter $s = x + iy = re^{i\theta}$, the equation of motion becomes

$$\ddot{s} = -\left(\Omega^2/\gamma^3\right)s + i\left(\Omega_0/\gamma\right)\dot{s} \tag{3.7}$$

which has the formal solution $s(t) = Ae^{i\omega_+ t} + Be^{i\omega_- t}$ where

$$\omega_\pm = \frac{\Omega_0}{2\gamma}\left[1 \pm \left(1 + \frac{4\Omega^2}{\gamma\Omega_0^2}\right)^{1/2}\right]$$

With the initial conditions $s(0) = s_0$, $\dot{s}(0) = 0$, in the limit of large Ω_0, the solution becomes

$$s(t) \approx s_0\left(e^{i\omega_- t} + \frac{\Omega^2}{\gamma\Omega_0^2}e^{i\omega_+ t}\right) \tag{3.8}$$

Hence, the single particle motion can be considered as the sum of two rotations: a large amplitude rotation at the slow $E \times B$ rotation frequency, and a small amplitude rotation at the fast cyclotron frequency. Hence, if $\Omega^2/\gamma\Omega_0^2 \ll 1$, then $r_L \ll r_b$ and the detailed fine structure of the particle orbits can be ignored. (As will be shown later there are also fast and slow rotation modes of the electron fluid equilibrium.)

The basic equations for examining the equilibrium are the fluid and Maxwell equations. For the case of an electron beam these are

$$\frac{\partial}{\partial t}n + \nabla\cdot(n\mathbf{v}) = 0 \tag{3.9}$$

$$\frac{\partial}{\partial t}\mathbf{p} + \mathbf{v}\cdot\nabla\mathbf{p} = -e\left(\mathbf{E} + \frac{1}{c}\mathbf{v}\times\mathbf{B}\right) \tag{3.10}$$

$$\nabla\times\mathbf{E} = -\frac{1}{c}\frac{\partial\mathbf{B}}{\partial t} \tag{3.11}$$

$$\nabla \times \mathbf{B} = -\frac{4\pi}{c} en\mathbf{v} + \frac{1}{c}\frac{\partial \mathbf{E}}{\partial t} \tag{3.12}$$

$$\nabla \cdot \mathbf{E} = -4\pi en \tag{3.13}$$

$$\nabla \cdot \mathbf{B} = 0 \tag{3.14}$$

In Eqs. (3.9)–(3.14) n, $\mathbf{v}$, and $\mathbf{p}$ represent the macroscopic equilibrium electron fluid density, velocity, and momentum, while $\mathbf{E}$ and $\mathbf{B}$ represent the self-consistent equilibrium electric and magnetic fields. Following the assumptions presented in the preceding paragraph and assuming the steady state ($\partial/\partial t = 0$), Eq. (3.10) yields the radial force balance given by

$$\frac{m\gamma(r)v_\theta^2(r)}{r} - e\left\{ E_r + \frac{1}{c}\left[v_\theta(B_0 + B_z^s) - v_z B_\theta \right] \right\} = 0 \tag{3.15}$$

where $\gamma = [1 - (v_\theta/c)^2 - (v_z/c)^2]^{-1/2}$. The equilibrium radial electric field is determined from the Poisson equation [Eq. (3.2) or Eq. (3.13)], and the equilibrium magnetic field is expressed as

$$\mathbf{B}(r) = (B_0 + B_z^s)\hat{e}_z + B_\theta \hat{e}_\theta \tag{3.16}$$

where B_0 is the externally applied uniform field. The azimuthal self-field B_θ is given by Eq. (3.3), while the axial diamagnetic self-field B_z^s is given by

$$B_z^s(r) = -\frac{4\pi e}{c}\int_r^R dr'\, n(r')v_\theta(r') + B_c \tag{3.17}$$

The constant B_c represents the uniform field due to the azimuthal image current in the drift tube wall; it is easily computed from flux conservation.

The first term in Eq. (3.15) is the outward centrifugal force, and the second term is the outward force of the electric field due to the beam space charge. The third and fourth terms represent the constraining forces of the magnetic fields. The third term also reflects the diamagnetic character of the beam, i.e., the reduction of the applied magnetic field strength interior to the beam.

Introducing the dimensionless variables $\beta_\theta = v_\theta/c$ and $\beta_z = v_z/c$ and using Eqs. (3.2), and (3.3) and (3.17) to eliminate E_r, B_z^s, and B_θ, Eq. (3.15) becomes

$$\frac{m\gamma(r)\left[\beta_\theta(r)c\right]^2}{r}$$

$$+\frac{4\pi e^2}{r}\int_0^r dr' r' n(r') - \beta_\theta(r)\left[e(B_0+B_c)-4\pi e^2\int_r^R dr' \beta_\theta(r')n(r')\right]$$

$$-\frac{4\pi\beta_z(r)e^2}{r}\int_0^r dr' r'\beta_z(r')n(r')=0 \qquad (3.18)$$

It is apparent from Eq. (3.18) that there is a certain arbitrariness in the macroscopic fluid treatment of the equilibrium; any two of the three variables $n(r)$, $\beta_\theta(r)$, and $\beta_z(r)$ can be specified independently.[1] In order to completely specify the equilibrium, it is necessary to choose two additional, physically realizable, constraints. A rather obvious first choice is conservation of energy: particle kinetic energy plus potential energy must be constant across the beam and equal to the applied diode potential according to

$$(\gamma-1)mc^2+e\phi(r)=-e\phi_c \qquad (3.19)$$

where ϕ_c is the (negative) cathode potential.

Following Reiser,[2] a second constraint is provided by the conservation law for canonical angular momentum given by

$$P_\theta(r)=\gamma(r)mrv_\theta(r)-\frac{er}{c}A_\theta(r)=-\frac{er_c}{c}A_{\theta s}(r_c) \qquad (3.20)$$

where $A_{\theta s}$ is given by the axial magnetic field B_z^c at the cathode according to

$$A_{\theta s}(r)=\frac{1}{r}\int B_z^c(r)r\,dr \qquad (3.21)$$

Under the assumption of laminar flow, Eq. (3.20) implies that all particles at a given radius r were emitted from the source at the same radius r_c, and, hence, have the same canonical angular momentum.

Equations (3.9)–(3.14), supplemented by Eqs. (3.19) and (3.20), form a complete set and uniquely specify the equilibrium in a self-consistent way, provided the source geometry is given.

There are three source conditions of particular practical importance.[2] The first is the nonimmersed source for which the cathode is shielded from the axial magnetic field in the downstream region. In this case $P_\theta = B_z^c = 0$, and Eq. (3.20) becomes simply

$$\gamma(r)mrv_\theta(r) = \frac{e}{c}\int_0^r B_z(r)r\,dr \qquad (3.22)$$

A second important case is that of an immersed source whose emitting area coincides with a flux surface; i.e., an annular cathode that produces a thin hollow beam. In this instance all particles have the same canonical angular momentum and Eq. (3.20) becomes

$$\gamma(r)mrv_\theta(r) - \frac{e}{c}\int_0^r B_z(r)r\,dr = -\frac{1}{2}\frac{e}{c}B_z^c(r_c)r_c^2 = P_\theta \qquad (3.23)$$

The final important case is that of an immersed source for which the axial magnetic field is uniform. In this instance P_θ varies across the beam. If it is also assumed that the beam expands or contracts in self-similar fashion, then the radii of particles in the downstream region must be related to their emission radii according to $r = \alpha r_c$, where α is a constant which depends on the actual system geometry. In this case $P_\theta(r) = -(eB_z^0/2c)(r^2/\alpha^2)$, and Eq. (3.20) becomes

$$\gamma(r)mrv_\theta(r) - \frac{e}{c}\int_0^r B_z(r)r\,dr = -\frac{eB_z^0}{2c}\left(\frac{r^2}{\alpha^2}\right) \qquad (3.24)$$

The primary distinction between the equilibria resulting from either shielded or immersed sources is provided by a comparison of Eq. (3.22) with Eq. (3.23) or Eq. (3.24). For the shielded source, as the electrons leave the cathode, v_θ is initially zero but rapidly increases as the electrons encounter the increasing external field (analogous to a magnetic mirror). The beam rotation generates an axial diamagnetic self-field, and equilibria for shielded sources are sometimes termed diamagnetic, or rotating beam, equilibria.

For immersed sources, on the other hand, beam rotation is generated as a result of the E_r and B_z field components, and v_θ is very small compared with v_z. Hence, a good first approximation is to ignore diamagnetic field contributions and set $B_z = B_0$ in Eq. (3.24). Solving for v_θ then yields

$$v_\theta = \frac{eB_0 r}{2\gamma mc}(1 - \alpha^{-2})$$

(3.25)

As $B_0 \to \infty$, the particles are trapped on field lines ($\alpha = r/r_s \to 1$), and it can be shown that the rotation rate drops to zero.

Before continuing with the development of laminar flow equilibria for finite external magnetic fields (Section 3.5), it is helpful to examine first the important concepts of space-charge-limiting current and virtual cathode formation. This analysis is most easily performed under the assumption of an infinite externally applied magnetic field.

3.3. Space-Charge-Limiting Current

The amount of beam current which can be propagated through an evacuated drift cavity is effectively limited by the electrostatic potential depression associated with the (unneutralized) beam space charge. To demonstrate this fact we examine beam propagation in a cylindrically symmetric drift space along an infinitely large, externally imposed longitudinal magnetic field. In this case the beam motion is one-dimensional ($v_r = v_\theta = 0$). Hence, in the steady state the set of fluid Maxwell equations reduces simply to the solution of Poisson's equation, Eq. (3.13), and the equation of continuity

$$\frac{\partial}{\partial z}(nv_z) = 0$$

(3.26)

subject to the conservation of energy constraint, Eq. (3.19). Combining Eqs. (3.13), (3.19), and (3.26), the Poisson equation reduces to the form

$$\frac{1}{r}\frac{\partial}{\partial r}\left(r\frac{\partial\phi}{\partial r}\right) + \frac{\partial^2\phi}{\partial z^2} = -\frac{4\pi j}{c}\left[1 - \left(\gamma_0 + \frac{e\,\phi}{mc^2}\right)^{-2}\right]^{-1/2}$$

(3.27)

where $\mathbf{E} = -\nabla\phi$, $(\gamma_0 - 1)mc^2 = |e\phi_c|$, and $j = -env_z$. The solution of Eq. (3.27) subject to the appropriate boundary conditions describes the electrostatic space charge potential (and, hence, the electron kinetic energy) as a function of position in the cavity.

There are a few important special cases for which analytical solutions to Eq. (3.27) are tractable. If the radius of the cavity becomes very much greater than the cavity length, then the geometry is essentially that of a gap between two infinite parallel planes perpendicular to the directed motion of the electron beam. This problem represents a generalization of the infinite planar diode. At the other extreme, when the length of the cavity is very much greater than the cavity radius, the geometry is that of a beam propagating in an infinitely long pipe.

3.3.1. Thin Annular Beam in an Infinitely Long Drift Space

As a first example consider an annular beam, whose thickness is much smaller than both the beam radius r_b and the distance between the beam and the chamber wall, propagating through a cylindrically symmetric drift space of radius R. In this case, the electrostatic potential is essentially constant across the thin beam and the problem reduces to the solution of the homogeneous equation

$$\frac{1}{r}\frac{\partial}{\partial r}\left(r\frac{\partial \phi}{\partial r}\right) = 0 \tag{3.28}$$

subject to the boundary conditions

$$\phi(R) = 0 \tag{3.29}$$

$$\left.\frac{\partial \phi}{\partial r}\right|_{r \downarrow r_b} = \frac{2I}{r_b v_b} \tag{3.30}$$

where I is the total beam current, and v_b is the longitudinal electron velocity (uniform across the thin beam). The solution is[3]

$$\phi(r) = -2\frac{I}{v_b}\begin{cases} \ln(R/r), & r_b < r < R \\ \ln(R/r_b), & r \le r_b \end{cases} \tag{3.31}$$

where v_b is given by

$$v_b = c\left[1 - \left(\gamma_0 + \frac{e\phi_b}{mc^2}\right)^{-2}\right]^{1/2} \tag{3.32}$$

Combining Eqs. (3.31) and (3.32) yields the following result for $\phi_b = \phi(r = r_b)$:

$$F(\phi_b) = \frac{-e\phi_b}{mc^2}\left[1 - \left(\gamma_0 + \frac{e\phi_b}{mc^2}\right)^{-2}\right]^{1/2} = \frac{2eI}{mc^3}\ln\left(\frac{R}{r_b}\right) \qquad (3.33)$$

A plot of $F(\phi_b)$ is shown in Fig. 3.2. It has a maximum at $|e\phi_b| = (\gamma_0 - \gamma_0^{1/3})mc^2$ given by

$$F_{\max}(\phi_b) = \left(\gamma_0^{2/3} - 1\right)^{3/2} \qquad (3.34)$$

Since maximizing F also maximizes the current allowed to propagate, the space-charge-limiting current for this system is given by

$$I_l = \frac{mc^3}{2e}\frac{\left(\gamma_0^{2/3} - 1\right)^{3/2}}{\ln(R/r_b)} \qquad (3.35)$$

For currents $I < I_l$, there are two possible solutions for ϕ_b according to Eq. (3.33): (1) $0 < |e\phi_b|/mc^2 < \gamma_0 - \gamma_0^{1/3}$, corresponding to particles of high velocity and low density; and (2) $\gamma_0 - \gamma_0^{1/3} < |e\phi_b|/mc^2 < \gamma_0 - 1$ corresponding to particles of lower velocity and higher density. For currents

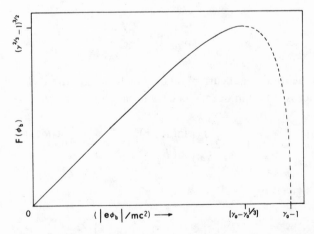

Figure 3.2. The function $F(\phi_b)$ for the case of a thin hollow beam illustrating the two possible solutions for beam currents $I < I_l$.

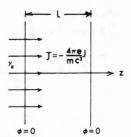

Figure 3.3. Schematic geometry of a beam injected into an infinite one-dimensional drift space.

$I > I_l$, there is no solution to Eq. (3.33), and the existence of a steady state flow of current in excess of the space charge limit is not possible.

3.3.2. Infinite One-Dimensional Drift Space

For a drift space of length L between two infinite parallel planes perpendicular to the direction of motion of a homogeneous beam which completely fills the space, Fig. 3.3., Eq. (3.27) becomes

$$\frac{d^2\gamma}{dz^2} = -\frac{J\gamma}{\left(\gamma^2 - 1\right)^{1/2}} \qquad (3.36)$$

where $J = -4\pi ej/(mc^3) = $ const.

When the boundary conditions have been specified, Eq. (3.36) describes the electron beam kinetic energy as a function of position between the parallel planes for a given current density.

The solution for Eq. (3.36), found using a procedure similar to that employed in Section (2.3.1), may be written as[4]

$$J^{1/2}|z - z_{min}| = f(\gamma, \gamma_{min}) \qquad (3.37)$$

where the constant γ_{min} is the minimum value of the electron beam kinetic energy which occurs at the position $z = z_{min}$ in the drift space $0 \leqslant z \leqslant L$. The function $f(\gamma, \gamma_{min})$ is defined by

$$f(\gamma, \gamma_{min}) = (2\gamma_{min})^{1/2}\left[\frac{\sin\phi}{1+\cos\phi}(1 - k\sin^2\phi)^{1/2}\right.$$

$$\left. + (1 - k^2)F(k, \phi) - E(k, \phi)\right] \qquad (3.38)$$

where $F(k, \phi)$ and $E(k, \phi)$ are the incomplete elliptic integrals of the first and second kind, and

$$k^2 = \frac{1}{2}\left[1 - \frac{(\gamma_{\min} - 1)^{1/2}}{\gamma_{\min}}\right] \tag{3.39}$$

$$\phi = \arccos\left\{\frac{\gamma_{\min} - \left[(\gamma^2 - 1)^{1/2} - (\gamma_{\min}^2 - 1)^{1/2}\right]}{\gamma_{\min} + \left[(\gamma^2 - 1)^{1/2} - (\gamma_{\min}^2 - 1)^{1/2}\right]}\right\} \tag{3.40}$$

An analysis of Eq. (3.37) indicates that a single solution exists for the current density range $0 \leqslant J \leqslant J_{l1}$ (see Fig. 3.4). For $J > J_{l2}$ there are no solutions, and the stationary state with a virtual cathode and partial reflection of the current is the only possibility. In the interval $J_{l1} \leqslant J \leqslant J_{l2}$, there are two different solutions for each value of J corresponding to the normal high-velocity, low-density flow and the low-velocity, high-density flow which is the relativistic analog of the "C overlap" regime.[5]

3.3.3. Iterative Procedure for the Limiting Current in a Long Drift Space

Consider next a solid electron beam of radius r_b propagating through an infinitely long cylindrically symmetric drift cavity of radius R. In this

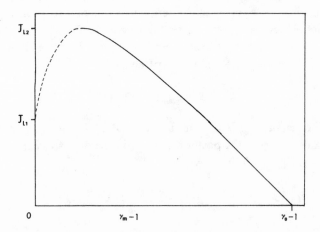

Figure 3.4. Electron beam current versus electron kinetic energy in the center of a planar gap. For $J_{l1} < J < J_{l2}$ there are two possible solutions.

case Eq. (3.27) may be expressed as

$$\frac{1}{r}\frac{d}{dr}\left(r\frac{d\gamma}{dr}\right) = \begin{cases} -\dfrac{J\gamma}{\left(\gamma^2-1\right)^{1/2}}, & 0 \le r \le r_b \\ 0, & r_b < r \le R \end{cases} \tag{3.41}$$

where now $J = [4eI/(m_0c^3r_b^2)] = $ const [see also Eq. (2.26)].

An equivalent integral equation representation of Eq. (3.41) is

$$J^{-1}\left[\gamma(r) - \gamma_0\right] = \left[\ln(r/R)\right]\int_0^r \frac{\gamma}{\left(\gamma^2-1\right)^{1/2}} r\,dr$$

$$+ \int_r^{r_b} \frac{\gamma}{\left(\gamma^2-1\right)^{1/2}} r\left[\ln(r/R)\right]\,dr \tag{3.42}$$

Approximations for the space-charge-limiting current can be obtained from Eq. (3.42) through an iteration process.[3,6,7] As a first approximation it is assumed that γ on the right-hand side of Eq. (3.42) is equal to the (as yet unknown) value on the drift space axis, i.e., $\gamma_a = \gamma(0)$.

Evaluation of Eq. (3.42) in this case yields the first iterative solution

$$\gamma(r) = \gamma_0 + \frac{(\gamma_0 - \gamma_a)}{1 + 2\ln(R/r_b)}\left\{(r/r_b)^2 - \left[1 + 2\ln(R/r_b)\right]\right\} \tag{3.43}$$

where

$$Jr_b^2\left[1 + 2\ln(R/r_b)\right] = (\gamma_0 - \gamma_a)\left(\gamma_a^2 - 1\right)^{1/2}\gamma_a^{-1} \tag{3.44}$$

Substitution of Eq. (3.43) into the right-hand side of Eq. (3.42) would yield a second, more accurate iterative solution, and the procedure could be repeated until the desired accuracy was achieved.

The iterative process produces a sequence of equations [of which Eq. (3.44) is the simplest] which relate the beam current (through the parameter J) to the value γ_a. Such equations represent better and better approximations for the space-charge-limiting current. For example, when $r_b/R \ll 1$, the variation in γ across the beam is small and Eqs. (3.43) and (3.44) resulting from the first iteration are expected to be a good approximation to the exact solution. In this case, the maximum value of J from Eq. (3.44) occurs when

$\gamma_a = \gamma_0^{1/3}$; hence, the limiting current for $r_b/R \ll 1$ is given approximately by

$$I_l \simeq \frac{\left(\gamma_0^{2/3} - 1\right)^{3/2} mc^3/e}{2\ln(R/r_b)} \tag{3.45}$$

which is the same result obtained in Section (3.3.1) for the case of an infinitely thin annular beam.

3.3.4. Upper Bound for the Limiting Current in Arbitrary Geometry

An upper bound for the limiting current of a drift space of arbitrary geometry can be found by considering Green's second identity

$$\int_V \left(\psi \nabla^2 \phi - \phi \nabla^2 \psi\right) dV = \int_S \left(\psi \frac{\partial \phi}{\partial n} - \phi \frac{\partial \psi}{\partial n}\right) dS \tag{3.46}$$

where S represents the bounding surface of the volume V and $\partial/\partial n$ is the derivative normal to the boundary. In Eq. (3.46) ϕ represents the solution of the nonlinear differential equation, Eq. (3.27)

$$\nabla^2 \phi = \frac{1}{r}\frac{\partial}{\partial r}\left(r\frac{\partial \phi}{\partial r}\right) + \frac{\partial^2 \phi}{\partial z^2} =$$

$$\begin{cases} -\dfrac{4\pi j}{c}\left[1 - \left(\gamma_0 + \dfrac{e\phi}{mc^2}\right)^{-2}\right]^{-1/2} = F(r, z, \phi) & \text{(in } V) \\ 0 & \text{(otherwise)} \end{cases} \tag{3.47}$$

subject to the appropriate boundary conditions. The function ψ is the solution to the associated eigenvalue equation

$$\nabla^2 \psi = \frac{1}{r}\frac{\partial}{\partial r}\left(r\frac{\partial \psi}{\partial r}\right) + \frac{\partial^2 \psi}{\partial z^2} = \Lambda\psi \tag{3.48}$$

By an appropriate choice of boundary conditions on ψ the right-hand side of Eq. (3.46) can be made to vanish yielding

$$\int \psi\left[F(\mathbf{r}, \phi) - \Lambda\phi\right] dV = 0 \tag{3.49}$$

The eigenvalue equation, Eq. (3.48), has an eigenfunction ψ_0, corresponding to the (first) eigenvalue $\Lambda_0 \neq 0$ which does not vanish in the volume V. Thus, if the quantity $[F(r, z, \phi) - \Lambda_0 \phi]$ does not change sign in V then a nontrivial function ϕ that solves Eq. (3.47) does not exist. Expressed in terms of $\gamma = \gamma_0 + e\phi/mc^2$, the criterion for the existence of a solution is

$$j = \frac{mc^3}{4\pi e} \Lambda_0 \frac{(\gamma_0 - \gamma)(\gamma^2 - 1)^{1/2}}{\gamma} \tag{3.50}$$

The upper bound estimate of the space-charge-limiting current density is then obtained as[4]

$$j_l = \frac{mc^3}{4\pi e} \Lambda_0 (\gamma_0^{2/3} - 1)^{3/2} \tag{3.51}$$

As an illustration of this formalism consider the geometry of Fig. 3.5 for the case in which the beam radius r_b is equal to the guide tube radius R. (This is a special case of the problem considered in Ref. 8.) For this example ϕ and ψ satisfy the same boundary conditions, and the eigenvalue Λ_0 is given by

$$\Lambda_0 = -\left[(\lambda_0/R)^2 + (\pi/L)^2\right] \tag{3.52}$$

where λ_0 is the first zero of J_0, the zeroth-order Bessel function of the first kind. In this instance Eq. (3.51) for the limiting current becomes

$$|I_l| = \frac{mc^3}{4e} \left[\lambda_0^2 + \left(\frac{\pi R}{L}\right)^2\right](\gamma_0^{2/3} - 1)^{3/2} \tag{3.53}$$

Eq. (3.53) is graphed as a function of (R/L) in Fig. 3.6 (see also Ref. 9). For comparison with an alternate solution method, see Problem 3.4.

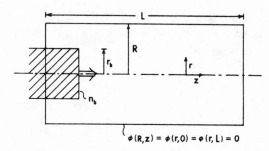

Figure 3.5. Schematic geometry of a beam injected into a cylindrically symmetric cavity of finite length.

$\phi(R,z) = \phi(r,0) = \phi(r,L) = 0$

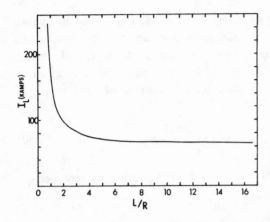

Figure 3.6. The limiting current as given by Eq. (3.53) for a beam completely filling a cylindrical drift tube of finite length ($\gamma_0 \approx 3$).

3.3.5. Wave Spectrum for a Nonrelativistic Beam in a One-Dimensional Drift Space

The dotted portions of the curves in Figs. 3.2 and 3.4 are the relativistic analogs of the "C-overlap" flow regime from Ref. 5. In this section we compute the wave frequency spectrum for the case of one-dimensional flow and show that this low-velocity, high-density flow is unstable.[10-12] (See also Problem 3.13.)

For nonrelativistic one-dimensional flow in the infinite guide field limit, Eqs. (3.9), (3.10), and (3.13) reduce to

$$\frac{\partial n}{\partial t} + \frac{\partial}{\partial z}(nv) = 0 \tag{3.54}$$

$$\frac{\partial v}{\partial t} + v\frac{\partial v}{\partial z} = \frac{e}{m}\frac{\partial \phi}{\partial z} \tag{3.55}$$

$$\frac{\partial^2 \phi}{\partial z^2} = 4\pi en \tag{3.56}$$

Assuming a linearization of the form $\psi(z, t) = \psi^0(z) + \delta\psi(z)e^{-i\omega t}$ for the quantities n, v, and ϕ, the Eqs. (3.54)–(3.56) yield the steady state problem characterized by

$$|en^0 v^0| = j = \text{const} \tag{3.57}$$

$$\tfrac{1}{2}mv^{0^2} - e\phi^0 = \tfrac{1}{2}mv_0^2 = \text{const} \tag{3.58}$$

$$\frac{\partial^2 \phi^0}{\partial z^2} = 4\pi en^0 \tag{3.59}$$

and the equations for the perturbed quantities, written as

$$-i\omega\delta n(z) + \frac{d}{dz}\left[v^0(z)\delta n(z)\right] + \frac{d}{dz}\left[n^0(z)\delta v(z)\right] = 0 \qquad (3.60)$$

$$-i\omega\delta v(z) + \frac{d}{dz}\left[v^0(z)\delta v(z)\right] = \frac{e}{m}\frac{d}{dz}\delta\phi(z) \quad (3.61)$$

$$\frac{d^2\delta\phi(z)}{dz^2} = 4\pi e\delta n(z) \quad (3.62)$$

The solution to the steady state problem is given by

$$\left(v^0 - v_m\right)\left(v^0 + 2v_m\right)^2 = \alpha(z - z_m)^2 \qquad (3.63)$$

where $\alpha = 18\pi ej/m$. The parameter v_m denotes the minimum velocity in the drift space which occurs at $z_m = L/2$. Setting $v^0 = v_0$ at $z = 0$ yields

$$(v_0 - v_m)(v_0 + 2v_m)^2 = \alpha L^2/4 \qquad (3.64)$$

The limiting current in this case is determined from $d\alpha/dv_m = 0$, which yields

$$\alpha_l = 8v_0^3/L^2 \qquad (3.65)$$

at $v_m = v_0/2$.

In general Eq. (3.64) has three possible solutions for the minimum electron velocity, one of which is negative and is discarded as unphysical. Of the two positive solutions, one corresponds to the low-density, high-velocity $(v_0/2 < v_m < v_0)$ regime, while the other corresponds to the low-velocity $(0 < v_m < v_0/2)$ regime.

At this point it is useful to define the Lagrangian time variable τ such that

$$\frac{dz}{d\tau} = v^0(\tau) \qquad (3.66)$$

Recognizing that $dv^0/d\tau = v^0 dv^0/dz$, differentiation of Eq. (3.63) yields

$$\frac{dv^0}{d\tau} = \pm\frac{2}{3}\alpha^{1/2}\left(v^0 - v_m\right)^{1/2} \qquad (3.67)$$

If t_0 denotes the electron transit time, then the appropriate sign convention for Eq. (3.67) is to use the negative sign for $0 < \tau < t_0/2$ and the positive sign for $t_0/2 < \tau < t_0$. A first integration of Eq. (3.67) yields

$$\left(v^0 - v_m\right)^{1/2} = \tfrac{1}{3}\alpha^{1/2}(t_0/2 - \tau) \tag{3.68}$$

since $v^0 = v_m$ at $\tau = t_0/2$. Recalling that $v^0 = dz/d\tau$, integration of Eq. (3.68) yields

$$z = v_m\tau - (\alpha/27)\left[(t_0/2 - \tau)^3 - (t_0/2)^3\right] \tag{3.69}$$

Setting $z = L/2$ in Eq. (3.69) yields the expression for the transit time given by

$$L = v_0 t_0 - k t_0^3/24 \tag{3.70}$$

where $k = \tfrac{4}{9}\alpha = 8\pi ej/m$. From Eq. (3.70) the transit time which occurs at the limiting current density is computed to be $t_l^3 = 12L/k_l$.

Returning now to the transient problem, Eqs. (3.60)–(3.62) can be combined to yield

$$\left[v^0(z)\right]^3 \frac{d^2y}{dz^2} + \frac{k}{2}y = C\left[v^0(z)\right]^2 \exp\left[-i\omega \int dz/v^0(z)\right] \tag{3.71}$$

where

$$y = v^0(z)\delta v(z)\exp\left[-i\omega \int dz/v^0(z)\right]$$

and C is a constant of integration. The solution of Eq. (3.71) is facilitated by transforming to the Lagrangian variable τ according to

$$v^0(\tau)\frac{d^2y}{d\tau^2} - \frac{dv^0}{d\tau}\frac{dy}{d\tau} + \frac{k}{2}y = C\left[v^0(\tau)\right]^2 e^{-i\omega\tau} \tag{3.72}$$

Writing the general solution of Eq. (3.72) as $y = y_h + y_p$, the solution of the homogeneous equation can be easily determined by the power series expansion method. This procedure yields

$$y_h = a_0 + a_1\tau - \frac{k}{4v_0}\left(a_0 + \frac{a_1 t_0}{2}\right)\tau^2 \tag{3.73}$$

The particular solution of Eq. (3.72) can be written as

$$y_p = \frac{C}{\omega^2}\left[-\left(v_0 + \frac{kt_0 i}{4\omega}\right) + \frac{k}{4}\left(t_0 + \frac{2i}{\omega}\right)\tau - \frac{k}{4}\tau^2\right]e^{-i\omega\tau} \qquad (3.74)$$

Combining Eqs. (3.73) and (3.74) the general solution for Eq. (3.72) becomes

$$y = a_0 + a_1\tau - \frac{k}{4v_0}\left(a_0 + \frac{a_1 t_0}{2}\right)\tau^2$$

$$+ \frac{C}{\omega^2}\left[-\left(v_0 + \frac{kt_0 i}{4\omega}\right) + \frac{k}{4}\left(t_0 + \frac{2i}{\omega}\right)\tau - \frac{k}{4}\tau^2\right]e^{-i\omega\tau} \qquad (3.75)$$

With Eq. (3.75) an expression for the perturbed velocity is easily obtained. The perturbed potential can be determined by an integration of Eq. (3.61) according to

$$\delta\phi = \frac{m}{e}\int_0^\tau \frac{dy}{d\tau}e^{i\omega\tau}d\tau + (\delta\phi)_0 \qquad (3.76)$$

and an expression for the perturbed density is calculated directly from Eq. (3.62)

$$\delta n = \frac{1}{4\pi e v^{0^2}}\left(\frac{d^2\delta\phi}{d\tau^2} - \frac{1}{v^0}\frac{dv^0}{d\tau}\frac{d\delta\phi}{d\tau}\right) \qquad (3.77)$$

The frequency spectrum of the oscillations can be determined by solving for the constants a_0, a_1, and $(\delta\phi)_0$ in terms of the appropriate boundary conditions:

$$\delta v(\tau)|_{\tau=0} = y(\tau)|_{\tau=0} = 0, \qquad \delta n(\tau)|_{\tau=0} = 0,$$

$$\delta\phi(\tau)|_{\tau=0} = \delta\phi(\tau)|_{\tau=t_0} = 0 \qquad (3.78)$$

The first and secondary boundary conditions yield

$$a_0 = C\frac{v_0}{\omega^2} + \frac{kt_0 i}{4\omega^3} \qquad (3.79)$$

$$a_1 = -\frac{Cki}{2\omega^3} \qquad (3.80)$$

while the third boundary condition implies that $(\delta\phi)_0 = 0$.

When Eqs. (3.79) and (3.80) are substituted into the final boundary conditions, the equation which results is

$$\frac{iL}{\omega} + \frac{k}{2}\left[\frac{it_0}{\omega^3}(1 + e^{i\omega t_0}) + \frac{2}{\omega^4}(1 - e^{i\omega t_0})\right] = 0 \qquad (3.81)$$

Equation (3.81) determines the frequency spectrum of the space charge waves which can appear in the drift space; it can be used to examine the stability of the various parameter regimes, with unstable solutions corresponding to $\text{Im}(\omega) > 0$.

Defining the variable $\beta = i\omega t_0$, Eq. (3.81) may be recast in the form

$$(2 - \beta)e^\beta = (2 + \beta) - \frac{2L\beta^3}{kt_0^3} \qquad (3.82)$$

with instability now corresponding to $R(\beta) < 0$. Note that from Eq. (3.82) if $k = k_l = 12L/t_l^3$, then $\beta = 0$.

Substituting $\beta = B + ib$ into Eq. (3.82) and separating real and imaginary parts yield

$$\frac{2L}{kt_0^3} = \frac{-e^B[(2 - B)\cos b + b\sin b] + B + 2}{B(B^2 - 3b^2)} \qquad (3.83)$$

$$\frac{2L}{kt_0^3} = \frac{-e^B[(2 - B)\sin b - b\cos b] + b}{b(3B^2 - b^2)} \qquad (3.84)$$

Equations (3.83) and (3.84) determine the functions B and b which are required to analyze the system stability. Rather than solve the general problem (as is done in Ref. 12) we shall restrict our analysis to the small-β limit, i.e., near the limiting current density. In this case the left-hand side of Eq. (3.82) can be expanded as

$$(2 - \beta)e^\beta \approx 2 + \beta - \frac{\beta^3}{6} - \frac{\beta^4}{6} + \cdots$$

and Eq. (3.82) becomes

$$\beta \approx \frac{12L}{kt_0^3} - 1 \qquad (3.85)$$

If $12L/kt_0^3 < 1$, Eq. (3.82) has a real negative root indicating instability.

We now reexamine Eq. (3.70), which is analogous to Eq. (3.64), but written in terms of the transit time t_0. Of the three roots of Eq. (3.70) there are two positive roots and one negative root with the sum of the roots being equal to zero. The smaller positive root represents the "normal" high-velocity flow, while the larger root describes the low-velocity "C-overlap" flow. Denoting the roots as t_n, t_{co}, and $-(t_n + t_{co})$, with $t_{co} > t_n > 0$, the product of the roots is

$$t_{co} t_n (t_{co} + t_n) = -\frac{24L}{k} \tag{3.86}$$

Hence, the normal and C-overlap transit times are characterized by

$$t_{co}^3 > \frac{12L}{k_l} \tag{3.87}$$

$$t_n^3 < \frac{12L}{k_l} \tag{3.88}$$

Equation (3.87) is recognized as just the condition for instability derived above; thus, the high-density, low-velocity flow regime is unstable to the growth of space charge waves for injected current densities near the space charge limit.

3.4. Virtual Cathode Formation

If the electron current injected into a drift cavity exceeds the space charge limit, total transmission of the injected current is not possible: a virtual cathode forms which reflects electrons back into the injector region. While it is possible to obtain a mathematical solution to the static problem of virtual cathode formation for simple geometries, such solutions are not physically realizable as demonstrated by numerous experiments and computer simulations. In the following sections the classical static solution is first presented for comparison purposes. The linear perturbation analysis of Section 3.3.5 is then used to predict unstable oscillations. As an introduction to the time-dependent phenomena associated with the self-electrostatic field effects, the motion of a single charge sheet in a one-dimensional drift cavity is studied in detail. Finally, the results of some numerical particle simulation code solutions are presented.

3.4.1. Classical Static Theory

For the one-dimensional geometry of Section 3.3.2, if the injected current density J exceeds the space-charge-limiting current density J_{l2}, there are no solutions for Eq. (3.37), and the stationary state with a virtual cathode ($\gamma_{min} = 1$) and partial reflection of the current is the only possibility. The virtual cathode may be considered to divide the gap into two regions. In the region beyond the virtual cathode ($z > z_{min}$), a certain transmitted current $J_2 \leqslant J$ flows and the potential distribution is described by

$$\left[J_2(z - z_0)^2 \right]^{1/2} = f(\gamma, 1) \tag{3.89}$$

In the region in front of the virtual cathode ($z < z_{min}$) two counterstreaming beams of particles, J and $J - J_2$, exist. Since the charge density is determined by the sum of the densities of these beams, the potential distribution is described by

$$\left[(2J - J_2)(z - z_0)^2 \right]^{1/2} = f(\gamma, 1) \tag{3.90}$$

For equal potentials at the boundaries, i.e.,

$$\gamma = \gamma_0, \qquad z = 0, L$$

the current J_2 and the position of the virtual cathode are determined from the following relations[4]:

$$(2J - J_2)^{-1/2} + J_2^{-1/2} = 2J_{l1}^{-1/2} \tag{3.91}$$

$$z_{min} = L \left[1 - \tfrac{1}{2}(J_{l1}/J_2)^{1/2} \right] \tag{3.92}$$

where J_{l1} is given by

$$J_{l1} = 4L^{-2}f^2(\gamma_0, 1) \tag{3.93}$$

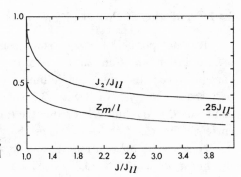

Figure 3.7. Transmitted current and location of the potential minimum according to the classical theory of one-dimensional space-charge-limited flow.

Solutions for Eq. (3.91) exist for injected currents in the range $J_{l1} \leqslant J \leqslant \infty$. For $J = J_{l1}$, all the charge is transmitted and the virtual cathode forms in the center of the gap [from Eq. (3.92)]. As the injected current increases, the transmitted current approaches $0.25J_{l1}$ and the location of the virtual cathode approaches the injection plane as shown in Fig. 3.7.

From Section 3.3.2 the range of currents $J_{l1} \leqslant J \leqslant \infty$ (with virtual cathode formation) overlaps the range of currents $0 \leqslant J \leqslant J_{l2}$ (with *no* virtual cathode). Hence, the classical quasistatic behavior for such a diode is expected to follow the description of Fig. 3.8, which presents the transmitted current as a function of the injected current. As J is increased from 0 to J_{l2} there is complete transmission of the current; however, for $J > J_{l2}$, J_2 drops to a fraction of J, and for very large J, $J_2 \approx \frac{1}{4}J_{l1}$. As the injected current is decreased from very large values, the transmitted current increases until $J = J_2 = J_{l1}$. However, the solution obtained from Eq. (3.91) differs from that obtained from Eq. (3.37) because a virtual cathode exists in one case but not in the other.

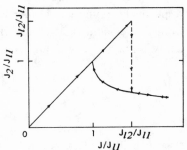

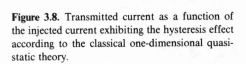

Figure 3.8. Transmitted current as a function of the injected current exhibiting the hysteresis effect according to the classical one-dimensional quasistatic theory.

3.4.2. Prediction of Unstable Flow

It is not possible from the steady state analysis of Sections 3.3.2 and 3.4.1 to predict the transient behavior of the current flow when the injected current exceeds the space charge limit. On the other hand, the small signal perturbation analysis of Section 3.3.5, although not rigorously applicable, can be used to qualitatively predict the transient behavior when the injected current is increased above the space charge limit.[15]

Near the limiting current density, $k_l = 8\pi e j_l / m$, β is small so that only the linear term of Eq. (3.85) need be retained. Defining the parameter $\xi = k/k_l$, the ratio of the injected current density to the limiting current density, Eq. (3.85) can be rewritten as

$$\beta = \frac{12L}{k_l}\left(\xi t_0^3\right)^{-1} - 1 \tag{3.94}$$

We have previously seen that under normal flow conditions if the injected current density is less than the limiting current ($\xi < 1$), then $\beta > 0$ corresponding to stability. When $\xi > 1$, the perturbation analysis is not rigorously valid and the transit time t_0 cannot be obtained. Nevertheless, if ξ is suddenly increased above unity, for a short time interval the transit time t_0 should not drastically change, becoming if anything, larger than $t_l = (12L/k_l)^{1/3}$. Under this assumption $\beta < 0$, and unstable flow is expected.

3.4.3. Single Charge Sheet Model*

As an introduction to time-dependent virtual cathode behavior consider the simple case of an infinitely thin sheet of charge density $(-\rho)$ that is positioned between two infinite, parallel, grounded conducting walls (Fig. 3.9). The sum of the charges induced in the walls is equal to the total charge in the sheet. Resolving this charge into two parts $-\rho_1$ and $-\rho_2$ which, respectively, represent the portions of the charge on which lines of force from the left and right boundaries terminate, from Gauss' law the electric fields in the two regions are given by

$$E_1 = 4\pi\rho_1 \tag{3.95}$$

$$E_2 = -4\pi\rho_2 \tag{3.96}$$

*Use of this model is described more fully in Ref. 16.

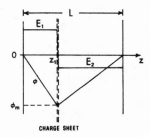

Figure 3.9. Schematic geometry of an infinitely thin sheet of charge positioned between two infinite conducting grounded parallel walls.

Since the fields are uniform, the electrostatic potential decreases linearly to a potential minimum ϕ_{min} at the sheet. Solving for ϕ_{min} yields

$$\phi_{min} = -4\pi\rho\,\frac{z_1(L-z_1)}{L} \tag{3.97}$$

and the electric fields, Eqs. (3.95) and (3.96), can be expressed as

$$E_1 = 4\pi\rho\left(\frac{L-z_1}{L}\right) \tag{3.98}$$

$$E_2 = -4\pi\rho\left(\frac{z_1}{L}\right) \tag{3.99}$$

It is easily shown that the net electric field acting to move the sheet is the average field $E_a = \tfrac{1}{2}(E_1 + E_2)$ or

$$E_a = 4\pi\rho\left(\tfrac{1}{2} - z_1/L\right) \tag{3.100}$$

If the mass per unit area of the sheet is designated as M, then the (nonrelativistic) equation of motion of the sheet is

$$\ddot{z}_1 = \frac{4\pi\rho^2}{ML}\left(z_1 - \frac{L}{2}\right) \tag{3.101}$$

With use of the initial conditions, $z_1 = 0$, $\dot{z}_1 = v_0$ at $t = 0$ the solution of Eq. (3.101) is

$$z_1 = (L/2)[1 - \cosh\omega t] + (v_0/\omega)\sinh\omega t \tag{3.102}$$

where ω is given by

$$\omega^2 = 4\pi\rho^2/ML \tag{3.103}$$

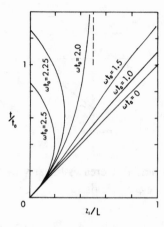

Figure 3.10. Typical charge sheet trajectories for various values of the parameter ω. Note trapping in the middle of the gap for $\omega t_0 = 2$, and reflection of the charge sheet for $\omega t_0 > 2$.

The characteristic behavior of Eq. (3.102) is shown in Fig. 3.10 for various values of the parameter ω. As $\omega \to 0$, Eq. (3.102) reduces to $z_1 = v_0 t$ and the sheet has a transit time given by

$$t_0 = L/v_0 \tag{3.104}$$

As the charge density increases the sheet can slow considerably in the center of the gap, but it still exits with velocity v_0 at $z_1 = L$. This behavior persists until the charge density reaches a limiting value determined by

$$\omega = 2/t_0 \tag{3.105}$$

In this case, Eq. (3.102) becomes

$$z_1 = (L/2)\left[1 - \exp(-2t/t_0)\right] \tag{3.106}$$

As $t \to \infty$, $z_1 \to (L/2)$ and the charge sheet is trapped in the center of the gap. For still larger charge densities the sheet is reflected back to the injection plane by its own self-field. For $\omega \gg v_0/L$, the position of zero velocity, $\dot{z}_1 = 0$, occurs at $z_r \approx 2v_0^2/\omega^2 L$ at the time $t_r = 2v_0/\omega^2 L$.

While the single charge sheet model would appear to be a drastic oversimplification, it serves as a good introduction to the multiparticle numerical simulation codes which have been used to study the time-dependent virtual cathode behavior. Further, as will be seen later it demonstrates many correct features of the simulations, including the existence of a limiting charge density (or current), and oscillation of the negative sheet charge with its multiple positive images when the limiting density is exceeded.

3.4.4. Time-Dependent Virtual Cathode Behavior

When the injected current exceeds the space-charge-limiting current there is no simple analytical framework that provides an adequate description of the complex time-dependent particle bunching and reflection that occur. As a result these phenomena have been extensively studied with numerical particle simulation codes. When the injected current is less than the space charge limit the simulation results agree well with the predictions of the perturbation theory. When the injected current exceeds the space charge limit, however, the most obvious feature of the simulations is that a steady state is never reached in which all particle trajectories are identical. A typical trajectory map (Fig. 3.11) for the case of a one-dimensional drift space illustrates this behavior.[15-19] As the injected current is increased above the limiting current, the magnitude of the electrostatic potential increases and the particles slow in the center of the space until some charges are reflected back to the injection plane (I). This reflection creates a charge bunch that further traps and reflects charge injected at later times (II). Meanwhile, charge that was injected prior to the first reflection continues to cross the gap (III). As charge exits the system in both directions the magnitude of the electrostatic potential decreases and charges injected at still later times are no longer trapped (IV). As these charges propagate across the gap, the magnitude of the potential again increases and the oscillation is repeated.[15] It is now clear why the single charge sheet gives such good results: the charge bunch formed as a result of the reflections is very similar to a single sheet of charge.

Such behavior is further illustrated in Fig. 3.12 which presents the amplitude of the potential minimum as a function of time. The oscillations are not sinusoidal; the rising slope approximately three times as steep as the falling slope is reminiscent of a relaxation oscillation. The position of the potential minimum also oscillates about the steady state value given from the analysis of Section 3.4.1. The frequency of the oscillations does not scale linearly with the beam plasma frequency, but tends to increase as the injected beam current increases above the space charge limit.

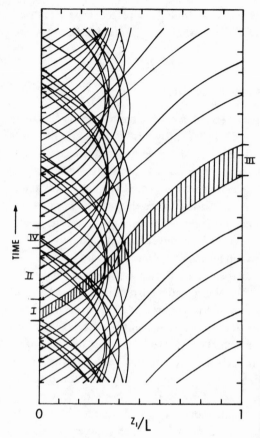

Figure 3.11. Typical particle (sheet) trajectories when the injected current exceeds the space charge limit. Note the time-dependent oscillatory behavior.

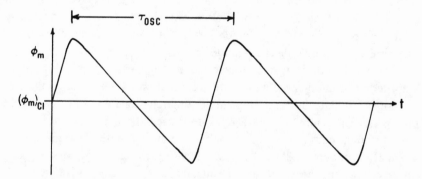

Figure 3.12. Typical oscillatory behavior of the amplitude of the potential minimum about the classical value as a function of time (one-dimensional drift space).

Once the oscillatory mode has been established the nonoscillatory mode may be recovered by reducing the injected current. As the charge in the system decreases the oscillation period lengthens until the steady state potential minimum solution is reestablished. The simulations indicate that recovery actually occurs somewhat sooner than predicted by the classical theory.[15]

While it is tempting to relate the existence of such an oscillatory mode to the appearance of an objectionable singularity in the charge density predicted by the stationary theory, such an interpretation must be rejected. Since actual beams are not monoenergetic, the stationary analysis has been extended to account for a small velocity spread. With this assumption the singularity in charge density is removed, although the solutions for current and potential are only slightly altered. On the other hand, when a velocity distribution is assumed for the numerical simulations, the oscillatory behavior, although modified somewhat, persists.[15]

Although the discussion in this section has largely been restricted to one-dimensional electrostatic flow, it should be mentioned that recent generation simulation codes are fully relativistic and fully electromagnetic, and permit two space and three velocity variables.[20-22] These newer codes have been applied to a wide variety of problems of intense beam physics. These less approximate calculations generally confirm the qualitative features of the one-dimensional models; e.g., for injected currents in excess of the space charge limit, the virtual cathode forms increasingly closer to the injection plane as I/I_l increases. Also, the oscillation frequency tends to increase as I/I_l increases. Hence, while the details of virtual cathode behavior are expected to vary on going from one- to two- and three-dimensional problems, the qualitative features of the one-dimensional models are expected to remain valid.

3.5. Laminar Flow Equilibria of Unneutralized Relativistic Electron Beams

After having considered the concepts of space-charge-limiting current and virtual cathode formation in some detail, we now return to the problem of laminar flow equilibria in finite external magnetic fields. Before turning to the general problem including the constraints of source geometry, we first consider two examples of equilibria analytically derivable from only the radial force equation, Eq. (3.18). Although not strictly correct, such equilibria are useful in practice as the basis for various stability calculations,[1] and as the starting point for iterative procedures which obtain more accurate equilibrium solutions.[23]

3.5.1. Relativistic Rigid Rotor Equilibrium[1]

If it is assumed that the azimuthal beam motion is very small compared to the axial motion, $v_\theta^2 \ll v_z^2$, then the axial diamagnetic field contribution can be neglected. In this case Eq. (3.18) can be written as

$$\gamma(r)\omega_\theta^2(r) + \frac{1}{r^2}\int_0^r dr' r' \omega_p^2(r') - \omega_\theta \Omega - \frac{\beta_z(r)}{r^2} \int_0^r dr'(r') \beta_z(r') \omega_p^2(r') = 0$$

(3.107)

where $\Omega = eB_0/mc$, $\omega_p^2 = 4\pi ne^2/m$, and $\omega_\theta(r) = \beta_\theta(r)c/r$.

If it is further assumed that the axial velocity profile of the electron beam is independent of radius r, i.e.,

$$\beta_z(r)c \equiv \beta_0 c = \text{const} \tag{3.108}$$

then

$$\gamma(r) \simeq \left[1 - \beta_z^2(r)\right]^{-1/2} \equiv \gamma_0 = \text{const} \tag{3.109}$$

With these simplifying assumptions, Eq. (3.107) reduces to

$$\gamma_0 \omega_\theta^2(r) + \frac{1}{r^2}\left(1 - \beta_0^2\right)\int_0^r dr' r' \omega_p^2(r') - \omega_\theta(r)\Omega = 0 \tag{3.110}$$

Solving Eq. (3.110) for the angular velocity $\omega_\theta(r)$ yields

$$\omega_\theta(r) = \omega_\theta^\pm(r) = \frac{\Omega}{2\gamma_0}\left\{1 \pm \left[1 - \frac{4}{\gamma_0 \Omega^2 r^2}\int_0^r dr' r' \omega_p^2(r')\right]^{1/2}\right\} \tag{3.111}$$

For a constant density profile

$$\omega_p^2(r) = \begin{cases} \dfrac{4\pi n_0 e^2}{m} = \omega_{p0}^2, & 0 < r < r_b \\ 0, & r > r_b \end{cases} \tag{3.112}$$

Eq. (3.111) reduces to

$$\omega_\theta(r) = \omega_\theta^\pm = \frac{\Omega}{2\gamma_0}\left[1 \pm \left(1 - \frac{2\omega_{p0}^2}{\gamma_0\Omega^2}\right)^{1/2}\right] \tag{3.113}$$

The condition for the radical to be real (i.e., the existence of an equilibrium) is

$$2\omega_{p0}^2 < \gamma_0\Omega^2 \tag{3.114}$$

The beam rotates as a rigid body with two allowable velocities of precession given approximately by

$$\omega_\theta^+ \simeq \frac{2\gamma_0\Omega^2 - \omega_{p0}^2}{2\gamma_0^2\Omega} \tag{3.115}$$

$$\omega_\theta^- \simeq \frac{\omega_{p0}^2}{2\gamma_0^2\Omega} \tag{3.116}$$

For the slow rotation mode the rotation velocity is $\omega_\theta^- \approx cE(r)/r\gamma^2 B_0$ which corresponds to a beam precession in θ at the $\mathbf{E} \times \mathbf{B}$ drift velocity (with relativistic corrections),[24] while the fast mode rotation corresponds to the cyclotron motion about the axis with $\omega_\theta^+ \simeq \Omega/\gamma_0$.

3.5.2. Relativistic Hollow Beam Equilibrium

As a second application of Eq. (3.107) consider the case of a hollow beam with density profile given by

$$\omega_p^2(r) = \begin{cases} 0, & 0 < r < r_0 \\ \dfrac{4\pi n_0 e^2}{m} = \omega_{p0}^2, & r_0 < r < r_b \\ 0, & r > r_b \end{cases} \tag{3.117}$$

Substituting Eq. (3.117) into Eq. (3.110) yields the solution for the angular velocity $\omega_\theta(r)$ given by

$$\omega_\theta(r) = \omega_\theta^\pm(r) = \frac{\Omega}{2\gamma_0}\left\{1 \pm \left[1 - \frac{2\omega_{p0}^2}{\gamma_0\Omega^2}\left(1 - \frac{r_0^2}{r^2}\right)\right]^{1/2}\right\}, \qquad r_0 < r < r_b$$

$$(3.118)$$

Since $\partial\omega_\theta/\partial r \neq 0$, the hollow beam no longer rotates as a rigid body, but has angular velocity shear. This difference in rotation rate across the beam provides a source of free energy that can drive an instability (see Section 3.7.2).

3.5.3. General Laminar Flow Beam Equilibria

For the equilibria of Sections 3.5.1 and 3.5.2 we were able to arbitrarily choose any two of the three quantities $n(r)$, $v_\theta(r)$, and $v_z(r)$. However, from conservation of energy, Eq. (3.19), the beam density is related to the beam kinetic energy through the Poisson equation. In addition, if the source geometry is known, then the variation of canonical angular momentum, $P_\theta(r)$, is specified through Eq. (3.20). These two additional constraints, together with the radial force equation, completely specify the allowed equilibrium states. In this and the following sections we follow Reiser and analyze the laminar flow equilibria corresponding to the three source geometries developed in Section 3.2. Alternate treatments have initially specified, e.g., $v_\theta(r)$, $v_z(r)$, or the current density[25-27]; in the case of the latter treatment, exact numerical solutions for the equilibrium have been obtained.[23] With these choices, however, it is necessary to determine $P_\theta(r)$, and then appropriately design the source region.

Beginning with the radial force balance, Eq. (3.15), we first eliminate E_r and B_z by using Eqs. (3.19) and (3.20) to obtain

$$\frac{mc}{e}r(\gamma v_z)' + \frac{c}{e}\frac{v_\theta}{v_z}v_\theta' + rB_\theta = 0 \qquad (3.119)$$

where the prime notation denotes differentiation with respect to the radial coordinate. After differentiating Eq. (3.119) with respect to r and employing the axial component of Eq. (3.12), some algebraic manipulation finally yields the first desired result:

$$\left[r(\gamma v_z)'\right]' + \frac{1}{m}\left(\frac{v_\theta}{v_z}P_\theta'\right)' - v_z(r\gamma')' = 0 \qquad (3.120)$$

To obtain a second independent relation involving the velocity components we differentiate Eq. (3.20) twice, and employ the azimuthal component of Eq. (3.12) to obtain

$$\gamma\left(\frac{v_\theta}{r}\right)' + (\gamma v_\theta')' + \gamma'v_\theta' - \frac{1}{m}\left(\frac{P_\theta'}{r}\right)' = 0 \qquad (3.121)$$

Before examining the detailed solutions to Eqs. (3.120) and (3.121) we first note that for the particular source geometries of Section 3.2 the equations can be considerably simplified since either $P_\theta = $ const or $P_\theta \propto r^2$. In the following sections these two cases are discussed in detail.

3.5.3.1. $P_\theta = $ const. For both the shielded source and the thin annular cathode it was shown in Section 3.2 that $P_\theta = $ const ($= 0$ for the shielded source). In this instance Eqs. (3.120) and (3.121) become simply

$$\left[r(\gamma v_z)'\right]' - v_z(r\gamma')' = 0 \qquad (3.122)$$

and

$$\gamma\left(\frac{v_\theta}{r}\right)' + (\gamma v_\theta')' + \gamma'v_\theta' = 0 \qquad (3.123)$$

When Eq. (3.122) is multiplied by γ, the result is that $(v_z'\gamma^2 r)' = 0$ or

$$v_z'\gamma^2 r = \kappa = \text{const} \qquad (3.124)$$

Elimination of γ from Eq. (3.123) via Eq. (3.124) yields

$$\frac{v_\theta}{r^2} + \frac{v_z''}{v_z'}v_\theta' - v_\theta'' = 0 \qquad (3.125)$$

Further, the relation $v_\theta = [c^2(1 - v_z'r/\kappa) - v_z^2]^{1/2}$ can be used to eliminate all v_θ terms from Eq. (3.125). The resulting complicated expression contains only v_z and its derivatives; it can be solved numerically to yield the allowed equilibrium states.

In order to proceed analytically we note that for the case of a shielded source and a solid beam the constant κ from Eq. (3.124) must equal zero, since $v_z'\gamma^2$ cannot become infinite at the origin. Hence, the solution must be

$$v_z = v_0 = \text{const} \tag{3.126}$$

where v_0 depends on the applied diode potential and the beam current.

With $v_z = v_0$, v_θ can be eliminated from Eq. (3.123) to yield

$$\gamma'' + \frac{\gamma'}{r} - \frac{\gamma(\gamma^2 - \gamma_0^2)}{r^2\gamma_0^2} - \frac{\gamma\gamma'^2}{\gamma^2 - \gamma_0^2} = 0 \tag{3.127}$$

The constant $\gamma_0 = (1 - v_0^2/c^2)^{-1/2}$ is the beam kinetic energy at the inner radius of the beam where $v_\theta = 0$ from Eq. (3.20). With the substitutions $y = \gamma/\gamma_0$ and $w = ry'$, Eq. (3.127) can be rewritten as

$$w\frac{dw}{dy} - y(y^2 - 1) - \frac{w^2 y}{y^2 - 1} = 0 \tag{3.128}$$

The solution of Eq. (3.128) is given by

$$w = ry' = \left[(y^2 + C)(y^2 - 1)\right]^{1/2} \tag{3.129}$$

where C is a constant of integration. A final integration of Eq. (3.129) yields

$$\frac{r_0}{r} = \exp\left[\int_0^\alpha \frac{d\alpha}{(1 - k^2\sin^2\alpha)^{1/2}}\right] = \exp F(\alpha, k) \tag{3.130}$$

where $F(\alpha, k)$ is the elliptic integral of the first kind, r_0 is a second constant of integration, and $k^2 = -C$ and $\sin\alpha = 1/y$.

Equation (3.130) is the general solution of the beam equilibrium for the case in which the canonical angular momentum and the axial velocity do not vary over the radial profile of the beam. As an application of Eq. (3.130) we examine in detail the case of a solid beam produced from a shielded source. At the origin y' must be finite so $C = -\gamma_m^2/\gamma_0^2$. However, since $v_\theta = 0$ at $r = 0$, then $\gamma_m = \gamma_0$ and $C = -1$, and Eq. (3.130) becomes

$$\gamma = \gamma_0(r_0^2 + r^2)/(r_0^2 - r^2) \tag{3.131}$$

The constant r_0 is determined from the beam drift tube geometry. If the beam completely fills the drift tube, then $\gamma = \gamma_i$ at $r = R$, where $(\gamma_i - 1)mc^2 = |e\phi_c|$. In this case

$$r_0 = R[(\gamma_i + \gamma_0)/(\gamma_i - \gamma_0)]^{1/2} \tag{3.132}$$

From Eq. (3.131) the rotation rate is easily found as

$$\omega(r) = \frac{2c}{\gamma_0} \frac{r_0}{r_0^2 + r^2} \tag{3.133}$$

while the beam density is calculated from Poisson's equation as

$$n(r) = \frac{2m_0 c^2}{\pi e^2} \gamma_0 r_0^2 \frac{r_0^2 + r^2}{\left(r_0^2 - r^2\right)^3} \tag{3.134}$$

As shown in Problem 3.10, $n(r)$ can be nearly constant for nonrelativistic beams, or markedly hollow for relativistic beams in the high-current limit. Integrating over the current density, $j_z = -env_0$, yields the total beam current given by

$$|I| = 2\beta_0 \gamma_0 \left(\frac{mc^3}{e}\right) \frac{R^2 r_0^2}{\left(r_0^2 - R^2\right)^2} = \frac{1}{2} \frac{mc^3}{e} \left(\gamma_0^2 - 1\right)^{1/2} \left(\frac{\gamma_i^2}{\gamma_0^2} - 1\right) \tag{3.135}$$

The value of γ_0 (and v_0) thus depends on the total beam current and the cathode potential.

The azimuthal self-magnetic field, computed from Eqs. (3.5) and (3.134), is given by

$$B_\theta(r) = -\frac{4mc^2 \beta_0 \gamma_0}{e} \left[\frac{rr_0^2}{\left(r_0^2 - r^2\right)^2}\right] \tag{3.136}$$

while the axial diamagnetic self-field, computed from Eqs. (3.17), (3.133), and (3.134) is given as

$$B_z^s(r) = -\frac{4mc^2 r_0^3}{e} \left[\frac{1}{\left(r_0^2 - R^2\right)^2} - \frac{1}{\left(r_0^2 - r^2\right)^2}\right] + B_c \tag{3.137}$$

From flux conservation the constant field B_c is easily computed as

$$B_c = \frac{4mc^2}{e} \frac{r_0 R^2}{\left(r_0^2 - R^2\right)^2}$$ (3.138)

and Eq. (3.137) for the axial self-field becomes

$$B_z^s(r) = -\frac{4mc^2 v_0^3}{e} \left[\frac{1}{r_0^2\left(r_0^2 - R^2\right)} - \frac{1}{\left(r_0^2 - r^2\right)^2} \right]$$ (3.139)

The required strength of the applied magnetic field B_0 is determined from Eq. (3.20) evaluated at the beam edge $r = R$. With $B_z(r) = B_z^s(r) + B_0$

$$P_\theta = \gamma_i mR v_\theta(r) - \frac{e}{c} \int_0^R B_z(r) r \, dr = 0$$

or

$$B_0 = \frac{4mc^2}{e} \frac{r_0}{r_0^2 - R^2}$$ (3.140)

The graph of beam current versus γ_0 is shown in Fig. 3.13. As in Section 3.3 the current is zero for $\gamma_0 = \gamma_i$ and $\gamma_0 = 1$, and has a maximum for

$$\gamma_{0m}^2 = \tfrac{1}{2}\gamma_i^2\left[\left(1 + 8/\gamma_i^2\right)^{1/2} - 1\right]$$ (3.141)

Substitution of Eq. (3.141) into Eq. (3.135) yields the space charge current

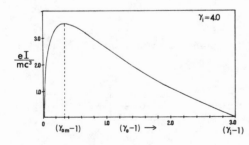

Figure 3.13. Equilibrium beam current for the case of a solid beam produced from a shielded source as a function of kinetic energy on the beam axis.

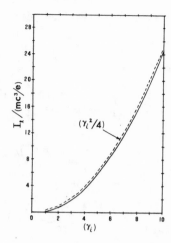

Figure 3.14. Variation of the space-charge-limiting current in the case of a solid beam produced from a shielded source as a function of the injected beam kinetic energy.

limit given by

$$|I_l| = \frac{mc^3}{2e} \left\{ \frac{\gamma_i^2}{2} \left[(1 + 8/\gamma_i^2)^{1/2} - 1 \right] - 1 \right\}^{1/2} \left[\frac{3 - (1 + 8/\gamma_i^2)^{1/2}}{(1 + 8/\gamma_i^2)^{1/2} - 1} \right]$$

(3.142)

$|I_l|$ is plotted versus γ_i in Fig. 3.14.

In the nonrelativistic limit ($\gamma_i \to 1$) Eq. (3.142) becomes

$$I_{lnr} \approx \left(\frac{2}{3} \right)^{3/2} \frac{mc^3}{e} (\gamma_i - 1)^{3/2}$$

(3.143)

while in the ultrarelativistic limit ($\gamma_i \gg 1$), the limit is given by

$$I_{lur} \approx \frac{mc^3}{4e} \gamma_i^2$$

(3.144)

Unlike the space charge limits calculated previously in Section 3.3, whose ultrarelativistic limit varied linearly with γ_i, Eq. (3.144) indicates that the limiting current for a solid beam produced by a shielded source varies quadratically with the applied diode potential when the beam completely fills the drift tube.[2]

When the beam does not completely fill the drift tube the situation is somewhat different. In this case the expression for the total beam current becomes (see Problem 3.11)

$$I = \left(\frac{mc^3}{2e} \right) \left(\gamma_0^2 - 1 \right)^{1/2} \left[\left(\gamma_{r_b} / \gamma_0 \right)^2 - 1 \right] \qquad (3.145)$$

where γ_{r_b}, the value of γ at the edge of the beam, is related to the injected kinetic energy according to

$$\gamma_i / \gamma_0 = \gamma_{r_b} / \gamma_0 + \left[\left(\gamma_{r_b} / \gamma_0 \right)^2 - 1 \right] \ln(R / r_b) \qquad (3.146)$$

Setting $R / r_b = e^n$ and eliminating γ_{r_b} from Eq. (3.145) yields

$$I = \left(mc^3 / e \right) \left[\left(\gamma_0^2 - 1 \right)^{1/2} / 4n^2 \right] \left(1 + 2\gamma_i / \gamma_0 - y \right) \qquad (3.147)$$

where $y = [1 + 4n(\gamma_i / \gamma_0 + n)]^{1/2}$. The limiting current for this case can be determined by setting the partial derivative $(\partial I / \partial \gamma_0)|_{R, r_b}$ equal to zero, which yields

$$y \left(1 + 2n\gamma_i / \gamma_0^3 \right) = 1 + 2n\gamma_i \left(\gamma_0^2 + 1 \right) / \gamma_0^3 + 4n^2 \qquad (3.148)$$

In the limit of small n (beam close to the drift tube wall), Eq. (3.148) yields the first-order correction to Eq. (3.141). For large n, however, the leading terms of Eq. (3.148) are[28]

$$2n \left(\gamma_i / \gamma_0^3 - 1 \right) + \left(1 + \gamma_i^2 / \gamma_0^4 - \gamma_i / \gamma_0 - \gamma_i / \gamma_0^3 \right) = 0 \qquad (3.149)$$

Hence, in this limit $\gamma_{0m} \approx \gamma_i^{1/3}$, and the space-charge-limiting current becomes

$$I_l \approx \frac{\left(\gamma_i^{2/3} - 1 \right)^{3/2} \left(mc^3 / e \right)}{1 + 2\ln(R / r_b)} \qquad (3.150)$$

which again scales linearly with γ_i in the ultrarelativistic limit.

3.5.3.2. Immersed Source with $P_\theta \propto r^2$. When the canonical angular momentum varies quadratically with radius, Eqs. (3.120) and (3.121) become

$$[r(\gamma v_z)']' + \frac{1}{m} \left(\frac{v_\theta}{v_z} \right)' P_\theta' - v_z [v\gamma']' = 0 \qquad (3.151)$$

$$\gamma(v_\theta / r)' + (\gamma v_\theta')' + \gamma' v_\theta' = 0 \qquad (3.152)$$

In general, it is not possible to obtain a single equation from Eqs. (3.151) and (3.152) that determines the equilibrium state, although it is easily shown that neither the rigid rotor nor the constant axial velocity case is a permissible solution.

Since beam rotation is generally small for immersed sources a reasonable first approximation is to neglect beam diamagnetism, which yields Eq. (3.25) for v_θ. Substituting this result into the nondiamagnetic form of Eq. (3.15) yields

$$-\frac{\Omega^2 r^2}{4\gamma v_z}(1-\alpha^{-4})+\frac{r}{v_z}\gamma'+\frac{e}{mc^3}rB_\theta=0 \qquad (3.153)$$

where $\Omega = eB/mc$. Differentiating Eq. (3.153) with respect to r and using Eqs. (3.12) and (3.13) gives

$$(\Omega/2)^2(1-\alpha^{-4})\left(\frac{r^2}{\gamma v_z}\right)'-\left(\frac{r\gamma'}{v_z}\right)'+\frac{v_z}{c^2}(r\gamma')'=0 \qquad (3.154)$$

Since $v_z^2 = c^2(1-\gamma^{-2}\{1-[\Omega r(1-\alpha^{-1})/2]^2\})$, Eq. (3.154) is a complicated expression for γ which can be solved numerically. Of course, in the limit that $B_0 \to \infty$, the rotation rate vanishes, and the current density must become uniform [Eq. (3.26)]. In this limit Eq. (3.154) reduces simply to Eq. (3.41).

It is of interest to compare the results of the limiting current calculations for both the immersed and shielded sources. For immersed sources, analytic expressions for the limiting current are only available in the very large magnetic field limit. Such expressions [e.g., Eqs. (3.35) and (3.53)] exhibit a linear dependence on the applied diode voltage in the relativistic limit. In contrast, the diode voltage dependence of the limiting current for a solid beam produced by a shielded source varies from linear (for the thin pencil beam limit) to quadratic (for the beam completely filling the drift tube). Although it is tempting to conclude that high current transport should be easier if the beam is produced by a shielded source, this result must be viewed with caution. The shielded source equilibrium configuration when the beam fills the drift tube is characterized by a weak external magnetic field and considerable beam rotation. This situation is difficult to realize in practice because slight misalignments and axial nonuniformities can cause a considerable loss of beam current. Increasing the magnetic field strength increases the beam separation from the drift tube wall, but the limiting current variation with the diode voltage is no longer quadratic. For more discussion of this point see Ref. 28.

3.6 Electron-Neutralized Transport of Intense Ion Beams in Vacuum

Although much of the previous analysis for intense electron beams can also, in principle, be applied to ion beam transport, there are obvious practical differences due primarily to the much larger ion rest mass. Consider, for example, propagation of an unneutralized annular proton beam in an evacuated drift tube assuming an infinitely strong axial magnetic field. For proton kinetic energies less than 100 MeV, the nonrelativistic form of Eq. (3.35) must be used to compute the space charge limit, i.e.,

$$I_l \approx \left(\frac{m_i c^3}{2e}\right)\left(\frac{2e\phi}{3m_i e}\right)^{3/2}\left[\ln(R/r_b)\right]^{-1} \qquad (3.155)$$

For proton currents in excess of I_l, a virtual anode would form and reflect protons back toward the injector.

In order to achieve a laminar flow ion beam equilibrium with an external field, additional field strength must be supplied because the self-magnetic pinch force provides significant cancellation of the repulsive space charge force only at relativistic energies (see Problem 3.1). A quick estimate of the necessary field strength can be obtained from Eq. (3.114), the radial force balance equilibrium condition. This criterion indicates that

$$B_i/B_e \propto \left(\gamma_e m_i/m_e\right)^{1/2} \qquad (3.156)$$

For example, while a 10-kA, 2-MeV electron beam can be controlled with a 10-kG field, a proton beam of the same parameters would require a field strength of ~ 1 MG. Because of such large strengths the most practical means for transporting large ion currents is neutralization of the beam space charge by low-energy electrons.[31]

3.6.1. Collinear Electron Neutralization

As a first example consider the case of active longitudinal injection of electrons into the ion beam. It is assumed that the ion beam has uniform

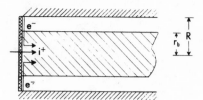

Figure 3.15. Ion beam space charge neutralization by collinear injection of electrons.

density and is of infinite transverse dimension (Fig. 3.15). It is further assumed that the electron current is emitted with uniform velocity v_{e0} in the direction of the ion beam, and is space charge limited. The latter condition implies that the electric field at the position of electron injection vanishes. Ignoring variations in the ion density, n_{i0}, Poisson's equation assumes the form

$$\frac{d^2\phi}{dz^2} = -4\pi e \left(n_{i0} - \frac{j_e}{ev_e} \right) \qquad (3.157)$$

where $j_e = ev_e n_e = \text{const.}$

A trivial solution, corresponding to ideal neutralization is realized when the electrons are injected with density and velocity equal to that of the ions. In practice, however, the injection velocity of the electrons will not be exactly equal to the ion velocity due to thermal effects. Further, such an active scheme must occupy discrete locations, thereby intercepting a portion of the ion beam. It is perhaps more practical to consider a distributed source of cold electrons which are drawn into the ion beam by the positive ion space charge.[29] Such a source might be represented by a plasma cloud, or a fine wire mesh with secondary electron production. In this case it is relatively easy to shown that there is no simple sheath solution that will permit an equal number of electrons to be drawn into the ion beam.[30] This situation can be resolved, however, by postulating the formation of a virtual cathode at the moving ion front and considering the effects of reflected electrons. The idealized solution has equal ion and electron densities, but half the electron distribution has zero velocity and half has a velocity equal to twice the ion velocity. Assuming that the electron source operates in a space-charge-limited fashion, the emerging ion beam must assume (in an average sense) the condition similar to a neutral plasma, at a potential slightly positive with respect to the electron source so as to draw off the required electron density.

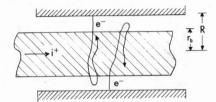

Figure 3.16. Ion beam space charge neutralization by transverse injection of electrons.

Of course such a static solution ignores the complex transient phenomena such as particle bunching and potential fluctuations, that we have come to expect in problems associated with space-charge-limited flow. For present purposes it is sufficient to note that in more physically realizable situations these results simply imply that the electrons would have a longitudinal velocity spread of the order of the ion velocity.

3.6.2. Transverse Electron Injection

A second class of ion beam neutralization schemes uses transverse injection of electrons from boundaries surrounding the ion beam volume. Consider the geometry of Fig. 3.16. An ion beam of uniform density and velocity and infinite depth passes through a planar region bounded by grounded conducting surfaces from which cold electrons are attracted into the beam by the positive ion space charge. The electrons could be produced as the result of secondary emission, or supplied by active sources. A simple equilibrium solution can be obtained if the electron emission is space charge limited. In this case the electric field vanishes at the boundaries, and also in the center of the beam from symmetry arguments. The equations which govern the steady state, in analogy with Section 3.3, are given by

$$\frac{d}{dy}(n_e v_e) = 0 \tag{3.158}$$

$$\frac{d^2\phi}{dy^2} = -4\pi e(n_{i0} - n_e) \tag{3.159}$$

$$\tfrac{1}{2} m_e v_e^2 = e\phi \tag{3.160}$$

Equations (3.158), (3.159), and (3.160) can be combined to yield

$$\frac{d^2\phi}{dy^2} = -4\pi e\left[n_{i0} - \frac{j_e}{e}\left(\frac{2e\phi}{m_c}\right)^{-1/2}\right] \tag{3.161}$$

where $j_e = en_ev_e = $ const. Making the substitution $u = d\phi/dy$, and recognizing that $d^2\phi/dy^2 = u\,du/d\phi$, Eq. (3.161) can be integrated to yield

$$\frac{d\phi}{dy} = (8\pi en_{i0})^{1/2}\left[(\phi\phi_{max})^{1/2} - \phi\right]^{1/2} \tag{3.162}$$

In writing Eq. (3.162) the boundary condition at the emission surface, $d\phi/dy = 0$, has been used to eliminate $j_e = en_{i0}(e\phi_{max}/2m_e)^{1/2}$, where ϕ_{max} is the value of the potential at the center of the beam. A final integration of Eq. (3.162) yields

$$y = \left(\frac{\phi_{max}}{2en_{i0}}\right)^{1/2}\left[\sin^{-1}\left(\frac{\phi}{\phi_{max}}\right)^{1/4} - \left(\frac{\phi}{\phi_{max}}\right)^{1/4}\left(1 - \frac{\phi}{\phi_{max}}\right)^{1/2} + \text{const}\right] \tag{3.163}$$

Evaluation of the constant follows from the condition that at $y = 0$, $\phi = 0$, which yields

$$\sin^{-1}(0) = -\pi l \tag{3.164}$$

where $l = 0, 1, 2, 3, \ldots$. At $y = \Delta/2$, the condition that $\phi = \phi_{max}$ indicates that

$$\phi_{max} = \frac{2en_{i0}\Delta^2}{\pi(2m + 1)^2} \tag{3.165}$$

where m is an integer corresponding to the number $(m + 1)$ of potential maxima across the ion beam.

If only the first mode develops, then the maximum potential at the center of the beam is reduced by a factor of only $(2/\pi)^2$ over the potential which would exist if no electrons were injected (see Problem 3.12).[32] This result is a consequence of the fact that the cold electrons cannot become trapped in the ion beam and must return to the emission plane; the electron space charge localized near the emission boundaries suppresses additional electron flow into the beam. This conclusion is overly pessimistic, however, because in any real system the electrons will acquire transverse velocity as the result of elastic deflections caused by transverse magnetic field components, longitudinal ion density fluctuations, wall irregularities, etc. Randomization of the electron distribution function decreases the electron density near the emission boundaries and permits enhanced neutralization in the center of the ion beam. Numerical particle simulations of this problem have indicated that the presence of an oblique magnetic field (components both normal and tangential to the ion flow) allows almost complete neutralization of the ion space charge on the time scale of the ion beam rise time.[33]

3.7. Electrostatic Stability of Intense Relativistic Electron Beams

In the preceding sections the equilibria of intense relativistic beams were considered; now, attention will be focused on the stability of small amplitude oscillations about various equilibrium configurations. The mean azimuthal motion of the electron beam is assumed to be nonrelativistic with

$$\frac{r^2 \omega_\theta^2(r)}{c^2} = \beta_\theta^2(r) \ll \beta_0^2 \tag{3.166}$$

where the unperturbed axial velocity $v_z = \beta_z c$ is assumed to be independent of radius. In this case the axial diamagnetic field contribution can be neglected and the radial force equation is approximated by [see Eqs. (3.18) and (3.110)]

$$\gamma_0 \omega_\theta^2(r) + \frac{1}{r^2}\left(1 - \beta_0^2\right)\int_0^r dr' \, r' \omega_p^2(r') - \omega_\theta(r)\Omega = 0 \tag{3.167}$$

To determine the stability of the various equilibrium configurations, the macroscopic fluid and field quantities are expressed as the sum of their equilibrium values plus a perturbation, i.e.,

$$n(\mathbf{r}, t) = n_0(r) + \delta n(\mathbf{r}, t)$$

$$\mathbf{v}(\mathbf{r}, t) = v_{\theta_0}(r)\hat{e}_\theta + v_{z0}\hat{e}_z + \delta\mathbf{v}(\mathbf{r}, t)$$

$$\mathbf{E}(\mathbf{r}, t) = E_{r_0}(r)\hat{e}_r + \delta\mathbf{E}(\mathbf{r}, t)$$

$$\mathbf{B}(\mathbf{r}, t) = B_0\hat{e}_z + \delta\mathbf{B}(\mathbf{r}, t)$$

The evolution of the small-amplitude perturbation is determined from Eqs. (3.9)–(3.14). In the electrostatic approximation the perturbed magnetic field, $\delta\mathbf{B}(\mathbf{r}, t)$, is assumed to remain negligibly small so that the perturbed electric field, $\delta\mathbf{E}(\mathbf{r}, t)$, can be expressed as the gradient of a scalar potential,

$$\delta\mathbf{E}(\mathbf{r}, t) = -\nabla\phi(\mathbf{r}, t) \tag{3.168}$$

When Eqs. (3.9), (3.10), (3.13), and (3.168) are combined and linearized (i.e., products of perturbed quantities are neglected), the macroscopic fluid–Poisson equations for the perturbed quantities become

$$\frac{\partial}{\partial t}\delta n(\mathbf{r}, t) + \nabla\cdot\left\{n_0(r)\,\delta\mathbf{v}(\mathbf{r}, t) + \delta n(\mathbf{r}, t)\left[v_{\theta_0}(r)\hat{e}_\theta + v_{z0}\hat{e}_z\right]\right\} = 0$$

$$\tag{3.169}$$

$$\nabla^2\delta\phi(\mathbf{r}, t) = 4\pi e\,\delta n(\mathbf{r}, t) \tag{3.170}$$

$$\frac{\partial}{\partial t}\delta\mathbf{p} + \mathbf{v}_0\cdot\nabla\,\delta\mathbf{p} + \delta\mathbf{v}\cdot\nabla\mathbf{p}_0 = -e\left(-\nabla\,\delta\phi + \frac{1}{c}\delta\mathbf{v}\times\mathbf{B}_0\right) \tag{3.171}$$

Since the axial motion of the beam may be relativistic, the two-mass approximation is assumed for the perturbed momentum according to

$$\delta p_r = \gamma_0 m\,\delta v_r$$

$$\delta p_\theta = \gamma_0 m\,\delta v_\theta$$

$$\delta p_z = \gamma_0^3 m\,\delta v_z$$

As a further simplification, all perturbed quantities are assumed to have harmonic dependence on time, the azimuthal coordinate, and the axial coordinate; i.e., $\delta\psi = \delta\psi(r)e^{-i\omega t}e^{il\theta}e^{ik_z z}$, where ω is the characteristic frequency of oscillations, l is the azimuthal harmonic number, and k_z is the axial wave number. With this assumption the linearized equations become[1]

$$-i\left(\omega - k_z v_{z0} - l\omega_{\theta_0}\right)\delta n + \left[\frac{1}{r}\frac{\partial}{\partial r}\left(r\delta v_r\right) + \frac{il\delta v_\theta}{r} + ik_z \delta v_z\right]n_0 = 0$$

$$\frac{1}{r}\frac{\partial}{\partial r}r\frac{\partial}{\partial r}\delta\phi - \frac{l^2}{r^2}\delta\phi - k_z^2\delta\phi = 4\pi e\,\delta n$$

$$-i\left(\omega - k_z v_{z0} - l\omega_{\theta_0}\right)\delta v_r + \left(\frac{\Omega}{\gamma_0} - 2\omega_{\theta_0}\right)\delta v_\theta = -\frac{e}{\gamma_0 m}\frac{\partial}{\partial r}\delta\phi$$

$$-i\left(\omega - k_z v_{z0} - l\omega_{\theta_0}\right)\delta v_\theta + \left[-\frac{\Omega}{\gamma_0} + \frac{1}{r}\frac{\partial}{\partial r}\left(r^2\omega_{\theta_0}\right)\right]\delta v_r = \frac{e}{\gamma_0 m}\frac{il\delta\phi}{r}$$

$$-i\left(\omega - k_z v_{z0} - l\omega_{\theta_0}\right)\delta v_z = \frac{e}{\gamma_0^3 m}ik_z\delta\phi$$

In these equations $\Omega = eB_0/mc$, and the equilibrium angular velocity profile, $\omega_{\theta_0} = v_{\theta_0}/r$, is related to the equilibrium density profile by the radial force equation. Solving for the density and fluid velocity perturbations in terms of the perturbed potential, Poisson's equation can be written as

$$\frac{1}{r}\frac{\partial}{\partial r}\left[r\left(1 - \frac{\omega_{p0}^2}{\gamma_0 v^2}\right)\frac{\partial}{\partial r}\delta\phi\right] - \frac{l^2}{r^2}\left(1 - \frac{\omega_{p0}^2}{\gamma_0 v^2}\right)\delta\phi$$

$$-k_z^2\left[1 - \frac{\omega_{p0}^2/\gamma_0^3}{\left(\omega - k_z v_{z0} - l\omega_{\theta_0}\right)^2}\right]\delta\phi \qquad (3.172)$$

$$= -\frac{l\delta\phi}{r}\frac{(\partial/\partial r)\left[\left(\omega_{p0}^2/v^2\right)\left(-\Omega/\gamma_0 + 2\omega_{\theta_0}\right)\right]}{\left(\omega - k_z v_{z_0} - l\omega_{\theta_0}\right)}$$

where

$$v^2(r) = \left(\omega - k_z v_{z_0} - l\omega_{\theta_0}\right)^2 - \left(-\frac{\Omega}{\gamma_0} + 2\omega_{\theta_0}\right)\left[-\frac{\Omega}{\gamma_0} + \frac{1}{r}\frac{\partial}{\partial r}\left(r^2\omega_{\theta_0}\right)\right]$$

and $\omega_{p0}^2(r) = 4\pi n_0(r)e^2/m$. The procedure for using Eq. (3.172) is to solve for $\delta\phi$ and ω as an eigenvalue problem.

3.7.1 Stability of the Rigid Rotor Equilibrium

Analytical solutions for Eq. (3.172) are tractable for only a few simple cases. As a first example of the use of Eq. (3.172) consider the case of constant beam density, corresponding to the rigid rotor relativistic electron beam equilibrium of Section 3.5.1. It is assumed that the beam radius is equal to the radius of a perfectly conducting wall, i.e., $[\delta\phi]_{r=r_b} = 0$.

Since the equilibrium rotation velocity $\omega_{\theta_0}(r)$, given by Eq. (3.113)

$$\omega_{\theta_0}(r) = \omega_{\theta_0}^\pm = \frac{\Omega}{2\gamma_0}\left[1 \pm \left(1 - \frac{2\omega_{p0}^2}{\gamma_0\Omega^2}\right)^{1/2}\right], \qquad 0 < r < r_b$$

is independent of radius, it follows that $v^2(r)$ is given by

$$v^2 = \left(\omega - k_z v_{z_0} - l\omega_{\theta_0}\right)^2 - \left(-\Omega/\gamma_0 + 2\omega_{\theta_0}\right)^2 \tag{3.173}$$

which is also independent of radius. Under these conditions Poisson's equation reduces to

$$\frac{1}{r}\frac{\partial}{\partial r}r\frac{\partial}{\partial r}\delta\phi - \frac{l^2}{r^2}\delta\phi + k_\perp^2\,\delta\phi = 0, \qquad 0 < r < r_b \tag{3.174}$$

where

$$k_\perp^2 = -k_z^2\frac{1 - \left[\dfrac{\omega_{p0}^2/\gamma_0^3}{\left(\omega - k_z v_{z0} - l\omega_{\theta_0}\right)^2}\right]}{1 - \omega_{p0}^2/\gamma_0 v^2} \tag{3.175}$$

Equation (3.174) is recognized as a form of Bessel's equation. The solution which remains finite at the origin is

$$\delta\phi = AJ_l(k_\perp r) \tag{3.176}$$

where A is a constant and J_l is the Bessel function of the first kind of order l. Enforcing the vanishing of the perturbed potential at the conducting wall leads to the condition

$$J_l(k_\perp r_b) = 0 \tag{3.177}$$

from which it follows that

$$k_\perp^2 r_b^2 = \rho_{lm}^2; \qquad m = 1, 2, \ldots \tag{3.178}$$

where ρ_{lm} is the mth zero of $J_l(x) = 0$, and $k_\perp$ is recognized as the effective perpendicular wave number quantized by the finite radial geometry.

Equation (3.178) is the dispersion relation for electrostatic waves in the electron-beam-filled guide tube. It relates the characteristic oscillation frequency ω to the azimuthal harmonic numbers l, the axial wave vector k_z, and the properties of the rigid rotor equilibrium configuration.

Equation (3.178) may be expressed in the equivalent form

$$1 - \frac{k_z^2}{k^2} \frac{\omega_{p0}^2/\gamma_0^3}{\left(\omega - k_z v_{z0} - l\omega_{\theta_0}\right)^2} - \frac{k_\perp^2}{k^2} \frac{\omega_{p0}^2/\gamma_0}{\left(\omega - k_z v_{z0} - l\omega_{\theta_0}\right)^2 - \omega_v^2} = 0 \tag{3.179}$$

where $k^2 = k_z^2 + k_\perp^2$, and ω_v is the vortex frequency defined by

$$\omega_v = -\left(-\Omega_0/\gamma_0 + 2\omega_{\theta_0}\right) \tag{3.180}$$

The solution to Eq. (3.179) can be written as[2]

$$\left(\omega - k_z v_{z0} - l\omega_{\theta_0}\right)^2 = \left[\omega_v^2 + \left(\omega_{p0}^2/\gamma_0\right)\left(\frac{k_z^2/\gamma_0^2 + k_\perp^2}{k^2}\right)\right]$$

$$\times \left\{ 1 \pm \left[1 - \frac{4\left(k_z^2/k^2\right)\omega_v^2 \omega_{p0}^2/\gamma_0^3}{\omega_v^2 + \left(\omega_{p0}^2/\gamma_0\right)\left(\dfrac{k_z^2/\gamma_0^2 + k_\perp^2}{k^2}\right)} \right]^{1/2} \right\} \tag{3.181}$$

For the analysis to be valid, the value of the beam density must be less than the maximum density for which equilibrium solutions exist. An upper bound on the density can be determined by requiring that the radical in Eq.

(3.181) be real; the condition that obtains (assuming the slow rotation mode) is also [see Eq. (3.114)]*

$$2\omega_{p0}^2/\gamma_0\Omega^2 \leqslant 1$$

Provided that the criterion of Eq. (3.114) is satisfied, $\text{Im}\,\omega_{\theta0} = 0$ (i.e., $\omega_{\theta0}$ contains no imaginary component). Hence, this condition implies the existence of undamped stable oscillations for the four characteristic eigenmodes of the dispersion relation. An examination of the character of the eigenmodes is simplified by restricting the analysis to the axisymmetric ($l = 0$) modes and assuming the slow rotation frequency. In this case the dispersion relation factors approximately as[35]

$$\left[(\omega - k_z v_{z0})^2 - \frac{k_z^2}{k^2}\frac{\omega_{p0}^2}{\gamma_0^3} \right] \left[(\omega - k_z v_{z0})^2 - \frac{\Omega^2}{\gamma_0^2} \right] \approx 0 \qquad (3.182)$$

and the eigenmodes can be identified as follows:

(1) Two Doppler-shifted plasma modes explicitly displaying the longitudinal electron mass effects:

$$\omega = k_z v_{z0} \pm \left(\frac{k_z^2}{k^2}\frac{\omega_{p0}^2}{\gamma_0^3} \right)^{1/2} \qquad (3.183)$$

(2) Two Doppler-shifted cyclotron modes:

$$\omega = k_z v_{z0} \pm \left(\frac{\Omega^2}{\gamma_0^2} \right)^{1/2} \qquad (3.184)$$

The plasma modes correspond to longitudinal bunching of the beam space charge, while the cyclotron waves are transverse modes corresponding to a traveling constriction in the beam. The various wave dispersion curves are graphed in Fig. 3.17. The variation of the wave phase velocity for the different modes is important for several collective ion acceleration concepts, and will be examined in more detail in Chapter 7.

---

*For a more detailed analysis of the stability of nonrelativistic rigid rotor equilibria, consult Ref. 34.

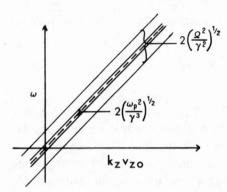

$2\left(\dfrac{\Omega^2}{\gamma^2}\right)^{1/2}$

ω

$2\left(\dfrac{\omega_p^2}{\gamma^3}\right)^{1/2}$

$k_z v_{zo}$

Figure 3.17. Wave dispersion curves for the case of the solid beam, rigid rotor equilibrium in the weak-beam–strong-field limit. ($\theta = 0°$).

Before leaving this section it should be emphasized that the relatively simple form of the dispersion relation, Eq. (3.179), could only be obtained by considering the radially homogeneous rigid rotor equilibrium. In general, radial inhomogeneities, particularly the variation of $\gamma(r)$ across the beam, strongly influence the structure of the wave spectra. Specific numerical solutions of the linear theory for more realistic solid beam equilibria indicate several significant differences.[23] For example, for any given pair of axial and azimuthal wave numbers, only a single slow cyclotron mode exists, and it is localized at the surface of the beam. Although distortion of the plasma modes is not so severe, the phase velocity of long-wavelength slow-space-charge waves does not vanish, even at the space-charge-limiting current, for finite strengths of the external magnetic field.

3.7.2. Stability of the Hollow Beam Equilibrium (Diocotron Instability)

For low-frequency perturbations of a tenuous beam in the long-wavelength limit ($k_z \to 0$), Eq. (3.172) can be expressed in the approximate form

$$\frac{1}{r}\frac{\partial}{\partial r}\left(r\frac{\partial}{\partial r}\delta\phi\right) - \frac{l^2}{r^2}\delta\phi = -\frac{l}{r}\frac{\delta\phi}{(\omega - l\omega_{\theta_0})}\frac{\gamma_0}{\Omega}\frac{\partial}{\partial r}\omega_{p0}^2 \qquad (3.185)$$

In this case it is straightforward to show that the equilibrium configuration is stable provided $(\partial/\partial r)(\omega_{p0}^2) \leqslant 0$ (see Problem 3.14); however, for the hollow beam equilibrium of Section 3.5.2 this criterion is obviously not satisfied since

$$\frac{\partial}{\partial r}\left[\omega_p^2(r)\right] = \omega_{p0}^2\left[\delta(r - r_0) - \delta(r - r_b)\right] \qquad (3.186)$$

Although induced oscillations in the body of the hollow beam are stable, with Eq. (3.186) the right-hand side of Eq. (3.185) corresponds to perturbations in charge density on the inner and outer edges of the beam. Because of the angular rotation shear of the hollow beam, the surface waves propagate relative to one another, and the motion of the charge perturbation of one wave can be modified by the electrostatic fields of the other. Under suitable conditions this interaction can become synchronized so as to produce a single exponentially growing wave mode. This type of instability has been termed "diocotron" to indicate that the charge perturbations on the two surfaces must slip parallel to each other to create the instability.[36]

In the beam interior the perturbed potential satisfies the homogeneous equation

$$\frac{1}{r}\frac{\partial}{\partial r}\left(r\frac{\partial}{\partial r}\delta\phi\right) - \frac{l^2}{r^2}\delta\phi = 0 \qquad (3.187)$$

with appropriate solutions given by

$$\delta\phi_{\mathrm{I}} = (A + Br_0^{-2l})r^l, \qquad\qquad 0 < r < r_0$$

$$\delta\phi_{\mathrm{II}} = (Ar^l + Br^{-l}), \qquad\qquad r_0 < r < r_b \qquad (3.188)$$

$$\delta\phi_{\mathrm{III}} = (Ar_b^{2l} + B)\left(\frac{R^{2l} - r^{2l}}{R^{2l} - r_b^{2l}}\right)r^{-l}, \qquad r_b < r < R$$

where the condition of continuity of $\delta\phi$ across the region boundaries has been enforced. Introducing the jump conditions at the discontinuities (to eliminate the constants A and B)

$$r_0\left[\frac{\partial}{\partial r}\delta\phi_{\mathrm{II}}\right]_{r=r_0} - r_0\left[\frac{\partial}{\partial r}\delta\phi_{\mathrm{I}}\right]_{r=r_0} = -l[\delta\phi]_{r=r_0}\left\{\frac{\gamma_0\omega_{p0}^2}{\Omega[\omega - l\omega_\theta(r_0)]}\right\}$$

$$(3.189)$$

$$r_b \left[\frac{\partial}{\partial r} \delta \phi_{III} \right]_{r=r_b} - r_b \left[\frac{\partial}{\partial r} \delta \phi_{II} \right]_{r=r_b} = l [\delta \phi]_{r=r_b} \left\{ \frac{\gamma_0 \omega_{p_0}^2}{\Omega [\omega - l \omega_\theta (r_b)]} \right\}$$

(3.190)

leads to an eigenvalue equation for ω given by[37]

$$(\omega / \omega_0)^2 - b(\omega / \omega_0) + c = 0 \tag{3.191}$$

where

$$b = l \left(1 - \frac{r_0^2}{r_b^2} \right) + \left(\frac{r_b^{2l}}{R^{2l}} - \frac{r_0^{2l}}{R^{2l}} \right) \tag{3.192}$$

$$c = l \left(1 - \frac{r_0^2}{r_b^2} \right) \left(1 - \frac{r_0^{2l}}{R^{2l}} \right) - \left(1 - \frac{r_0^{2l}}{r_b^{2l}} \right) \left(1 - \frac{r_b^{2l}}{R^{2l}} \right) \tag{3.193}$$

$$\omega_0 = \omega_{p0}^2 / (2\gamma_0^2 \Omega) \tag{3.194}$$

If $b^2 \geqslant 4c$, then $\mathrm{Im}(\omega) = 0$, and the equilibrium configuration is stable; if $b^2 < 4c$, the system is unstable with the characteristic growth rate being given by

$$\omega_i = \frac{\omega_0}{2} (4c - b^2)^{1/2} \tag{3.195}$$

Employing Eqs. (3.192) and (3.193) the stability criterion can be explicitly expressed as

$$\left[-l(1 - \beta^2) + 2 - \alpha^{2l} - (\alpha\beta)^{2l} \right]^2 \geqslant 4\beta^{2l} (1 - \alpha^{2l})^2 \tag{3.196}$$

where $\alpha = r_b / R$ and $\beta = r_0 / r_b$. It is seen by inspection that choosing $\beta = 0$ (solid beam) or $\alpha = 1$ (outer edge of the beam extending to the conducting wall) trivially satisfies the stability criterion. Physically, these cases correspond to the elimination of one of the beam surfaces which supports wave propagation.

For $l = 1$, Eq. (3.196) reduces to

$$(R^2 - r_b^2)^2 (r_b^2 - r_0^2)^2 > 0 \tag{3.197}$$

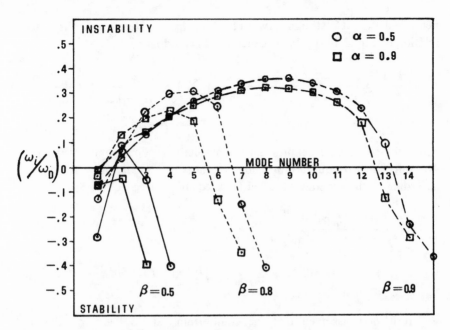

Figure 3.18. Diocotron instability growth rate versus mode number for various values of the beam thickness and separation from the drift tube wall. The most unstable mode number scales approximately as $(1 - \beta)^{-1}$

which is always satisfied. In this case the two surface waves cannot become synchronized due to insufficient rotational shear, and the system does not become unstable.[38]

In Fig. 3.18 the growth of the diocotron instability is presented as a function of mode number for various values of the parameters α and β. It is apparent that the instability is most serious for very hollow beams ($\beta \rightarrow 1$) with radii substantially smaller than the drift tube radius. It is also interesting to note that the most unstable mode number qualitatively scales as $(1 - \beta)^{-1}$. This leads to a quick approximate criterion for α to ensure stability. For example, with $\beta = 0.5$, $l = 2$ is the most unstable mode number. Substituting these values into Eq. (3.196) indicates that the diocotron mode is stabilized for $\beta = 0.5$ provided that $\alpha^2 = (r_b/R)^2 \geqslant 0.8$.

3.7.3. An Electron–Electron Two-Stream Instability

For a multicomponent plasma whose components completely fill the waveguide and may have relativistic axial motions along an external magnetic field $B_0\hat{e}_z$, it is straightforward to show that the electrostatic dispersion relation of Eq. (3.179) may be generalized according to

$$1-\left(\frac{k_z}{k}\right)^2\sum_i\frac{\omega_{pi}^2/\gamma_0^3}{\left(\omega-l\omega_i-k_zv_{iz}\right)^2}-\left(\frac{k_\perp}{k}\right)^2\sum_i\frac{\omega_{pi}^2/\gamma_0}{\left(\omega-l\omega_i-k_zv_{iz}\right)^2-\omega_{iv}^2}=0$$

(3.198)

where the subscript i denotes the ith plasma component and $k_\perp^2 r_b^2=\rho_{lm}^2$. The unperturbed equilibrium values of the various quantities are related according to the generalized radial force equations [see Eq. (3.18)]:

$$-\frac{m_i\gamma_i(\beta_{i\theta}c)^2}{r}=e_i\left[\frac{4\pi}{r}\sum_j e_j\int_0^r dr'r'n_j(r')\right.$$

$$\left.+\beta_{i\theta}B_0-4\pi\frac{\beta_{iz}}{r}\sum_j e_j\int_0^r dr'r'\beta_{jz}n_j\right]$$

(3.199)

It is now assumed that the equilibrium axial velocity profiles are independent of radius and that $\beta_{1z}=\beta_0=-\beta_{2z}$. It is further assumed that the constant densities of the counterstreaming electron beams are related according to $n_2=\alpha n_1$, where α is a constant. In this case the dispersion relation becomes

$$1-\frac{k_z^2\omega_{p1}^2/\gamma_0^3}{k^2}\left[\frac{1}{\left(\omega-\omega_{1\theta}l-k_zv_{1_z}\right)^2}+\frac{\alpha}{\left(\omega-\omega_{2\theta}l+k_zv_{1_z}\right)^2}\right]-\frac{k_\perp^2\omega_{p1}^2/\gamma_0}{k^2}$$

$$\times\left[\frac{1}{\left(\omega-\omega_{1\theta}l-k_zv_{1_z}\right)^2-\Omega^2/\gamma^2}+\frac{\alpha}{\left(\omega-\omega_{2\theta}l+k_zv_{1_z}\right)^2-\Omega^2/\gamma^2}\right]=0$$

(3.200)

where the rotation rates are given by

$$\omega_{1\theta}=\frac{\Omega}{2\gamma_0}\left\{1\pm\left[1-\frac{2\gamma_0\omega_{p1}^2}{\Omega^2}\left(1-\beta_0^2\right)+\alpha\left(1+\beta_0^2\right)\right]^{1/2}\right\}$$

$$\omega_{2\theta}=\frac{\Omega}{2\gamma_0}\left\{1\pm\left[1-\frac{2\gamma_0\omega_{p1}^2}{\Omega^2}\left(1+\beta_0^2\right)+\alpha\left(1+\beta_0^2\right)\right]^{1/2}\right\}$$

(3.201)

If the magnetic field is sufficiently strong $(\Omega^2 \gg \omega_{p1}^2 \gamma_0)$ then a good approximation for the dispersion relation is to neglect the beam rotation altogether (assuming slow rotation modes), in which case Eq. (3.200) becomes

$$1 - \frac{k_z^2 \omega_{p1}^2 / \gamma_0^3}{k^2} \left[\frac{1}{(\omega - k_z v_{1z})^2} + \frac{\alpha}{(\omega + k_z v_{1z})^2} \right] \simeq 0 \qquad (3.202)$$

Although the natural oscillation modes of the separate beams are stable, in the case of the counterstreaming beams the motions of the charge perturbations of one beam are altered by the electrostatic fields of the charge perturbations of the other beam. Under certain conditions, this coupling tends to further bunch the beam electrons, which then increases the wave amplitudes and causes even stronger bunching.

It can be shown from Eq. (3.202) that instability exists for axial wave vectors k_z that satisfy[39]

$$4k_z^2 v_{1z}^2 < \omega_{p1}^2 \gamma_0^{-3} \left(\frac{k_z^2}{k^2} \right) (1 + \alpha^{1/3})^3 \qquad (3.203)$$

Since $k_\perp^2 = \rho_{lm}^2 / r_b^2$, an equivalent expression for Eq. (3.203) is

$$4k_z^2 v_{1z}^2 < \omega_{p1}^2 \gamma_0^{-3} (1 + \alpha^{1/3})^3 - \frac{\rho_{lm}^2}{r_b^2} v_{1z}^2 \qquad (3.204)$$

Since k_z is real by definition, a necessary condition for instability is that the right-hand side of Eq. (3.204) be positive, i.e.,

$$\omega_{p1}^2 > \frac{(\rho_{lm}/r_b)^2 \gamma_0^3 v_{1z}^2}{(1 + \alpha^{1/3})^3} \qquad (3.205)$$

For the special case of symmetric equidensity counterstreaming electron beams $(\alpha = 1)$, it is straightforward to show that the maximum growth rate obtained can be approximated by[1]

$$[\Gamma_i]_{max} \simeq \tfrac{1}{2} \omega_{p1} \gamma_0^{-3/2} \qquad (3.206)$$

and the corresponding axial wave vector in this instance is

$$[k_z]_{max} \simeq \frac{3}{4} \frac{\omega_{p1}^2 / \gamma_0^3}{v_{1z}^2} \qquad (3.207)$$

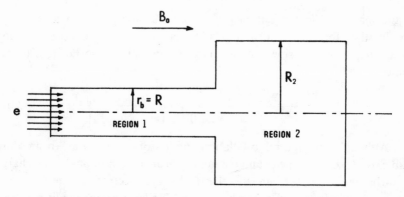

Figure 3.19. Schematic geometry of an electron beam injected along an applied magnetic field into a waveguide which has a sharp discontinuity in the boundary.

The motivation for this particular problem can be established as follows. Suppose an intense electron beam is injected into a waveguide along a strong magnetic field. The waveguide is assumed to have a sharp discontinuity in the boundary (see Fig. 3.19). In region 1, the beam completely fills the waveguide ($r_b = R$), while in region 2, $r_b \ll R_2$. Under appropriate conditions the possibility arises that while the beam current may be less than the space charge limit for region 1, it may substantially exceed the space charge limit for region 2. In this case a virtual cathode will form in the region of discontinuity and reflect some of the injected beam particles.[40] The difference between the streaming velocities of the injected and reflected particles can provide a source of free energy to drive a strong two-stream instability.

3.8. Summary

When an intense relativistic electron beam propagates through a less dense stationary ion background the amount of fractional beam space charge neutralization required for the existence of a radial force equilibrium is $f_e = n_i/n_e = \gamma^{-2}$. Hence, for propagation of intense beams in vacuum ($f_e = 0$) an external means of preventing beam expansion is necessary, e.g., a strong longitudinal magnetic field. However, even under the assumption of an infinitely strong magnetic field, the amount of current which can be propagated though an evacuated drift cavity is effectively limited by the electrostatic potential depression associated with the unneutralized beam space charge; if the beam current exceeds the space-charge-limiting current, a virtual cathode forms and reflects a fraction of the beam current back toward the injector.

While it is possible to obtain mathematical solutions to the steady state problem of virtual cathode formation in simple geometries, such solutions are not physically realizable. Indeed, the most obvious feature of numerical simulations of this problem is that the solutions are oscillatory and do not approach a steady state in which all trajectories are identical. Such unsteady transient behavior can be qualitatively predicted using a small signal perturbation analysis, and the essential features of the oscillations can be understood on the basis of a simple model which uses a single charge sheet in a one-dimensional diode.

If the beam current does not exceed the space charge limit the most important problem is to determine the requirements for the existence of a stable, unneutralized beam equilibrium. A discussion of several possible beam equilibria proceeds from the steady state radial force balance equation supplemented by the two additional constraints of energy conservation and conservation of canonical angular momentum. The allowed equilibrium states depend greatly on the conditions at the beam source. Three source geometries of practical importance are (1) the magnetically shielded source for which the canonical angular momentum vanishes, (2) the immersed thin annular cathode for which P_θ is a constant, and (3) the source immersed in a uniform axial magnetic field for which $P_\theta \propto r^2$. For immersed sources beam rotation is generally small. In contrast, as the beam from a shielded source encounters the increasing external magnetic field, the beam rotation rapidly increases producing significant beam diamagnetic fields.

Much of the analysis developed for intense electron beams also applies, in principle, to the case of ion beams; however, the required magnetic field strength for an ion beam laminar flow equilibrium is impractically large, and ion beam transport is most easily realized by neutralization of the ion space charge by low-energy electrons. In this chapter emphasis was focused on the mechanisms of charge neutralization by either collinear or transverse introduction of electrons into the drift space.

Once the existence of several beam equilibria has been established it is necessary to examine the transient behavior of various configurations when subjected to small-amplitude oscillations; i.e., linearization of the equations to examine system stability. Analyses of several electron beam instabilities were performed in the electrostatic approximation in which the perturbed magnetic field was assumed to remain negligibly small. As a further simplification, all perturbed quantities were assumed to have a harmonic dependence on time, and the azimuthal and axial coordinates. As a result of this procedure, dispersion relations (relating the characteristic frequency of oscillation to the azimuthal harmonic number, the axial wave vector, and the properties of the specific equilibria) were obtained for a few important situations. In particular, the normal modes of the solid relativistic beam rigid rotor equilibrium were identified and found to be stable, while the hollow relativistic beam equilibrium was shown to be diocotron unstable due to the radial dependence (shear) of the angular velocity profile.

Problems

3.1. (a) Compare the expansion of a 10-kA, 10-MeV electron beam and a 10-kA, 10-MeV proton beam propagating in an evacuated drift space. Assume that the current (or charge) density is a constant function of radius out to the beam edge, r_b.

  (b) Estimate the strength of an axial magnetic field required to maintain radial force balance for the two cases.

3.2. (a) Show that the nonrelativistic limit of Eq. (3.35) is given by

$$I_l \approx \frac{(e/m)^{1/2}(2\phi/3)^{3/2}}{2\ln(R/r_b)}$$

  where $|e\phi| = (\gamma_0 - 1)mc^2$.

  (b) Assuming $R = 2r_b$, compare the space charge current limit for 1-, 10-, and 100-MeV electron and proton beams.

3.3. Calculate the space-charge-limiting current for a cylindrically symmetric, uniform density, hollow electron beam of inner and outer radii r_1 and r_2 propagating in an evacuated drift tube of radius R along a very strong longitudinal magnetic field.

  (a) Use Gauss' law to show that the radial electric field is given by

$$E_r = -\frac{2\pi e n_0}{r} \times \begin{cases} 0, & r < r_1 \\ (r^2 - r_1^2), & r_1 < r < r_2 \\ (r_2^2 - r_1^2), & r_2 < r < R \end{cases}$$

(b) Using (a) show that the electrostatic potential in the drift tube is given by

$$\phi(r) = -2\pi e n_0$$

$$\times \begin{cases} (r_2^2 - r_1^2)\ln(R/r_2) - r_1^2\ln(r_2/r_1) + (r_2^2 - r_1^2)/2, & 0 < r < r_1 \\ (r_2^2 - r_1^2)\ln(R/r_2) - r_1^2\ln(r_2/r) + (r_2^2 - r^2)/2, & r_1 < r < r_2 \\ (r_2^2 - r_1^2)\ln(R/r), & r_2 < r < R \end{cases}$$

(c) Obtain the approximate expression for the space charge limit, given below, by setting the magnitude of the electrostatic potential energy, obtained from (b), equal to the injected beam kinetic energy $(\gamma_0 - 1)mc^2$:[41]

$$I_l \approx \beta(\gamma_0 - 1)\left(\frac{mc^3}{e}\right)\left[1 - \frac{2r_1^2}{r_2^2 - r_1^2}\ln\left(\frac{r_2}{r_1}\right) + 2\ln\left(\frac{R}{r_2}\right)\right]^{-1}$$

(d) The expression found in (c) is only correct in the limit of ultrarelativistic beam kinetic energies; its range of validity is extended by replacing the factor $\beta(\gamma_0 - 1)$ by $(\gamma_0^{2/3} - 1)^{3/2}$ obtained in Section 3.3.1. With this *ad hoc* assumption, show that the space charge limit for a solid beam is given by[42]

$$I_l \approx \frac{(\gamma_0^{2/3} - 1)^{3/2}(mc^3/e)}{1 + 2\ln(R/r_0)}$$

while the limit for a thin hollow beam is [see Eq. (3.45)]

$$I_l \approx \frac{(\gamma_0^{2/3} - 1)^{3/2}(mc^3/e)}{2\ln(R/r_b)}$$

3.4. Calculate the space-charge-limiting current for a cylindrically symmetric solid electron beam of radius r_b, propagating in an evacuated drift tube of radius R and length L along a very strong longitudinal magnetic field.

(a) Show that the cylindrically symmetric Green's function is given by

$$G(r, z'; r', z') = \frac{8\pi}{R} \sum_{n=1}^{\infty} \frac{J_0(\lambda_n r/R)J_0(\lambda_n r'/R)}{\lambda_n[J_1(\lambda_n)]^2\sinh(\lambda_n L/R)}$$

$$\times \begin{cases} \sinh(\lambda_n z/R)\sinh[\lambda_n(L-z')/R], & z < z' \\ \sinh[\lambda_n(L-z)/R]\sinh(\lambda_n z'/R), & z > z' \end{cases}$$

(b) Using the Green's function of (a), show that the electrostatic potential in the drift space is given by

$$\phi(r,z) = -8\pi e n_0 r_b R \sum_{n=1}^{\infty} \frac{J_0(\lambda_n r/R) J_1(\lambda_n r_b/R)}{\lambda_n^3 [J_1(\lambda_n)]^2 \sinh(\lambda_n L/R)}$$

$$\times \left\{ \begin{array}{l} \sinh[\lambda_n(L-z)/R][\cosh(\lambda_n z/R)-1] \\ +\sinh(\lambda_n z/R)[\cosh[\lambda_n(L-z)/R]-1] \end{array} \right\}$$

(c) Using the procedures of Problems 3.1(c) and 3.1(d), show that the space-charge-limiting current is given approximately by [9]

$$I_l^L = \frac{(\gamma_0^{2/3}-1)^{3/2}(mc^3/e)}{8} \left(\frac{r_b}{R}\right)$$

$$\times \left\{ \sum_{n=1}^{\infty} \frac{J_1(\lambda_n r_b/R)}{\lambda_n^3 [J_1(\lambda_n)]^2} = \left[1 - \mathrm{sech}\left(\frac{\lambda_n L}{2R}\right)\right] \right\}^{-1}$$

(d) Show that the result of (c) above reduces to the result obtained in Problem 3.1(d) in the limit of $L \gg R$.

(e) Obtain an appropriate criterion for the cavity length such that the simple formula of Problem 3.1(d) is a good approximation for the space charge limit of the cavity.

3.5. (a) Derive Eq. (3.63) using Eqs. (3.57)–(3.59).

(b) Using Eq. (3.63) compute the three possible solutions for the minimum electron velocity for the case of nonrelativistic flow in the one-dimensional drift space.

(c) Plot the beam current density as a function of the minimum electron velocity for several values of v_0.

3.6. Show that the first two boundary conditions of Eq. (3.78) yield Eqs. (3.79) and (3.80).

3.7. (a) For the single charge sheet model, Section 3.4.3, show that the electrostatic potential in the drift space is given by

$$\phi(z) = \left\{ \begin{array}{ll} \phi_{\min}(z/z_1), & 0 < z < z_1 \\ \phi_{\min}\left(\dfrac{L-z}{L-z_1}\right), & z_1 < z < L \end{array} \right.$$

where $\phi_{\min}$ is given by Eq. (3.97).

(b) The result of (a) is essentially the Green's function for the given geometry. Now assume that a constant density beam in excess of the space charge limit is injected into the drift space. Neglecting beam slowing effects, show that the virtual cathode ($|e\phi| = (\gamma_0 - 1)mc^2$) first forms at the position[43]

$$z_{vc} \approx [2(\gamma_0 - 1)]^{1/2}(c/\omega_p)$$

where $\omega_p^2 = (4\pi e^2 n_0/m)$.

3.8. If the axial motion of a particle beam is very nonrelativistic, i.e., $\beta_z^2 \ll 1$, the self-magnetic pinching force may be neglected.

(a) In this instance show that the radial force equation can be expressed as

$$\gamma(r)\omega_\theta^2(r) + \frac{1}{r^2}\int_0^r dr' \, r'\omega_p^2(r') - \omega_\theta\Omega + \frac{\omega_\theta}{c^2}\int_r^\infty dr' \, \omega_\theta(r')\omega_p^2(r') = 0$$

(b) For the case of a rigid rotor, $\omega_\theta = \omega_0$, show that the density profile is given by

$$\omega_p^2(r) = \frac{\omega_p^2(0)}{1 - r^2\omega_0^2/c^2}\left\{1 + \frac{2\omega_0^2}{\omega_p^2(0)}\left[1 - \frac{1 + r^2\omega_0^2/2c^2}{\left(1 - r^2\omega_0^2/c^2\right)^{1/2}}\right]\right\}$$

where $\omega_p(0)$ is given by

$$\omega_p^2(0) = 2(\omega_0\Omega - \omega_0^2)\left\{1 - \frac{r_b^2\omega_0^2/c^2}{\omega_0\Omega - \omega_0^2}\left[\omega_0\Omega - \frac{\omega_0^2}{\left(1 - r_b^2\omega_0^2/c^2\right)^{1/2}}\right]\right\}$$

(c) Show that when the azimuthal motion is also nonrelativistic, $r_b^2\omega_0^2/c^2 \ll 1$, the result of (b) reduces to a constant density profile.

3.9. (a) Show that rigid rotor equilibria are not allowed for the cases of $P_\theta = \text{const}$ or $P_\theta \propto r^2$.

(b) Show that constant axial velocity equilibria are not permissible solutions when $P_\theta \propto r^2$.

3.10. Investigate the radial profile of the beam density for the immersed source as a function of both total beam current and injected beam kinetic energy.

3.11. (a) Derive Eqs. (3.143) and (3.144).

(b) Derive Eqs. (3.145) and (3.146) when the beam does not fill the drift tube.

3.12. Show that the maximum potential at the center of an unneutralized ion beam for the geometry of Fig. 3.16 is given by

$$\phi_{\max} = 2\pi e n_{i0}(\Delta/2)^2$$

3.13. Compute the wave phase velocity for both the slow space charge wave and the slow cyclotron wave [Eqs. (3.183) and (3.184)].

3.14. Using Eq. (3.185) show that the equilibrium configuration is stable provided that ω_p^2 is a monotonically decreasing function of r.

3.15. From Eq. (3.202) derive the condition for instability, Eq. (3.203), for the case of the counterstreaming beams.

3.16. For the special case of symmetric equidensity counterstreaming electron beams, show that the maximum growth rate for the two-stream instability is given by

$$[\Gamma_i]_{\max} \approx \tfrac{1}{2}\omega_p\gamma_0^{-3/2}$$

with the corresponding axial wave vector

$$[k_z]_{\max} \approx \frac{3}{4}\frac{\omega_p^2/\gamma_0^3}{v_z^2}$$

4

Propagation of Intense Beams in Plasma

4.1. Introduction

When an intense charged particle beam is injected into a dense plasma, the beam space charge can be effectively neutralized by a redistribution of the plasma particles. For example, during injection of an intense electron beam, plasma electrons will move out of the beam region during the characteristic time $(4\pi\sigma)^{-1}$, which is typically quite short ($\ll 10^{-9}$ sec). Hence, net space charge effects (e.g., radial expansion or virtual cathode formation) do not pose serious limitations to beam transport in a plasma. Another general limitation arises, however, from the constriction of the beam due to its azimuthal self-magnetic field. In the original analyses,[1] an upper limit for current propagation was derived as

$$I_A \simeq \beta\gamma mc^3/e \qquad (4.1)$$

where, as usual, β is the particle stream velocity divided by the velocity of light and $\gamma = (1 - \beta^2)^{-1/2}$. I_A, the Alfven current limit, physically corresponds to the situation in which the beam self-magnetic field is sufficient to reverse the direction of the electron trajectories at the outer edge of the beam. In principle, this limitation can be avoided if the induced emf associated with the rising beam current is sufficient to drive a plasma current directed opposite to the beam current. Depending upon the plasma conductivity, the induction electric field can effectively cause the net current, and hence the magnetic field in the system, to vanish.

The concept of current neutralization can be introduced qualitatively by modifying Eq. (4.1) as

$$I_A^* = I_A(1 - f_m)^{-1} \qquad (4.2)$$

where f_m is the fractional magnetic neutralization due to the counterstreaming plasma current.[2] When $f_m = 1$, no net fields act on the beam particles and they propagate with linear trajectories. In this case the limitations to beam transport are due to the development of various instabilities produced by the motion of the beam through the plasma. Such unstable oscillations are usually subdivided into two classes: large-scale (macroscopic), with characteristic scale lengths of the order of the beam radius or larger, and small-scale (microscopic) for perturbations of length scale smaller than the beam radius. Examples of the macroscopic instabilities are the sausage and kink distortions, which can produce considerable deformation of the equilibrium beam configuration; examples of microscopic instabilities include the two-stream, cyclotron, and the parametric instabilities that are mainly responsible for beam–plasma heating and particle diffusion.

In this chapter important problems associated with the propagation of intense beams in plasmas are considered. The existence of the stationary state of an electron beam with counterstreaming plasma return currents is demonstrated, and the development of various beam–plasma instabilities is examined, taking into account both charge and current neutralization.

4.2. Current Neutralization

When a high-current particle beam is injected into a plasma, plasma electron currents may be induced to flow in such a manner as to cancel the self-magnetic field of the primary beam.

To illustrate the current neutralization phenomenon consider a relativistic electron beam–plasma system which is cylindrically symmetric about the z axis, the direction of beam propagation (Fig. 4.1). As the beam propagates, it perturbs the plasma inducing charges, currents, and fields as described by the Maxwell equations

$$\nabla \cdot \mathbf{E} = -4\pi e (n_p + n_b) \tag{4.3}$$

$$\nabla \cdot \mathbf{B} = 0 \tag{4.4}$$

$$\nabla \times \mathbf{B} = \frac{4\pi}{c} (\mathbf{j}_p + \mathbf{j}_b) + \frac{1}{c} \frac{\partial \mathbf{E}}{\partial t} \tag{4.5}$$

$$\nabla \times \mathbf{E} = -\frac{1}{c} \frac{\partial \mathbf{B}}{\partial t} \tag{4.6}$$

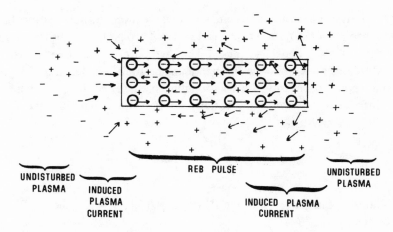

Figure 4.1. Schematic diagram of the cylindrically symmetric beam–plasma system.

The subscripts p and b denote plasma and beam quantities, respectively, and $\mathbf{j} = -en\mathbf{v}$. The plasma is assumed to consist of mobile electrons and immobile ions. It is further assumed that the plasma current j_p is related to the electric field $\mathbf{E}$ by Ohm's law

$$\mathbf{j}_p = \sigma \mathbf{E} \tag{4.7}$$

where σ is the plasma conductivity (assumed for the purpose of this calculation to be a scalar quantity).

If there are no beam or plasma θ currents, then the Maxwell equations are satisfied using field quantities of the form

$$\mathbf{E} = E_r \hat{r} + E_z \hat{z} \tag{4.8}$$

$$\mathbf{B} = B_\theta \hat{\theta} \tag{4.9}$$

In this case Eqs. (4.6) and (4.5) become (in component form)

$$\frac{\partial E_r}{\partial z} - \frac{\partial E_z}{\partial r} = -\frac{1}{c}\frac{\partial B_\theta}{\partial t} \tag{4.10}$$

$$\frac{1}{r}\frac{\partial}{\partial r}(rB_\theta) = \frac{4\pi}{c}\sigma E_z + \frac{4\pi}{c}j_{b_z} + \frac{1}{c}\frac{\partial E_z}{\partial t} \tag{4.11}$$

$$-\frac{\partial B_\theta}{\partial z} = \frac{4\pi}{c}\sigma E_r + \frac{4\pi}{c}j_{b_r} + \frac{1}{c}\frac{\partial E_r}{\partial t} \tag{4.12}$$

The current neutralization problem is thus reduced to solving the field equations [Eqs. (4.10), (4.11), and (4.12)] for a given electron beam configuration $\rho_b = -en_b$ and $\mathbf{j}_b = -en_b\mathbf{v}_b$. The method of solution employed by several treatments utilizes the Fourier–Laplace transform technique.[2-4] An alternative approach using Green's function methods is illustrated here.[5, 6]

Introducing the vector and scalar potentials $\mathbf{A}$ and ϕ defined by

$$\mathbf{B} = \nabla \times \mathbf{A}$$

$$\mathbf{E} = -\nabla \phi - \frac{1}{c}\frac{\partial \mathbf{A}}{\partial t}$$

and choosing the gauge condition $A_r = 0$, Eqs. (4.10), (4.11), and (4.12) can be written as

$$\frac{\partial^2 A_z}{\partial z\,\partial r} = \frac{4\pi}{c}j_{b_r} - \frac{\partial}{\partial r}\left(\frac{4\pi\sigma}{c} + \frac{1}{c}\frac{\partial}{\partial t}\right)\phi + \frac{4\pi}{c}\frac{\partial\sigma}{\partial r}\phi \qquad (4.13)$$

$$-\frac{1}{r}\frac{\partial}{\partial r}\left(r\frac{\partial A_z}{\partial r}\right) + \left(\frac{4\pi\sigma}{c} + \frac{1}{c}\frac{\partial}{\partial t}\right)\frac{1}{c}\frac{\partial A_z}{\partial t}$$

$$= \frac{4\pi}{c}j_{b_z} - \frac{\partial}{\partial z}\left(\frac{4\pi\sigma}{c} + \frac{1}{c}\frac{\partial}{\partial t}\right)\phi + \frac{4\pi}{c}\frac{\partial\sigma}{\partial z}\phi \qquad (4.14)$$

(A_θ has been neglected since only B_θ, E_r, and E_z are of interest.) Integrating Eq. (4.13) over r (assuming σ is not a function of r) gives

$$-\frac{\partial A_z}{\partial z} + \frac{4\pi}{c}\int_0^r j_{b_r}\,dr' + f(z,t) = \left(\frac{4\pi\sigma}{c} + \frac{1}{c}\frac{\partial}{\partial t}\right)\phi \qquad (4.15)$$

where $f(z,t)$ is an arbitrary function. Substitution of Eq. (4.15) into Eq. (4.14) yields

$$-\frac{\partial^2 A_z}{\partial z^2} - \frac{1}{r}\frac{\partial}{\partial r}\left(r\frac{\partial A_z}{\partial r}\right) + \frac{4\pi\sigma}{c^2}\frac{\partial A_z}{\partial t} + \frac{1}{c^2}\frac{\partial^2 A_z}{\partial t^2} = \frac{4\pi}{c}j_{b_z} - \frac{\partial}{\partial z}Q = S$$

$$(4.16)$$

where

$$Q = \frac{4\pi}{c}\int_0^r j_{b_r}\,dr' + f(z,t)$$

Equation (4.16) is an equation for A_z which only involves source functions. The scalar potential ϕ may be expressed in terms of A_z as [from Eq. (4.15)]

$$\phi = c \exp\left(-\int 4\pi\sigma \, dt'\right)\left[\int dt' \exp\left(4\pi \int \sigma \, dt''\right)\left(-\frac{\partial A_z}{\partial z} + Q\right) + H(r, z)\right]$$

(4.17)

where $H(r, z)$ is an arbitrary function.

If it is assumed that the beam travels at constant velocity v_b in the positive z direction, and the plasma is uniform ($\sigma = \text{const}$), Eq. (4.16) can be conveniently rewritten in terms of the beam front variable $u = \gamma(v_b t - z)$ as

$$\left(-\nabla_r^2 - \frac{\partial^2}{\partial u^2} + 2k\frac{\partial}{\partial u}\right)A(u, r) = S(u, r)$$

(4.18)

where $k = 2\pi\sigma\gamma_0 v_b / c^2$. Setting $A = We^{ku}$ yields

$$\left(\nabla_r^2 + \frac{\partial^2}{\partial u^2} - k^2\right)W = -Se^{-ku} = S'(r, u)$$

(4.19)

The function $W(r, u)$ may be expressed in terms of the Green's function $G(r, u; r', u')$ according to

$$W = \int_0^\infty r' \, dr' \int_{-\infty}^\infty du' \, G(r, u; r', u')S'(r', u')$$

(4.20)

where

$$\left(\nabla_r^2 + \frac{\partial^2}{\partial u^2} - k^2\right)G = \frac{\delta(r - r')}{r}\delta(u - u')$$

(4.21)

The explicit form of the Green's function depends on the boundary and initial conditions. Two examples of practical importance that may be easily obtained using standard methods are the following:

Case 1. Green's function for a beam in an infinite homogeneous plasma:

$$G^\infty = -\frac{1}{2}\int_0^\infty \frac{\exp\left[-(\lambda^2 + k^2)^{1/2}|u - u'|\right]}{(\lambda^2 + k^2)^{1/2}}\lambda J_0(\lambda r)J_0(\lambda r') \, d\lambda$$

(4.22)

Case 2. Green's function for a beam propagating in an infinitely long metallic waveguide of radius R filled with a homogeneous plasma:

$$G = -\frac{1}{R^2} \sum_{i=1}^{\infty} \frac{\exp\left\{-\left[(\lambda_i/R)^2 + k^2\right]^{1/2}|u-u'|\right\}}{\left[(\lambda_i/R)^2 + k^2\right]^{1/2}\left[J_1(\lambda_i)\right]^2} J_0\left(\frac{\lambda_i r}{R}\right) J_0\left(\frac{\lambda_i r'}{R}\right)$$

(4.23)

where the λ_i are the zeros of J_0.

As an application of the Green's function formalism the perturbation fields arising from the propagation of a rigid, blunt beam through a plasma-filled waveguide will be calculated (see also Ref. 7). The beam current density is specified as

$$j_{b_z} = -env_b J_0\left(\frac{\lambda_1 r}{R}\right) g(u)$$

(4.24)

$$g(u) = \begin{cases} 1, & u > 0 \\ 0, & u < 0 \end{cases}$$

The zero-order Bessel function has been chosen for the radial profile because it gives closed expressions for the fields.

Since there are no radial beam currents, $Q = 0$, and the source function S and the potential Φ are given by

$$S = \frac{4\pi}{c} j_{b_z}$$

$$\phi = ce^{-4\pi\sigma t}\left[\int dt'\, e^{4\pi\sigma t'}\left(-\frac{\partial A}{\partial z}\right)\right]$$

(4.25)

or, in terms of the parameter $u = \gamma(vt - z)$

$$\phi = \frac{1}{\beta} e^{-\varepsilon u}\int_{-\infty}^{u} du'\, e^{\varepsilon u'} \frac{\partial A}{\partial u'}$$

(4.26)

where $\varepsilon = 4\pi\sigma/\gamma v_b = 2k/\gamma^2\beta^2$.

Using the appropriate Green's function of Eq. (4.23), the function W is given by

$$W = -\frac{4\pi env_b}{R^2 c} \sum_{i=1}^{\infty} \frac{J_0(\lambda_i r/R)}{\left[(\lambda_i/R)^2 + k^2\right]^{1/2}\left[J_1(\lambda_i)\right]^2}$$

$$\times \int_{-\infty}^{\infty} du' \exp\left\{-\left[(\lambda_i/R)^2 + k^2\right]^{1/2}|u - u'|\right\}$$

$$\times g(u')e^{-ku'}\int_0^R r'\,dr'\,J_0\left(\frac{\lambda_i r'}{R}\right)J_0\left(\frac{\lambda_i r'}{R}\right) \tag{4.27}$$

Performing the integrations over r' and u' yields

$$W = -2\pi en\beta\frac{J_0(\lambda_1 r/R)}{\xi^{1/2}}$$

$$\times \begin{cases} \dfrac{e^{\sqrt{\xi}\,u}}{\eta_2}, & u < 0 \\[2ex] \dfrac{e^{-\sqrt{\xi}\,u}}{\eta_1}(e^{\eta_1 u} - 1) + \dfrac{e^{\sqrt{\xi}\,u}}{\eta_2}e^{-\eta_2 u}, & u > 0 \end{cases} \tag{4.28}$$

where the functions η_1, η_2, and ξ are defined according to

$$\xi = (\lambda_1/R)^2 + k^2 \tag{4.29}$$

$$\eta_1 = \sqrt{\xi} - k$$
$$\eta_2 = \sqrt{\xi} + k \tag{4.30}$$

Using the above expression the perturbation fields can be computed for each of the two regions of the plasma.

Region 1: $u < 0$ (region ahead of the beam)

$$B_\theta = -2\pi en\beta\left(\frac{\lambda_1}{R}\right)J_1\left(\frac{\lambda_1 r}{R}\right)\frac{e^{\eta_2 u}}{\eta_2\sqrt{\xi}} \tag{4.31}$$

$$E_r = -2\pi en\left(\frac{\lambda_1}{R}\right)J_1\left(\frac{\lambda_1 r}{R}\right)\frac{e^{\eta_2 u}}{(\varepsilon + \eta_2)\sqrt{\xi}} \tag{4.32}$$

$$E_z = -2\pi en\gamma J_0\left(\frac{\lambda_1 r}{R}\right)\frac{e^{\eta_2 u}}{\sqrt{\xi}}\left(\frac{\eta_2}{\varepsilon + \eta_2} - \beta^2\right) \tag{4.33}$$

Region 2: $u > 0$ (region behind the beam front)

$$B_\theta = -2\pi e n \beta \left(\frac{\lambda_1}{R} \right) J_1 \left(\frac{\lambda_1 r}{R} \right) \left[\frac{(1 - e^{-\eta_1 u})}{\eta_1 \sqrt{\xi}} + \frac{1}{\eta_2 \sqrt{\xi}} \right] \qquad (4.34)$$

$$E_r = \frac{-2\pi e n}{\sqrt{\xi}} \left(\frac{\lambda_1}{R} \right) J_1 \left(\frac{\lambda_1 r}{R} \right) \left(\frac{e^{-\epsilon u}}{\eta_2 + \epsilon} + \frac{e^{-\eta_1 u} - e^{-\epsilon u}}{\epsilon - \eta_1} \right) \qquad (4.35)$$

$$E_z = \frac{-2\pi e n \gamma}{\sqrt{\xi}} J_0 \left(\frac{\lambda_1 r}{R} \right)$$

$$\times \left(\frac{-\epsilon e^{-\epsilon u}}{\eta_2 + \epsilon} + \frac{-\eta_1 e^{-\eta_1 u} + \epsilon e^{-\epsilon u}}{\epsilon - \eta_1} - \beta^2 e^{-\eta_1 u} \right) \qquad (4.36)$$

At this point several properties of the field expressions can be discussed. As required, $E_r = B_\theta = 0$ at $r = 0$ while E_z is a maximum at $r = 0$ and vanishes at the waveguide radius. The fields extend ahead of the beam and decay exponentially with a characteristic length $1/\gamma \eta_2 = (1/\gamma)[(k^2 + (\lambda_1/R)^2)^{1/2} + k]^{-1}$. Behind the beam front the longitudinal electric field E_z changes sign (so as to accelerate plasma electrons in a direction opposite the primary beam electrons).

In the limit of high conductivity ($k \gg \lambda_1/R$), $\eta_2 \simeq 2k$, $\eta_1 \simeq \frac{1}{2}k(\lambda_1/Rk)^2$, and the field expressions become

$u < 0$:

$$B_\theta = -2\pi e n \beta \left(\frac{\lambda_1}{R} \right) J_1 \left(\frac{\lambda_1 r}{R} \right) \frac{e^{2ku}}{2k^2} \qquad (4.37)$$

$$E_r = -2\pi e n \left(\frac{\lambda_1}{R} \right) J_1 \left(\frac{\lambda_1 r}{R} \right) \left(\frac{\gamma^2 - 1}{2k^2 \gamma^2} \right) e^{2ku} \qquad (4.38)$$

$$E_z \simeq 0 \qquad (4.39)$$

$u > 0$:

$$B_\theta = -2\pi e n \left(\frac{\lambda_1}{R} \right) J_1 \left(\frac{\lambda_1 r}{R} \right) \beta \left\{ 2 \left\{ 1 - \exp \left[-\left(\frac{\lambda_1}{Rk} \right)^2 \frac{ku}{2} \right] \right\} + \frac{1}{2} \left(\frac{\lambda_1}{Rk} \right)^2 \right\}$$

$$(4.40)$$

$$E_r = -2\pi e n \left(\frac{\lambda_1}{R}\right) J_1 \left(\frac{\lambda_1 r}{R}\right) \frac{\gamma^2 - 1}{2k^2} \left\{ \exp\left[-\left(\frac{\lambda_1}{Rk}\right)^2 \frac{ku}{2} \right] - \beta^2 \exp\left(\frac{-2ku}{\gamma^2 - 1} \right) \right\}$$

(4.41)

$$E_z = 2\pi e n J_0 \left(\frac{\lambda_1 r}{R}\right) \left(\frac{\gamma^2 - 1}{k\gamma}\right) \left\{ \exp\left[-\left(\frac{\lambda_1}{Rk}\right)^2 \frac{ku}{2} \right] - \exp\left(\frac{-2ku}{\gamma^2 - 1} \right) \right\} \quad (4.42)$$

In this case the net longitudinal current $j_{z_{net}}$ due to both the primary beam current and the plasma return current is given by

$$j_{z_{net}} = j_b \left\{ 1 - \left\{ \exp\left[-\left(\frac{\lambda_1}{R}\right)^2 \frac{c^2}{4\pi\sigma} \left(t - \frac{z}{v_b} \right) \right] - \exp\left[-4\pi\sigma \left(t - \frac{z}{v_b} \right) \right] \right\} \right\}$$

(4.43)

Hence, the return current increases rapidly with distance from the beam front. [$(4\pi\sigma)^{-1}$ is recognized as the characteristic charge neutralization time, i.e., the time required for the plasma charge to redistribute so as to nullify the electric field produced by the beam.] Note also that the radial dependence of the plasma current is identical to that of the primary current; the return current is induced to flow in the region of the beam. Decay of the plasma current due to the finite plasma conductivity occurs over the characteristic decay time τ_c, which is given by

$$\tau_c = \frac{4\pi\sigma}{c^2} \left(\frac{R}{\lambda_1}\right)^2 = \frac{\tau_B}{\lambda_1^2} \quad (4.44)$$

where τ_B is the characteristic magnetic diffusion time. The qualitative behavior of the current density (net) is indicated in Fig. 4.2.

For the case in which the conductivity is low ($k \ll \lambda_1/R$) geometric effects are predominant. In this limit the net longitudinal current becomes

$$j_{z_{net}} \approx j_b \left\{ 1 - \frac{1}{\gamma^2 - 1} \left(\frac{Rk}{\lambda_1}\right) \exp\left[-\frac{\lambda_1 u}{R} \left(1 - \frac{kR}{\lambda_1} \right) \right] \right\} \quad (4.45)$$

When the conductivity decreases to zero, the return current flows entirely in the drift tube wall.

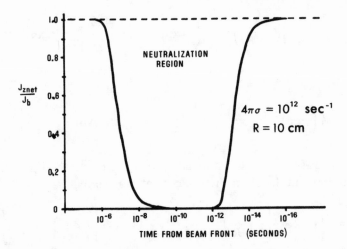

Figure 4.2. Variation of the net current density with time (distance) from the beam front in the limit of high plasma conductivity. Current neutralization is achieved rapidly (10^{-13} sec), and slowly decays over hundreds of nanoseconds.

Current neutralization calculations have also been performed assuming the presence of a magnetic field applied either parallel or transverse to the motion of the beam [8, 9] In the first case if the beam plasma frequency exceeds the electron cyclotron frequency, $\Omega/\omega_\rho \lesssim 1$, then the current neutralization process is essentially the same as for the case of zero applied field discussed above. If $\Omega/\omega_\rho \gg 1$, however, the flow of plasma electrons out of the beam region is impeded and charge neutralization cannot occur. In this case the electrostatic component of the electric field parallel to the beam motion is larger than the inductive component. The plasma electrons are accelerated forward at the beam front and the net current increases. When the magnetic field is applied perpendicular to the beam motion a return current can still be induced to flow across the field. In this case the charge neutralization process leads to a nonaxisymmetric plasma electron distribution, and the resulting electric field cancels the magnetic force on the return current electrons.

4.3. Macroscopic Beam–Plasma Equilibria

In the previous section the response of the plasma to the injection of a charged particle beam was examined. In this section we wish to couple the beam and plasma fluid components in order to develop various equilibrium configurations. To proceed, the fluid Maxwell equations of Section 3.2 are generalized to include the effects of finite beam and plasma temperatures. Assuming that the plasma ions form a stationary background, these equations become

$$\frac{1}{r}\frac{d}{dr}(rE_r) = 4\pi e(n_i - n_e - n_b) \tag{4.46}$$

$$\frac{1}{r}\frac{d}{dr}(rB_\theta) = -\frac{4\pi e}{c}(n_e v_{ez} + n_b v_{bz}) \tag{4.47}$$

$$-\frac{m\gamma_b v_{b\theta}^2}{r} + \frac{kT_b}{n_b}\frac{d}{dr}n_b = -e\left[E_r + \frac{1}{c}(v_{b\theta}B_0 - v_{bz}B_\theta)\right] \tag{4.48}$$

$$-\frac{m\gamma_e v_{e\theta}^2}{r} + \frac{kT_e}{n_e}\frac{d}{dr}n_e = -e\left[E_r + \frac{1}{c}(v_{e\theta}B_0 - v_{ez}B_\theta)\right] \tag{4.49}$$

In writing Eqs. (4.48) and (4.49) it has been assumed that components of the pressure tensors perpendicular to the z axis are given by $p_\alpha = n_\alpha kT_\alpha$, where k is the Boltzmann constant, and T_e and T_b are the isotropic plasma electron and beam electron kinetic temperatures. The constant B_0 corresponds to the application of a uniform axial magnetic field.

4.3.1. Warm Fluid Equilibria with $B_0 = 0$

As a first application of Eqs. (4.46)–(4.49), it is assumed that $B_0 = 0$, and that beam and plasma centrifugal effects are negligible. In this case Eqs. (4.48) and (4.49) become

$$\frac{kT_b}{n_b}\frac{dn_b}{dr} = -e\left\{E_r - \frac{v_{bz}}{c}B_\theta\right\} \tag{4.50}$$

$$\frac{kT_e}{n_e}\frac{dn_e}{dr} = -e\left\{E_r - \frac{v_{ez}}{c}B_\theta\right\} \tag{4.51}$$

Further consideration of these equations leads to specific cases of practical importance. Historically, the first example of a high-current, warm-fluid equilibrium was the Bennett pinch condition.[10, 11] It can be

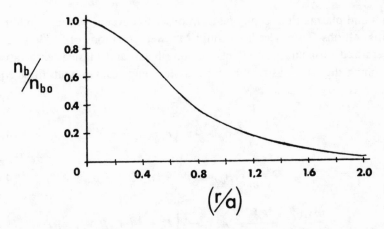

Figure 4.3. Beam density variation as given by the Bennett distribution.

derived by assuming that $n_e = 0$, $v_{bz}/c = \beta_0 = \text{const}$, and $f_e = n_i/n_b = \text{const}$. In this case Eq. (4.50) becomes

$$\frac{kT_b}{n_b}\frac{dn_b}{dr} = \frac{4\pi e^2}{r}\left(1 - f_e - \beta_0^2\right)\int_0^r r\,dr\,n_b \qquad (4.52)$$

If n_b is a monotonically decreasing function of radius ($dn_b/dr \leqslant 0$), and if $\beta_0^2 > 1 - f_e$, then an equilibrium configuration exists in which the outward forces of the beam fluid pressure and the electrostatic repulsion are balanced by the magnetic pinching forces. Differentiation of Eq. (4.52) with respect to r yields

$$\frac{d^2 n_b}{dr^2} + \frac{1}{r}\frac{dn_b}{dr} - \frac{1}{n_b}\left(\frac{dn_b}{dr}\right)^2 + Kn_b^2 = 0 \qquad (4.53)$$

where the constant $K = (4\pi e^2/kT_b)[\beta_0^2 - (1 - f_e)]$. The solution of Eq. (4.53) is given by the Bennett profile (Fig. 4.3)

$$n_b = n_{b0}\left[1 + (r/a)^2\right]^{-2} \qquad (4.54)$$

where n_{b0} denotes the beam density on axis, and a, the Bennett radius, is given by

$$a^2 = \frac{8}{Kn_{b0}} = \frac{8\lambda_D^2}{\beta_0^2 - (1 - f_e)} \qquad (4.55)$$

$\lambda_D^2 = kT_b/4\pi n_{b0}e^2$ denotes the beam electron Debye length at $r = 0$. Note that a^2 is a minimum for $f_e \approx 1$, and becomes large if $f_e \approx \gamma_0^{-2}$.

Generalized Bennett relations can also be obtained for the case of space charge neutral equilibria ($E_r = 0$), which are assumed by setting $n_e + n_b = n_{i0} = $ const. In this case it is easy to show that[12]

$$B_\theta^2 = 8\pi k(T_e - T_b)\left(n_b - \frac{2}{r^2}\int_0^r r\,dr\, n_b\right) \qquad (4.56)$$

$$v_{bz} = \frac{ckT_b}{eB_\theta}\left(\frac{1}{n_b}\frac{dn_b}{dr}\right) \qquad (4.57)$$

$$v_{ez} = -\frac{ckT_e}{eB_\theta}\left(\frac{1}{n_{i0} - n_b}\frac{dn_b}{dr}\right) \qquad (4.58)$$

If it is further assumed that the beam velocity is constant across the beam radial profile ($v_{bz} = v_{b0}$), then Eqs. (4.56) and (4.57) yield

$$\frac{d^2 n_b}{dr^2} + \frac{1}{r}\frac{dn_b}{dr} - \frac{1}{n_b}\left(\frac{dn_b}{dr}\right)^2 + K^1 n_b^2 = 0 \qquad (4.59)$$

where the constant

$$K^1 = \left(\frac{e\beta_0}{T_b}\right)^2 \frac{4\pi(T_b - T_e)}{k}$$

with $\beta_0 = v_{b0}/c$. In analogy with Eq. (4.53), the solution to Eq. (4.59) is given by Eq. (4.54), with the Bennett radius specified as

$$a^2 = \frac{8}{K^1 n_{b0}} = \frac{8\lambda_D T_b}{\beta_0^2(T_b - T_e)} \qquad (4.60)$$

Note that physically acceptable solutions are obtained only if the beam temperature exceeds the plasma electron temperature. In this case the degree of current neutralization is easily shown to be $f_m = T_e/T_b$ from Eqs. (4.57) and (4.58).

4.3.2. Warm Fluid Equilibria in a Discharge Channel

For certain applications requiring the transport of beam currents in excess of the Alfven limit it is desirable to inject the beam into a current-carrying plasma discharge channel.[13-16] The plasma provides both charge and current neutralization, while the azimuthal magnetic field of the discharge current provides radial confinement of the beam electrons. In this case Eq. (4.47) becomes

$$\frac{1}{r}\frac{d}{dr}(rB_\theta) = -\frac{4\pi j_d}{c}$$ (4.61)

where j_d is the discharge current density. Assuming that the equilibrium is electrically neutral ($E_r = 0$), and that the beam temperature and axial velocity are constant, Eqs. (4.50) and (4.61) yield

$$\frac{d^2 n_b}{dr^2} + \frac{1}{r}\left(\frac{dn_b}{dr}\right) - \frac{1}{n_b}\left(\frac{dn_b}{dr}\right)^2 + K^* j_d n_b = 0$$ (4.62)

where $K^* = 4\pi e\beta_0/ckT_b$. Eq. (4.62) has Bennett profile solutions only if $j_d(r) \propto n_b(r)$.

4.3.3. Cold Fluid Equilibria with an Axial Magnetic Field

Equations (4.46)–(4.49) can also be used to obtain cold-beam–cold-plasma fluid equilibria following the development contained in Section 3.5.1. Dropping the pressure terms from Eqs. (4.48) and (4.49) yields

$$\gamma_0 \omega_{b\theta}^2 + \Omega\omega_{b\theta} - \frac{e}{mr}(E_r - \beta_{b0}B_\theta) = 0$$ (4.63)

$$\omega_{e\theta}^2 + \Omega\omega_{e\theta} - \frac{e}{mr}(E_r - \beta_{ez}B_\theta) = 0$$ (4.64)

where $\beta_\alpha = v_\alpha/c$, $\Omega = eB_0/mc$, and $\omega_\alpha = v_\alpha/r$. In writing these equations it has been assumed that $\gamma_b \approx (1 - \beta_{b0}^2)^{-1/2} = \gamma_0 = $ const, and that B_z contributions from beam and plasma electron rotations are negligible. Solving Eq. (4.63) for $\omega_{b\theta}$ yields

$$\omega_{b\theta}^\pm = \frac{\Omega}{2\gamma_0}\left\{1 \pm \left[1 + \frac{4e\gamma_0(E_r - \beta_{b0}B_\theta)}{mr\Omega^2}\right]^{1/2}\right\}$$ (4.65)

It is further assumed for simplicity that the beam density is constant ($n_b = n_{b0}$), and that there are constant charge and current neutralization fractions defined by

$$f_e = \frac{n_{i0} - n_e}{n_b} \qquad (4.66)$$

$$f_m = -\frac{n_e v_{ez}}{n_{b0} v_{b0}} \qquad (4.67)$$

In this case Eq. (4.65) becomes simply

$$\omega_{b\theta}^{\pm} = \frac{\Omega}{2\gamma_0} \left\{ 1 \pm \left[1 - \frac{2\omega_{b0}^2 \gamma_0}{\Omega^2} \left((1 - f_e) - \beta_{b0}^2 (1 - f_m) \right) \right]^{1/2} \right\} \qquad (4.68)$$

and it is observed that these are rigid rotor equilibria.

If there is no external guide field ($\Omega = 0$), then the rotational velocities are given by

$$\omega_{b\theta}^{\pm} = \pm \left[\omega_{b0} / (2\gamma_0)^{1/2} \right] \left[\beta_{b0}^2 (1 - f_m) - 1 + f_e \right]^{1/2} \qquad (4.69)$$

The condition for radial confinement of the beam is that the radical be real, i.e.,

$$\beta_{b0}^2 (1 - f_m) \geqslant 1 - f_e \qquad (4.70)$$

Equation (4.70) simply states that the electrostatic repulsive force must be smaller than the magnetic pinching force resulting from the interaction of the beam velocity with the magnetic field of the net current.

If there is an external magnetic field present the equilibrium criterion is somewhat less stringent. The condition for the radical to be real in Eq. (4.68) is

$$\beta_0^2 (1 - f_m) + \Omega^2 / 2\gamma_0 \omega_{b0}^2 \geqslant 1 - f_e \qquad (4.71)$$

From Section 4.2, if $\Omega/\omega_{b0} \gg 1$, then both the charge and current neutralization processes will be impeded, yet an equilibrium is possible provided that $\Omega^2 \geqslant 2\omega_{b0}^2/\gamma_0$. If $\Omega/\omega_{b0} \lesssim 1$, then $f_e \sim 1$ and $f_m \sim 1$, and Eq.

(4.68) yields

$$\omega_{b\theta}^{+} \simeq \frac{\Omega}{2\gamma_0} \left\{ 2 - \frac{\gamma_0 \omega_{b0}^2}{\Omega^2} \left[(1 - f_e) - \beta_{b0}^2 (1 - f_m) \right] \right\} \qquad (4.72)$$

$$\omega_{b\theta}^{-} \simeq \frac{\omega_{b0}^2}{2\Omega} \left[(1 - f_e) - \beta_{b0}^2 (1 - f_m) \right] \qquad (4.73)$$

In the limit $f_e = 1$, the slow mode rotation direction is always opposite that of the fast mode, regardless of the degree of current neutralization. In this case consideration of Eq. (4.64) indicates that the plasma electron rotation rate is given by

$$\omega_{e\theta}^{-} = \frac{\omega_{b0}^2}{2\Omega} \frac{\beta_{b0}^2 f_m (1 - f_m)}{(n_{i0}/n_{b0} - 1)} > 0 \qquad (4.74)$$

Combining Eqs. (4.73) and (4.74), the net azimuthal current for the slow rotation mode is given by

$$j_{n\theta} = j_{b\theta} + j_{e\theta} = - e n_{b0} r \frac{\omega_{b0}^2 \beta_{b0}^2}{2\Omega} (1 - f_m)^2 \qquad (4.75)$$

Hence, if the axial beam current is completely neutralized, then the slow mode azimuthal beam current is also neutralized because both the beam and plasma rotations vanish.

While the calculations of this section have assumed that the plasma is unbounded, it should be mentioned that many of the results are also applicable to intense beam transport in plasma channels. In general, if the channel radius exceeds the beam radius, and the skin depth of the beam is less than half the beam radius, then steady state equilibria can be found for which the beam current exceeds the Alfven limit.[17]

4.4. Macroscopic Beam–Plasma Instabilities

Having analyzed a variety of steady-state beam plasma equilibria it is now necessary to examine the system stability with respect to small-scale perturbations. The macroscopic hose (kink) and sausage instabilities are examined first. In the case of the hose instability the perturbation is a transverse displacement of the beam; for the sausage mode the perturbation is an azimuthally symmetric change in beam radius.

4.4.1. Resistive Hose Instability

An intense beam propagating through a plasma can be hydromagnetically unstable to transverse perturbations because of the finite plasma conductivity.[18-23] When the beam axis undergoes a small displacement, the magnetic field axis lags behind for times on the order of the magnetic diffusion time $\tau_B = 4\pi\sigma r_b^2/c^2$, and a restoring force, due to the interaction between the longitudinal beam current and the perpendicular component of the residual magnetic field, tends to push the beam back to its original position. In contrast to the kink instability of a self-pinched discharge, the resistive hose instability is essentially controlled by the plasma electron collision rate (which determines the plasma conductivity). Under collisionless conditions the plasma is a perfect conductor ($\tau_B \to \infty$), and the magnetic field of the beam is "frozen" in the plasma. In this limit, beam displacements simply result in betatron oscillations. If σ is finite however, the beam displacements and the restoring forces can become out of phase leading to instability (Fig. 4.4).

As an example of hose instability analysis consider a highly relativistic electron beam with radius r_b propagating with constant velocity $\beta_0 c$ along the axis of an infinitely conducting guide tube of radius R filled with plasma characterized by scalar conductivity σ.* Transverse velocities are assumed to be nonrelativistic. In addition, the beam is assumed to be completely charge neutralized, while the plasma electron currents are taken into account through the simple Ohm's law, $j_p = \sigma E$. Since it is the interaction between j_{bz} and $B_\perp$ that is of interest, the relevant Maxwell equation is (neglecting the displacement current)

$$\frac{1}{r}\frac{\partial}{\partial r}(rB_\theta) - \frac{1}{r}\frac{\partial B_r}{\partial \theta} = \frac{4\pi}{c}(j_{bz} + \sigma E_z) \qquad (4.76)$$

---

*Mjolsness et al. have considered the tensor character of the plasma (due to the beam self field) and have found the effects to be small.[24]

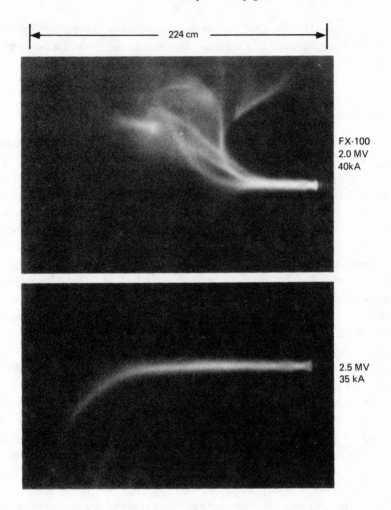

Figure 4.4. Hose instability in a relativistic electron beam (anode: 0.0127 cm Titanium). Note the sausage mode oscillations near the injection plane.

Introducing the vector potential according to $\mathbf{B} = \nabla \times \mathbf{A}$ Eq. (4.76) may be expressed in terms of A_z as

$$\frac{1}{r}\frac{\partial}{\partial r}r\frac{\partial A_z}{\partial r} + \frac{1}{r^2}\frac{\partial^2 A_z}{\partial \theta^2} = -\frac{4\pi}{c}j_{bz} + \frac{4\pi\sigma}{c}\frac{\partial A_z}{dt} \qquad (4.77)$$

with choice of gauge condition $E_z = -(1/c)\partial A_z/\partial t$.

If the beam axis undergoes a small transverse displacement, y, each beam electron experiences a restoring force given by

$$F_\perp = \gamma_0 m \frac{dv_\perp}{dt} = -e(\boldsymbol{\beta}_0 \times \mathbf{B})_\perp \qquad (4.78)$$

where $v_\perp = (d/dt)y$, and it has been assumed that $\boldsymbol{\beta}_0 = (v_{z0}/c)\hat{e}_z$ and $\gamma_0 = (1 - \beta_0^2)^{-1/2}$.

If it is further assumed that the transverse beam displacement occurs without deforming the beam cross section (rigid beam approximation),[18] then the force on the beam is the average

$$\gamma_0 m \frac{d^2 y}{dt^2} = \frac{e^2 \beta^2 c}{I_b} \int_0^{2\pi} d\theta \int_0^R r n_b (\hat{e}_z \times \mathbf{B}) \, dr \qquad (4.79)$$

where n_b denotes the beam density and I_b the total beam current.

In the rigid beam approximation the longitudinal current density is given by

$$j_{bz}(r) \simeq j_{b0} - \mathbf{y} \cdot \nabla j_{b0} \qquad (4.80)$$

where j_{b0} is the unperturbed current density and only the linear term of the expansion has been retained. For a uniform current density out to a maximum beam radius r_b, Eq. (4.80) becomes

$$j_{bz} = j_{b0} + \mathbf{y} \cdot \hat{e}_r j_{b0} \delta(r - r_b) \qquad (4.81)$$

Without loss of generality it may be taken that

$$\mathbf{y} = y\hat{e}_y \qquad \text{and} \qquad j_{bz} = j_{b0} + \delta j_b$$

where

$$\delta j_b = y(\sin\theta) j_{b0} \delta(r - r_b).$$

Assuming a linearization of A_z according to*

$$A_z = A_{z0} + \delta A_z \sin\theta \qquad (4.82)$$

---

*Equation (4.82) corresponds to taking the first term of a multipole expansion.

Eq. (4.77) can be written as

$$\frac{1}{r}\frac{\partial}{\partial r}\left(r\frac{\partial A_{z0}}{\partial r}\right) = -\frac{4\pi}{c}j_{b0} \tag{4.83}$$

$$\frac{1}{r}\frac{\partial}{\partial r}\left(r\frac{\partial}{\partial r}\delta A_z\right) - \frac{\delta A_z}{r} - \frac{4\pi\sigma}{c^2}\frac{\partial}{\partial t}\delta A_z = -\frac{4\pi}{c}j_{b0}y\delta(r-r_b) \tag{4.84}$$

With the specific form of the assumed displacement Eq. (4.79) becomes

$$\left(m\gamma_0\pi r_b^2 n_b\right)\left(\frac{\partial}{\partial t} + \beta_0 c\frac{\partial}{\partial z}\right)^2 y = \frac{1}{c}\int_0^{2\pi}d\theta\int_0^R dr\, r\left(B_{0x}\delta j_b + j_{b0}\delta B_x\right) \tag{4.85}$$

where

$$B_x = (\cos\theta)B_r - (\sin\theta)B_\theta = (\cos\theta)\left(\frac{1}{r}\frac{\partial A_z}{\partial\theta}\right) + (\sin\theta)\left(\frac{\partial A_z}{\partial r}\right) \tag{4.86}$$

Hence,

$$B_{0x} = (\sin\theta)\left(\frac{\partial A_{z0}}{\partial r}\right)$$

$$\delta B_x = (\cos^2\theta)\left(\frac{\delta A_z}{r}\right) + (\sin^2\theta)\left(\frac{\partial}{\partial r}\delta A_z\right)$$

and Eq. (4.85), after performing the integrations, can finally be written as

$$\left(\frac{\partial}{\partial t} + \beta_0 c\frac{\partial}{\partial z}\right)^2 y = \frac{\pi e\beta_0 j_{b0}r_b}{\gamma_0 m I_b}\left(y\frac{\partial A_{z0}}{\partial r}\bigg|_{r_b} + \delta A_z\big|_{r_b}\right) \tag{4.87}$$

To analyze the system stability Eqs. (4.83), (4.84), and (4.87) must be solved subject to the appropriate boundary conditions. Equation (4.83) may be integrated to yield

$$\frac{\partial A_{z0}}{\partial r}\bigg|_{r_b} = \frac{1}{r_b}\int_0^{r_b}dr\, r\left(-\frac{4\pi j_{b0}}{c}\right) = -\frac{2I_b}{r_b c} \tag{4.88}$$

Assuming for the perturbed quantities solutions of the form $y = ye^{i(k_z z - \omega t)}$, $\delta A_z = \delta A_z e^{i(k_z z - \omega t)}$, substitution of Eq. (4.88) into Eq. (4.87) yields the beam response given by

$$y = \frac{\omega_\beta^2}{\omega_\beta^2 - (\omega - \beta_0 c k_z)^2} \frac{r_b c}{2I_b} \delta A_z \bigg|_{r_b} \qquad (4.89)$$

and Eq. (4.84) becomes

$$\left(\frac{1}{r} \frac{\partial}{\partial r} r \frac{\partial}{\partial r} - \frac{1}{r^2} + \frac{4\pi \sigma i \omega}{c^2} \right) \delta A_z = \left(-\frac{2\omega_\beta^2}{\omega_\beta^2 - \Omega^2} \right) \delta A_z(r_b) \frac{\delta(r - r_b)}{r_b}$$

$$(4.90)$$

where ω_β, the betatron frequency, is given by

$$\omega_\beta = \left(\frac{e\beta_0}{\gamma_0 m} \frac{2I_b}{r_b^2 c} \right)^{1/2}$$

and $\Omega = \omega - \beta_0 c k_z$ is the wave frequency Doppler shifted to the beam frame.

The appropriate boundary conditions are that A_z is continuous at the beam boundary ($r = r_b$), remains finite at the origin ($r = 0$), and vanishes at the conducting wall ($r = R$). Hence, $r\delta A_z|_{r=0} = 0$, $\delta A_z(r_{b-}) = \delta A_z(r_{b+})$, $\delta A_z(R) = 0$.

In addition, δA_z must satisfy the jump condition at the beam radius given from Eq. (4.90) as

$$\left(\frac{\partial \delta A_z}{\partial r} \right)_{r_{b+}} - \left(\frac{\partial \delta A_z}{\partial r} \right)_{r_{b-}} = -\frac{2\omega_\beta^2}{\omega_\beta^2 - \Omega^2} \frac{\delta A_z(r_b)}{r_b} \qquad (4.91)$$

Defining the quantities x and g by $x = r/r_b$, $g^2 = -(4\pi\sigma r_b^2/c^2)i\omega = -i\omega\tau_B$, Eq. (4.90) can be rewritten as

$$\frac{1}{x} \frac{d}{dx} x \frac{d}{dx} \delta A_z - \frac{\delta A_z}{x^2} - g^2 \delta A_z = -\frac{2\omega_\beta^2}{\omega_\beta^2 - \Omega^2} \delta(x - 1) \delta A_z \qquad (4.92)$$

Equation (4.92) is recognized as a modified Bessel equation except at the beam radius ($x = 1$). Hence, the correct solutions for the regions interior and exterior to the beam must be matched appropriately at $x = 1$.

For $0 < x < 1$:

$$\delta A_z = \alpha I_1(gx) + \alpha' K_1(gx)$$

and for $1 < x < R/r_b$

$$\delta A_z = \beta I_1(gx) + \beta' K_1(gx)$$

Since δA_z must remain finite at the origin, $\alpha' = 0$. The vanishing of δA_z at the conducting boundary requires the ratio β'/β to be

$$\frac{\beta'}{\beta} = -\frac{I_1(gR/r_b)}{K_1(gR/r_b)} \tag{4.93}$$

Hence the solution is of the form

$$\delta A_z = \begin{cases} \alpha I_1(gx), & 0 < x < 1 \\ \beta \left[I_1(gx) - K(gx)\dfrac{I_1(gR/r_b)}{K_1(gR/r_b)} \right], & 1 < x < R/r_b \end{cases} \tag{4.94}$$

When the continuity condition on δA_z at the beam radius and the jump condition [Eq. (4.91)] are applied to the solution, Eq. (4.94), the following relations are obtained:

$$\alpha I_1(g) = \beta \left[I_1(g) - \frac{I_1(gR/r_b)}{K_1(gR/r_b)} K_1(g) \right] \tag{4.95}$$

$$\beta g \left[I_1'(g) - \frac{I_1(gR/r_b)}{K_1(gR/r_b)} K_1'(g) \right] - \alpha g I_1'(g) = -\frac{2\omega_\beta^2}{\omega_\beta^2 - \Omega^2} I_1(g) \tag{4.96}$$

Equations (4.95) and (4.96) are linear, homogeneous equations. For the existence of a nontrivial solution the determinant of the coefficients of α and β must vanish. This condition determines the dispersion relation. Use of Bessel function identities and some algebra yields

$$1 = \frac{\omega_\beta^2}{\omega_\beta^2 - \Omega^2} 2 I_1^2(g) \left[\frac{K_1(g)}{I_1(g)} - \frac{K_1(gR/r_b)}{I_1(gR/r_b)} \right] \tag{4.97}$$

Inspection of the dispersion relation [Eq. (4.97)] indicates that if the beam radius is equal to the radius of the conducting wall, $r_b = R$, only stable oscillations are obtained with $\Omega = \omega - \beta_0 c k_z = \pm \omega_\beta$. For $R > r_b$, however, the possibility exists for instability.

Equation (4.97) may be written as

$$1 = \frac{\omega_\beta^2}{\omega_\beta^2 - \Omega^2} F(g) \tag{4.98}$$

where

$$F(g) = 2I_1^2(g)\left[\frac{K_1(g)}{I_1(g)} - \frac{K_1(gR/r_b)}{I_1(gR/r_b)}\right] \tag{4.99}$$

In the low-frequency limit ($g \ll 1$) the function F can be expanded in powers of g^2. In the limit of small argument x the modified Bessel functions may be written as

$$I_1(x) = \frac{x}{2} + \frac{x^3}{16} + \cdots$$

$$K_1(x) = \left(\gamma + \ln\frac{x}{2}\right)\frac{x}{2} + \frac{1}{x} - \frac{1}{4} + \cdots$$

where $\gamma = 0.577\ldots$. Substituting these expansions into Eq. (4.99) and retaining only zeroth- and first-order terms yields

$$F(g) \simeq \left[1 - (r_b/R)^2\right]\left\{1 - \frac{g^2}{4}\left[\frac{2\ln(R/r_b)}{1 - (r_b/R)^2} - 1\right]\right\} \tag{4.100}$$

Inserting Eq. (4.100) into Eq. (4.98) gives (to lowest order)

$$\frac{g^2}{4}\left[\frac{2\ln(R/r_b)}{1 - (r_b/R)^2} - 1\right] = \frac{\Omega^2 - (r_b\omega_\beta/R)^2}{\omega_\beta^2 - \Omega^2} \tag{4.101}$$

The frequency $\omega_\beta' = r_b\omega_\beta/R$ may be thought of as the reduced betatron frequency associated with wall focusing of the displaced beam. (Note that ω_β' is independent of the beam radius.) Defining the low-frequency decay constant

$$\alpha^{-1} = \frac{\tau_B}{4}\left[\frac{2\ln(R/r_b)}{1 - (r_b/R)^2} - 1\right] \tag{4.102}$$

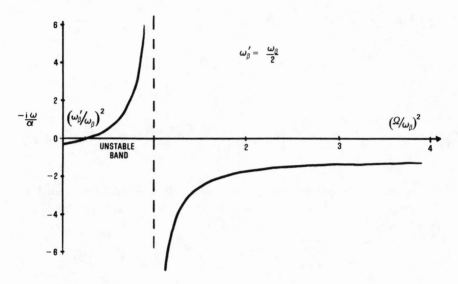

Figure 4.5. Variation of the hose instability growth rate with frequency. Instability exists over the frequency range of $\omega_\beta'^2 < \Omega^2 < \omega_\beta^2$.

Eq. (4.101) can be rewritten as

$$-i\omega = \alpha \frac{\Omega^2 - \omega_\beta'^2}{\omega_\beta^2 - \Omega^2} \qquad (4.103)$$

The growth rate is plotted against real Ω in Fig. 4.5. Instability occurs when $\Omega^2 > \omega_\beta'^2$; for $\Omega^2 < \omega_\beta'^2$, the dangerous low-frequency range, instability does not occur. In essence, the conducting wall provides a stabilizing restoring force on the beam; because of the very high wall conductivity the induced image currents in the wall cannot get out of phase with the beam perturbations.

Depending on how the stability problem is posed, either Ω or ω may be regarded as the independent variable, and taken to be real. For Eq. (4.103) if a beam segment is taken to oscillate at a fixed real frequency Ω, then ω is pure imaginary and represents the growth rate of the instability as a function of distance from the beam front. On the other hand, in experimental studies of hose growth[25-26] the perturbation occurs at a fixed position; ω is thus the real frequency of the deflection, and Ω (complex in general) represents the oscillation and growth of the wave as a particular beam segment propagates. In this case the appropriate form of Eq. (4.103) is

$$\beta_0 c k_z = \omega \pm \left(\frac{- i \omega \omega_\beta^2 + \alpha \omega_\beta'^2}{\alpha - i \omega} \right)^{1/2} \qquad (4.104)$$

where the square root is taken such that $\text{Im}(k_z) < 0$; i.e., growth occurs in the $+z$ direction.

The rigid beam model is valid for the low-frequency behavior ($\Omega \ll \omega_\beta$). As Ω approaches the betatron resonance, however, the beam displacement becomes dominated by groups of resonant particles. In this frequency range the beam motion is not rigid and a Vlasov kinetic model of the beam must be used to investigate the linear stability. The results of this approach indicate that the rigid-beam analysis can be retained provided that the quantity $(\Omega^2 - \omega_\beta'^2)/(\omega_\beta^2 - \Omega^2)$ is replaced by the appropriate integral over the distribution of betatron frequencies, $f(\omega_\beta)$. In the limit that $R/r_b \gg 1$, Eq. (4.103) then becomes

$$- i \omega = \alpha \int_0^\infty d\omega_\beta f(\omega_\beta) \left(\frac{\Omega^2}{\omega_\beta^2 - \Omega^2} \right) \qquad (4.105)$$

and it is observed that the betatron frequency spread detunes the resonance and limits the growth rate. If $2\Delta\omega_\beta$ is the width of that spread, then

$$- i \omega \lesssim \frac{\alpha \omega_\beta}{2 \Delta \omega_\beta} \qquad (4.106)$$

Such betatron frequency spreading can result from γ variations, rounded radial beam profiles, etc.

Although both the rigid beam and the distributed frequency models give comparable growth rates, the rigid beam model predicts that the hose instability is unstable in the absolute sense, as a consequence of the infinite resonance in the dispersion relation. In contrast, the kinetic model predicts a cutoff frequency for unstable growth, and indicates that the instability is convective, moving from the head of the pulse to the tail when viewed in the rest frame of the beam.[27]

The effect of a longitudinal magnetic field on the hose instability growth rate can easily be incorporated into Eq. (4.103) by making the substitution $\Omega^2 \to \bar{\Omega}^2 = \Omega(\Omega + \Omega_c)$, where Ω_c is the beam electron cyclotron frequency $(eB_z/\gamma mc)$.[23] For $R/r_b \gg 1$, Eq. (4.103) becomes

$$-i\omega = \alpha \frac{\Omega^2 + \Omega\Omega_c}{\omega_\beta^2 - \left(\Omega^2 + \Omega\Omega_c\right)} \tag{4.107}$$

and the unstable frequencies are shifted to different values of Ω. Solving Eq. (4.107) for $\beta_0 ck_z$ yields

$$\beta_0 ck_z = \omega + \frac{\Omega_c}{2}\left\{1 \pm \left[1 - \frac{4\omega_\beta^2}{\Omega_c^2}\left(\frac{i\omega}{\alpha - i\omega}\right)\right]^{1/2}\right\} \tag{4.108}$$

In the limit of large Ω_c/ω_β, $\mathrm{Re}(k_z) \approx (\omega + \Omega_c)/2$ and (for growth in the $+z$ direction)

$$\mathrm{Im}(k_z) \approx -\frac{1}{\beta_0 c}\frac{\omega_\beta^2}{\Omega_c}\frac{\alpha\omega}{(\alpha^2 + \omega^2)} \tag{4.109}$$

In this limit the peak growth rate is

$$\mathrm{Max}\left[-\mathrm{Im}(k_z)\right] \approx \frac{1}{2\beta_0 c}\frac{\omega_\beta^2}{\Omega_c} \tag{4.110}$$

which occurs when $\omega = \alpha$. In Problem 4.5 the peak growth rate for $B_z = 0$ is shown to be approximately $0.3\ \omega_\beta/\beta_0 c$. Hence, application of the axial magnetic field reduces the peak growth rate by a factor of $\sim \Omega_c/\omega_\beta$.

As a concluding remark it should be mentioned that the above calculations have assumed that neutralization of the primary beam current was small. In this case the instability is driven by the phase lag between the beam displcements and the restoring forces. For large f_m a second mechanism, due to the repulsive interaction between the beam current and the return current, becomes important.[28] The effect is particularly strong when the beam is injected into a very narrow, highly conducting channel.

4.4.2. Sausage Instability

Radially symmetric perturbations of the beam envelope can also be destabilized by the finite conductivity of a background plasma leading to growth of the macroscopic "sausage" instability.[23] In this case it is the plasma response to beam current density variations resulting from perturbations of the equilibrium beam radius that is of interest. Because of the finite plasma conductivity the return current cannot respond immediately to beam density perturbations, lagging behind for times on the order of the magnetic diffusion time, τ_B. Hence, induced variations in the plasma current density can become phased so as to exert forces on the beam which cause the initial perturbation to grow.

Assuming that the beam is very energetic with $v_z/c = \beta_0 \approx 1$, $\gamma_0 \simeq (1 - \beta_0^2)^{-1/2}$, and $\mathbf{B} = B_\theta \hat{e}_\theta$, the perpendicular components of the equation of motion are given by

$$\frac{d^2\mathbf{r}}{dt^2} = \hat{e}_r \left[\frac{d^2r}{dt^2} - r\left(\frac{d\theta}{dt}\right)^2 \right] + \hat{e}_\theta \left(2\frac{dr}{dt}\frac{d\theta}{dt} + r\frac{d^2\theta}{dt^2} \right) = \frac{eB_\theta}{\gamma_0 m}\hat{e}_r \quad (4.111)$$

Integration of the theta component yields the angular momentum constant

$$L = r^2 \frac{d\theta}{dt} = \text{const} \quad (4.112)$$

and substitution for $d\theta/dt$ in the radial component of Eq. (4.111) gives

$$\frac{d^2r}{dt^2} - \frac{L^2}{r^3} = \frac{e}{\gamma_0 m}B_\theta \quad (4.113)$$

The azimuthal magnetic field arises from the beam current, I_b, and plasma currents, I_p, according to

$$B_\theta(r) = \frac{2}{rc}\left[I_b(r) + I_p(r) \right] \quad (4.114)$$

Rather than using Maxwell's equation to compute the plasma response to the injected beam (as was done in Section 4.4.1) the results of Section 4.2 are used to represent the plasma current in phenomenological fashion as

$$\frac{dI_p}{du} = -\frac{dI_b}{du} - \alpha I_p \quad (4.115)$$

where u represents the distance from the beam front ($u = \beta_0 ct - z$). Equation (4.115) implies that the plasma current results from a changing beam current and decays owing to finite plasma conductivity with the decay constant α given by

$$\alpha = (c\tau_B)^{-1} \tag{4.116}$$

Because of the finite conductivity the plasma current cannot respond immediately to perturbations in the beam current. Hence, the magnetic field strength also lags behind the beam perturbations, and it is expected that instability should result under appropriate circumstances.

Equation (4.115) is recognized as a first-order linear differential equation which can be made exact with the integrating factor $\exp(\int \alpha \, du)$. The solution is given by

$$I_p(r) = \int_{-\infty}^{u} du' \, e^{\alpha(u'-u)} \left(\frac{\partial I_b}{\partial u'} \right)_r \tag{4.117}$$

If it is assumed that the current density is uniform out to the beam radius r_b, i.e.,

$$I_b(r) = \begin{cases} I_{b0}(r/r_b)^2, & 0 < r < r_b \\ 0, & r > r_b \end{cases} \tag{4.118}$$

then Eq. (4.117) becomes

$$I_p(r) = I_{b0} \int_{-\infty}^{u} du' \, e^{\alpha(u'-u)} \left(\frac{2}{r_b} \frac{\partial r_b}{\partial u'} \right) \left(\frac{r}{r_b} \right)^2 \tag{4.119}$$

(It should be noted that in this instance it is the plasma response to beam current density variations resulting from perturbations in the beam radius that is of interest.)

Substitution of Eqs. (4.114) and (4.119) into Eq. (4.113) yields

$$\frac{d^2 r}{dt^2} - \frac{L^2}{r^2} = \frac{e}{\gamma_0 m} \frac{2}{rc} \left\{ I_{b0} \left(\frac{r}{r_b} \right)^2 + I_{b0} \int_{-\infty}^{u} du' \, e^{\alpha(u'-u)} \left(\frac{2}{r_b} \frac{\partial r_b}{\partial u'} \right) \left(\frac{r}{r_b} \right)^2 \right\}$$

$$\tag{4.120}$$

An envelope equation for the beam radius can be obtained from Eq. (4.120) by setting $r = r_b$, in which case

$$\frac{d^2 r_b}{dt^2} - \frac{L^2}{r_b^3} = \frac{\omega_{\beta 0}^2 r_{b0}^2}{r_b} \left(1 + 2 \int_{-\infty}^{u} du' \, e^{\alpha(u'-u)} \frac{1}{r_b} \frac{\partial r_b}{\partial u'}\right) \qquad (4.121)$$

and Eq. (4.112), for the angular momentum becomes

$$L = r_b^2 \frac{d\theta}{dt}(r_b) = r_{b0}^2 \omega_{\beta 0} = r_{b0}^2 c k_{\beta 0} = \text{const} \qquad (4.122)$$

where r_{b0}, $\omega_{\beta 0}$, and $k_{\beta 0}$ are the equilibrium radius, and the equilibrium betatron frequency and wave number, respectively.

Observing that

$$\frac{d}{dt} \equiv \left(\frac{\partial}{\partial t}\right)_z + \beta_0 c \left(\frac{\partial}{\partial z}\right)_t = \left(\frac{\partial}{\partial t}\right)_u = \beta_0 c \left(\frac{\partial}{\partial z}\right)_u$$

and

$$\left(\frac{\partial}{\partial t}\right)_z \to \beta_0 c \left(\frac{\partial}{\partial u}\right)_z$$

Eq. (4.121) may be rewritten as

$$\left(\frac{\partial^2 r_b}{\partial z^2}\right)_u - \frac{k_{\beta 0}^2 r_{b0}^4}{r_b^3} = -\frac{k_{\beta 0}^2 r_{b0}^2}{r_b} \left[1 + 2 \int_{-\infty}^{u} du' \, e^{\alpha(u'-u)} \frac{1}{r_b} \frac{\partial r_b}{\partial u'}\right] \qquad (4.123)$$

which is a nonlinear equation for the beam envelope.

To examine the question of beam stability small perturbations of the beam radius are assumed about the equilibrium, i.e., $r_b = r_{b0} + \delta r_b$ with $\delta r_b \ll r_{b0}$. Substituting into Eq. (4.123) and retaining only terms linear in δr_b gives

$$\frac{\partial^2 \delta r_b}{\partial z^2} + 2 k_{\beta 0}^2 \delta r_b = -2 k_{\beta 0}^2 \int_{-\infty}^{u} du' \, e^{\alpha(u'-u)} \frac{\partial \delta r_b}{\partial u'} \qquad (4.124)$$

An integration by parts yields the alternative form

$$\frac{\partial^2 \delta r_b}{\partial z^2} + 4k_{\beta 0}^2 \delta r_b = 2k_{\beta 0}^2 \int_{-\infty}^{u} du' \, \alpha e^{\alpha(u'-u)} \delta r_b \qquad (4.125)$$

Taking the derivative of Eq. (4.125) with respect to u gives the differential equation

$$\frac{\partial}{\partial u} \left(\frac{\partial^2 \delta r_b}{\partial z^2} + 4k_{\beta 0}^2 \delta r_b \right) = 2k_{\beta 0}^2 \alpha \, \delta r_b - 2k_{\beta 0}^2 \alpha \int_{-\infty}^{u} e^{\alpha(u'-u)} \delta r_b \, du'$$

or

$$\left(\alpha + \frac{\partial}{\partial u} \right) \left(\frac{\partial^2 \delta r_b}{\partial z^2} + 4k_{\beta 0}^2 \delta r_b \right) = 2k_{\beta 0}^2 \alpha \, \delta r_b \qquad (4.126)$$

Assuming that the beam envelope perturbations are of the form

$$\delta r_b \sim e^{-i(k_u u + k_z z)} \qquad (4.127)$$

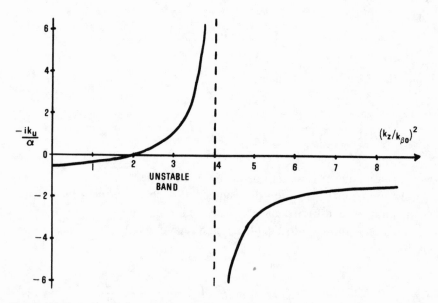

Figure 4.6. Variation of the sausage instability growth rate with longitudinal wave number. Instability exists over the wave number range of $2k_{\beta 0}^2 < k_z^2 < 4k_{\beta 0}^2$.

the dispersion relation for the beam sausage mode becomes

$$(\alpha - ik_u)(-k_z^2 + 4k_{\beta 0}^2) = 2\alpha k_{\beta 0}^2 \qquad (4.128)$$

and the growth rate in u for real k_z is given by

$$-ik_u = \alpha \frac{k_z^2 - 2k_{\beta 0}^2}{4k_{\beta 0}^2 - k_z^2} \qquad (4.129)$$

Since instability exists only in the narrow band $2k_{\beta 0}^2 < k_z^2 < 4k_{\beta 0}^2$ (Fig. 4.6), it is expected that a relatively small betatron frequency spread can effectively "detune" the sausage mode and suppress the instability (see Problem 4.6).

4.5. Microscopic Instabilities

In addition to the macroscopic instabilities, which can greatly affect the shape of a particle beam propagating in a dense plasma, there is also the possibility for development of microscopic instabilities. Such instabilities can lead to plasma heating and particle diffusion, and can effectively limit the amount of beam energy which can be transported through a system.

To examine the question of microscopic stability consider a uniform electron beam propagating along the longitudinal (z) axis of an infinitely long plasma-filled metallic waveguide of radius R. To prevent the growth of the macroscopic instabilities it is assumed that the system is placed in a strong longitudinal magnetic field. It is further assumed that the electron beam is both charge and current neutralized, i.e., the equilibrium radial electric and azimuthal magnetic fields may be neglected in the analysis. Finally, it is assumed that the equilibrium azimuthal motions of the beam and the plasma components are also negligible so that diamagnetic effects are unimportant. For the case of the beam and plasma components completely filling the waveguide, the electrostatic dispersion relation reduces to (see Section 3.7.1)

$$1 - \left(\frac{k_z}{k}\right)^2 \sum_j \frac{\omega_j^2/\gamma_j^3}{(\omega - k_z v_{jz})^2} - \left(\frac{k_\perp}{k}\right)^2 \sum_j \frac{\omega_j^2/\gamma_j}{(\omega - k_z v_{jz})^2 - (\Omega_j/\gamma_j)^2} = 0$$

$$(4.130)$$

where ω_j^2 denotes $(4\pi n_j e^2/m_j)$, $k_\perp^2 R^2 = \rho_{lm}^2$, and ρ_{lm} denotes the mth zero of the Bessel function $J_l(x)$. The summations extend over all species of charged particles in the system (beam electrons, and plasma electrons and ions).

On the basis of Eq. (4.130) we can expect the development of various streaming instabilities because there are natural plasma modes with phase velocities very near the beam velocity. Hence, in the beam reference frame, the wave electric field can couple strongly to the beam electrons causing beam bunching. As the bunches are decelerated by the wave, the wave extracts energy from the electron bunches. This feedback mechanism results in exponential growth of the wave amplitude until nonlinear wave saturation occurs. The source of the free energy required to drive the instabilities is the differential stream velocities of the various beam and plasma components.

4.5.1. Stability of a Charge-Neutralized Electron Beam (Buneman Instability)

Depending on the frequency and parameter ranges of interest Eq. (4.130) may be considerably simplified. As a first example consider the case of an electron beam propagating in an ionic plasma with $n_b = n_i = n_0$. In this instance the assumption of current neutralization is not valid, and a magnetic field whose strength is large compared to the beam self-magnetic field must be applied to prevent beam constrictions. In the following, the investigation is restricted to the frequency region $(\omega - k_z v_{bz})^2 \ll \Omega_e^2/\gamma_e^2$. It is further assumed that the ions are stationary with $v_{iz} \simeq 0$ and $\gamma_i \simeq 1$. In this case the dispersion relation Eq. (4.130) reduces to[29]

$$k_z^2 \left[1 - \frac{\omega_b^2}{\gamma_b^3(\omega - k_z v_{bz})^2} - \frac{\omega_i^2}{\omega^2} \right] + k_\perp^2 \left(1 - \frac{\omega_i^2}{\omega^2} \right) = 0 \qquad (4.131)$$

In writing Eq. (4.131) the effect of the external magnetic field on the ion motion has been neglected ($\omega > \Omega_i$). Equation (4.131) is recognized as the generalization of Buneman's equation to the case of a relativistic beam.[30]

In the long-wavelength limit ($k_z \to 0$) the unstable solutions lie in the region $\omega \simeq k_z v_{bz}$. Expanding ω about $k_z v_{bz}$ in terms of the small increment

δ_0 as $\omega \simeq k_z v_{bz} + \delta_0$, and substituting this expression into Eq. (4.131) yields

$$\delta_0^2 = \frac{k_z^2}{k_\perp^2} \frac{\omega_b^2}{\gamma_b^3} \left[1 - \frac{\omega_i^2}{\left(k_z v_{bz} \right)^2} \right]^{-1} \tag{4.132}$$

Unstable oscillations corresponding to $\delta_0^2 < 0$ occur for

$$\omega_i^2 > k_z^2 v_{bz}^2 \tag{4.133}$$

4.5.2. Electron Beam Stability in a Dense Plasma (Two-Stream and Cyclotron Instabilities)

As a second application of Eq. (4.130) consider the case in which the plasma density greatly exceeds the beam density, i.e., $n_e \gg n_b / \gamma_b$. In this case the beam represents a perturbation which can excite natural plasma oscillations and it is expected that the character of the oscillations will be primarily determined by the plasma parameters.

It is assumed that the motions of both the plasma electrons and plasma ions are nonrelativistic ($\gamma_i \simeq \gamma_e \simeq 1$). For simplicity it is further assumed that the ions are only weakly magnetized ($\Omega_i^2 \ll \omega_i^2$), and that the oscillation frequencies of interest are well removed from the ion resonance ($\omega^2 \gg \omega_i^2$). Under these conditions the dispersion relation reduces to[31]

$$k_z^2 \left[1 - \frac{\omega_e^2}{\omega^2} - \frac{\omega_b^2/\gamma_b^3}{\left(\omega - k_z v_{bz} \right)^2} \right] + k_\perp^2 \left[1 - \frac{\omega_e^2}{\omega^2 - \Omega_e^2} - \frac{\omega_b^2/\gamma_b}{\left(\omega - k_z v_{bz} \right)^2 - \Omega_e^2/\gamma_b^2} \right] = 0$$

$$\tag{4.134}$$

An examination of Eq. (4.134) indicates that the importance of electron beam induced effects is maximized for the frequency ranges

$$\omega \simeq k_z v_{bz} \tag{4.135}$$

$$\omega \simeq k_z v_{bz} \pm \Omega_e / \gamma_b \tag{4.136}$$

Neglecting beam effects entirely yields the plasma dispersion relation

$$1 - \frac{\omega_e^2}{\omega^2} \frac{k_z^2}{k^2} - \frac{\omega_e^2}{\omega^2 - \Omega_e^2} \frac{k_\perp^2}{k^2} = 0 \qquad (4.137)$$

with natural oscillation frequencies given by

$$\omega_\pm^2 = \frac{1}{2} \left(\Omega_e^2 + \omega_e^2 \right) \left\{ 1 \pm \left[1 - \frac{4\omega_e^2 \Omega_e^2 k_z^2 / k^2}{\left(\omega_e^2 + \Omega_e^2 \right)^2} \right]^{1/2} \right\} \qquad (4.138)$$

The instability associated with the intersections of the line $\omega = k_z v_{bz}$ with the branches of the natural oscillations $\omega_\pm$ is called two-stream, or Čerenkov, instability while the instability associated with the intersections of $\omega = k_z v_{bz} \pm \Omega_e \gamma_b^{-1}$ is termed cyclotron instability.

To analyze the two-stream instability ω is expanded as

$$\omega \simeq k_z v_{bz} + \delta_0 = \omega_0 + \delta_0 \qquad (4.139)$$

where ω_0 coincides with one of the natural oscillation frequencies given by Eq. (4.139). In this case the δ_0 satisfies the relation

$$2 \frac{\delta_0}{\omega_0} \left[\frac{\omega_e^2}{\omega_0^2} k_z^2 + \frac{\omega_e^2 \omega_0^2}{\left(\omega_0^2 - \Omega_e^2 \right)^2} k_\perp^2 \right] - \frac{k_z^2 \omega_b^2 / \gamma_b^3}{\delta_0^2} - \frac{k_\perp^2 \omega_b^2 / \gamma_b}{\delta_0^2 - \Omega_e^2 / \gamma_b^2} = 0$$

$$(4.140)$$

To examine the cyclotron instability ω is expanded as

$$\omega = k_z v_{bz} \pm \Omega_e / \gamma_b + \delta_0 = \omega_0 + \delta_0 \qquad (4.141)$$

in which case

$$2 \frac{\delta_0}{\omega_0} \left[\frac{\omega_e^2}{\omega_0^2} k_z^2 + \frac{\omega_e^2 \omega_0^2}{\left(\omega_0^2 - \Omega_e^2 \right)^2} k_\perp^2 \right] - \frac{k_z^2 \omega_b^2 / \gamma_b^3}{\left(\delta_0 \pm \Omega_e / \gamma_b \right)^2} - \frac{k_\perp^2 \omega_b^2 / \gamma_b}{\delta_0 (\delta_0 \pm 2\Omega_e / \gamma_b)} = 0$$

$$(4.142)$$

In the limit $\Omega_e/\gamma_b \ll \delta_0$, Eqs. (4.140) and (4.142), both have the same approximate solution given by

$$\delta_0^3 \simeq \frac{\omega_0}{2} \frac{k_z^2\omega_b^2/\gamma_b^3 + k_\perp^2\omega_b^2/\gamma_b}{\dfrac{\omega_e^2}{\omega_0^2}k_z^2 + \dfrac{\omega_e^2\omega_0^2}{\left(\omega_0^2-\Omega_e^2\right)^2}k_\perp^2} \tag{4.143}$$

On the other hand, when $\delta_0 \ll \Omega_e/\gamma_b$, for the two-stream instability

$$\delta_0^3 \simeq \frac{\omega_0}{2} \frac{k_z^2\omega_b^2/\gamma_b^3}{\dfrac{\omega_e^2}{\omega_0^2}k_z^2 + \dfrac{\omega_e^2\omega_0^2}{\left(\omega_0^2-\Omega_e^2\right)^2}k_\perp^2} \tag{4.144}$$

whereas for the cyclotron instability

$$\delta_0^2 \approx -\frac{\omega_0}{4} \frac{k_\perp^2\omega_b^2/\Omega_e}{\dfrac{\omega_e^2}{\omega_0^2}k_z^2 + \dfrac{\omega_e^2\omega_0^2}{\left(\omega_0^2-\Omega_e^2\right)^2}k_\perp^2} \tag{4.145}$$

The instabilities determined from Eqs. (4.143)–(4.145) divide naturally into three regions depending upon the ratio of the electron plasma frequency to the electron cyclotron frequency, ω_e/Ω_e.[31] (See Figs. 4.7–4.9.)

Figure 4.7. Region 1 $[0 < \Omega_e/\omega_e < \frac{1}{2}\gamma_b(n_b/2n_e\gamma_b)^{1/3}]$ dispersion characteristic for an electron beam in a dense plasma, illustrating the degeneracy of the two-stream and cyclotron instabilities. This dispersion characteristic is also applicable for the two-stream instability for region 2, $[\frac{1}{2}\gamma_b(n_b/2n_e\gamma_b)^{1/3} < (\omega_e/\Omega_e) \approx 1]$ at small angles.

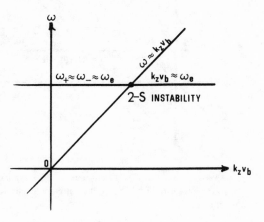

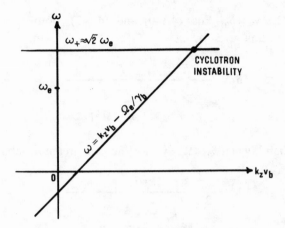

Figure 4.8. Region 2 dispersion characteristic illustrating the cyclotron interaction at large angles.

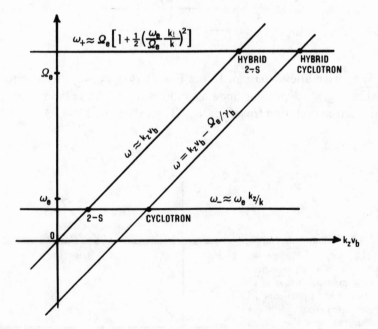

Figure 4.9. Region 3 ($\omega_e/\Omega_e > 1$) dispersion characteristic illustrating four distinct instabilities: two-stream instability, hybrid two-stream instability, cyclotron instability, and hybrid cyclotron instability.

Region 1: $0 < \Omega_e/\omega_e < \gamma_b(n_b/2n_e\gamma_b)^{1/3}$

$$k_z v_{bz} = \omega_e\left(1 + \frac{\Omega_e^2}{2\omega_e^2}\frac{k_\perp^2}{k^2}\right) \simeq \omega_e \tag{4.146}$$

$$\mathrm{Im}(\delta_0) \simeq \frac{\sqrt{3}}{2}\omega_e\left(\frac{n_b}{2\gamma_b n_e}\right)^{1/3}\left(\frac{k_\perp^2}{k^2} + \frac{k_z^2}{k^2\gamma_b^2}\right)^{1/3} \tag{4.147}$$

Region 2: $\gamma_b(n_b/2n_e\gamma_b)^{1/3} < \Omega_e/\omega_e \approx 1$
two-stream:

$$k_z v_{bz} \simeq \omega_e \tag{4.148}$$

$$\mathrm{Im}(\delta_0) \simeq \frac{\sqrt{3}}{2}\frac{\omega_e}{\gamma_b}\left(\frac{n_b}{2n_e}\right)^{1/3}\left(\frac{k_z^2}{k^2}\right)^{2/3} \tag{4.149}$$

cyclotron:

$$k_z v_{bz} - \Omega_e/\gamma_b \simeq 2\omega_e \tag{4.150}$$

$$\mathrm{Im}(\delta_0) \simeq \frac{\omega_e}{2}\left(\frac{n_b}{n_e}\right)^{1/2}\frac{k_\perp}{k} \tag{4.151}$$

Region 3: $\Omega_e/\omega_e > 1.$
two-stream:

$$(k_z v_{bz})_- = \omega_e\frac{k_z}{k} \tag{4.152}$$

$$(k_z v_{bz})_+ = \Omega_e\left(1 + \frac{\omega_e^2}{2\Omega_e^2}\frac{k_\perp^2}{k^2}\right) \tag{4.153}$$

$$\mathrm{Im}(\delta_0)_- = \frac{\sqrt{3}}{2}\frac{\omega_e}{\gamma_b}\left(\frac{n_b}{2n_e}\right)^{1/3}\frac{k_z}{k} \tag{4.154}$$

$$\mathrm{Im}(\delta_0)_+ = \frac{\sqrt{3}}{2}\frac{\omega_e}{\gamma_b}\left(\frac{n_b}{2n_e}\right)^{1/3}\left(\frac{\omega_e}{\Omega_e}\right)^{1/3}\left(\frac{k_z k_\perp}{k^2}\right)^{2/3} \tag{4.155}$$

cyclotron:

$$\left(k_z v_{bz} - \frac{\Omega_e}{\gamma_b} \right)_- = \omega_e \frac{k_z}{k} \tag{4.156}$$

$$\left(k_z v_{bz} - \frac{\Omega_e}{\gamma_b} \right)_+ = \Omega_e \left(1 + \frac{\omega_e^2}{2\Omega_e^2} \frac{k_\perp^2}{k^2} \right) \tag{4.157}$$

$$\mathrm{Im}(\delta_0)_- = \frac{\omega_e}{2} \left(\frac{n_b}{n_e} \right)^{1/2} \left(\frac{\omega_e}{\Omega_e} \right)^{1/2} \frac{k_\perp}{k} \left(\frac{k_z}{k} \right)^{1/2} \tag{4.158}$$

$$\mathrm{Im}(\delta_0)_+ = \frac{\omega_e}{2} \left(\frac{n_b}{n_e} \right)^{1/2} \left(\frac{\omega_e}{\Omega_e} \right) \frac{k_\perp^2}{k^2} \tag{4.159}$$

In Eqs. (4.146)–(4.159) the subscript refers to the $\pm$ sign associated with the natural oscillation frequencies of Eq. (4.138). The lower branch is excited only in the case of large external magnetic fields. There is either no instability, or else the growth rate is very small.

In addition to the three regions described above, there is also a transition interval in which an upper hybrid oscillation mode is driven unstable by the beam cyclotron resonance at large angles. An estimate for the growth rate in this interval is[32]

$$\delta_T \simeq \frac{\sqrt{3}}{2} \omega_e \left(\frac{n_b}{2\gamma_b n_e} \right)^{1/3} - \frac{\Omega_e}{\gamma_b} \left[\frac{\sqrt{3}}{2} - \frac{(n_b/n_e)^{1/2}(\omega_e/\Omega_e)^{1/2}}{2(n_b/2n_e\gamma_b)^{1/3}} \right] \tag{4.160}$$

It is of interest to compare the relative growth rates for the various parameter regimes. For the case of a dense plasma, $\omega_e \gg \Omega_e$, the cyclotron instability degenerates into the two-stream instability and the growth rate is peaked at large angles. For regions 2 and 3, however, the cyclotron modes remain essentially radial modes while the two-stream modes becomes longitudinal. For region 2 a comparison of Eqs. (4.149) and (4.151) indicates that under the condition

$$\frac{k_z^4}{kk_\perp^3} \gtrsim \gamma_b^3 \left(\frac{n_b}{n_e}\right)^{1/2} \tag{4.161}$$

two-stream instability should predominate over cyclotron instability; in the opposite limit cyclotron instability predominates.

For region 3 (rarefied plasma with $\omega_e < \Omega_e$) there are four distinct instabilities possible: (i) two-stream instability, (ii) hybrid two-stream instability, (iii) electron cyclotron instability, and (iv) upper hybrid instability, in order of increasing angle of propagation. The growth rate of the (hybrid) two-stream instability always predominates over the (hybrid) cyclotron instability on the upper branch [compare Eqs. (4.155) and (4.159)]. On the lower branch the two-stream growth increment is even larger [Eq. (4.154)], while the cyclotron instability growth rate is relatively smaller [Eq. (4.158)]. Hence, for region 3 parameters the two-stream instability should practically always predominate.

4.5.3. Electromagnetic Filamentation (Weibel) Instability[33]

In addition to the electrostatic instabilities already considered, current neutralized intense electron beams can also be unstable to transverse electromagnetic modes. The physical mechanism of the filamentation instability is due to the attractive force between currents of like sign. For example, if there is a local increase in the plasma electron return current, the resulting uncompensated magnetic field further pinches off the plasma electron current. A similar interaction can also occur between beam particles, although on a slower time scale if the beam is relativistic. The interaction acts transversely to the beam velocity and "tears" the beam into filaments.

The electromagnetic dispersion relation which describes the filamentation mode can be determined by generalizing Eq. (3.173) to include plasma species and retaining the perturbed magnetic field $\delta\mathbf{B}(\mathbf{r}, t)$ in the analysis. The result can be written in matrix notation as[32]

$$\mathbf{A} \cdot \mathbf{E} = \begin{pmatrix} A_{11} & A_{12} & A_{13} \\ A_{12}^* & A_{22} & A_{23} \\ A_{13}^* & A_{23}^* & A_{33} \end{pmatrix} \begin{pmatrix} E_x \\ E_y \\ E_z \end{pmatrix} = 0 \tag{4.162}$$

where

$$A_{11} = k_\perp^2 c^2 - \omega^2 + \omega_e^2 + \frac{\omega^2 \omega_b^2 / \gamma_b^3}{\left(\omega - k_z v_{bz}\right)^2} + \frac{k_\perp^2 v_{bz}^2 \omega_b^2 / \gamma_b}{\left(\omega - k_z v_{bz}\right)^2 - \Omega_e^2 / \gamma_b^2}$$

$$A_{12} = -k_z k_y c^2 + \frac{\omega_b^2 / \gamma_b}{\left(\omega - k_z v_{bz}\right)^2 - \Omega_e^2 / \gamma_b^2} \left[\left(\omega - k_z v_{bz}\right) k_y v_{bz} + i\varepsilon_b \frac{\Omega_e}{\gamma_b} k_x v_{bz} \right]$$

$$A_{13} = -k_z k_x c^2 + \frac{\omega_b^2 / \gamma_b}{\left(\omega - k_z v_{bz}\right)^2 - \Omega_e^2 / \gamma_b^2} \left[\left(\omega - k_z v_{bz}\right) k_x v_{bz} - i\varepsilon_b \frac{\Omega_e}{\gamma_b} k_y v_{bz} \right]$$

$$A_{22} = k_\perp^2 c^2 - \omega^2 + \frac{\omega_e^2 \omega^2}{\omega^2 - \Omega_e^2} + \frac{\left(\omega_b^2 / \gamma_b\right)\left(\omega - k_z v_{bz}\right)^2}{\left(\omega - k_z v_{bz}\right)^2 - \Omega_e^2 / \gamma_b^2}$$

$$A_{23} = -k_y k_x c^2 - \frac{i\varepsilon_b \left(\omega_b^2 / \gamma_b\right)\left(\omega - k_z v_{bz}\right)\Omega_e / \gamma_b}{\left(\omega - k_z v_{bz}\right)^2 - \Omega_e^2 / \gamma_b^2} - \frac{i\varepsilon_e \omega_e \omega \Omega_e}{\omega^2 - \Omega_e^2}$$

$$A_{33} = k_\perp^2 c^2 - \omega^2 + \frac{\left(\omega_b^2 / \gamma_b\right)\left(\omega - k_z v_{bz}\right)^2}{\left(\omega - k_z v_{bz}\right)^2 - \Omega_e^2 / \gamma_b^2} + \frac{\omega_e^2 \omega^2}{\omega^2 - \Omega_e^2}$$

The matrix components A_{12}^*, A_{13}^*, and A_{23}^* in Eq. (4.162) are the complex conjugates of A_{12}, A_{13}, and A_{23}. Also, $k_\perp^2 = (k_x^2 + k_y^2)$ and ε_α is the sign of the charge. The off-axis terms can be simplified by specifying a particular perpendicular direction.

In the zero magnetic guide field case, $\Omega_e = 0$, $A_{23} = A_{23}^* = 0$, and the determinant of Eq. (4.162) yields

$$\omega^2 = k^2 c^2 + \omega_e^2 + \omega_b^2 / \gamma_b \tag{4.163}$$

from $A_{22} = 0$, and

$$\left(k^2 c^2 - \omega^2 + \omega_e^2 + \omega_b^2 / \gamma_b\right)\left(1 - \frac{\omega_e^2}{\omega^2} - \frac{\omega_b^2 / \gamma_b^3}{\left(\omega - k_z v_{bz}\right)^2}\right) - \frac{\omega_e^2 k_\perp^2 v_{bz}^2 \omega_b^2 / \gamma_b}{\omega^2 \left(\omega - k_z v b_z\right)^2} = 0$$

$$\tag{4.164}$$

from $A_{11} A_{33} - A_{13} A_{13}^* = 0$.

In the weak-beam limit ($\omega_e \gg \omega_b$) at large angles ($k_z \to 0$), the electromagnetic dispersion relation Eq. (4.164), predicts a purely growing mode with growth increment $\delta_0 \sim \omega$ given by

$$\delta_w = \omega_e \left(\frac{n_b}{\gamma_b n_e} \right)^{1/2} \left(\frac{k^2 v_{bz}^2}{k^2 c^2 + \omega_e^2} \right)^{1/2} \tag{4.165}$$

Equation (4.165) is the growth rate for the so-called Weibel, or tearing mode, instability for the case of zero magnetic field. Equation (4.165) may be simply generalized to incorporate the effect of nonzero magnetic field as[32, 34]

$$\delta_w = \left[\left(\frac{n_b}{n_e \gamma_b} \right) \left(\frac{k^2 v_{bz}^2}{k^2 c^2 + \omega_e^2} \right) \omega_e^2 - \frac{\Omega_e^2}{\gamma_b^2} \right]^{1/2} \tag{4.166}$$

While the magnetic field has a stabilizing effect the instability is still purely growing. The ratio of the Weibel instability growth rate to that of the two-stream instability in the weak magnetic field limit at large angles is approximately given by[35]

$$\frac{\delta_w}{\delta_{ts}} \sim \left(\frac{n_b}{2\gamma_b n_e} \right)^{1/6} \tag{4.167}$$

The growth rate for the Weibel instability is plotted vs k in Fig. 4.10.

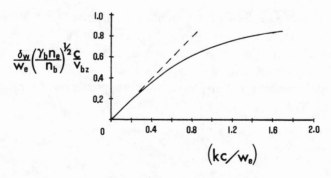

Figure 4.10. Variation of the Weibel instability growth rate as a function of wave number.

Early analyses suggested that the instability could be suppressed by transverse temperature[36, 37]; unfortunately, the results of experimental studies did not support this conclusion.[35] More recent calculations accounting for beam temperature and plasma collisional effects indicate that the instability can only be stabilized by an applied magnetic field. Transverse temperature stabilization cannot occur because of the nonzero resistivity of the background plasma, although the growth rate (proportional to the collision frequency in the plasma) is typically 1–3 orders of magnitude smaller than the cold fluid result.[38]

Nonlinear evolution of the filamentation instability has been studied numerically.[34] Linear mode growth is observed to cease when the electron density in the filaments exceeds the charge neutrality condition. However, mutually attractive Lorentz forces still exist between the beam filaments, although the magnitude of the force is reduced by the plasma currents which flow in a thin sheath ($\sim c/\omega_p$) outside each filament. As a result, the individual filaments gradually coalesce, until the beam eventually becomes a single large filament, free of plasma return current and self-pinched to the extent of maintaining charge neutrality.

It has also been suggested that the filamentation instability may nonlinearly couple to the hose instability with the beam spraying outward into several filaments. In addition, the formation of such high-current structures should appreciably alter the plasma heating properties of high-current beams.[37]

4.6. Plasma Heating by Linear Relativistic Electron Beams

The possible exploitation of an intense electron beam as an energy source for heating a dense plasma to thermonuclear temperatures has received much attention. This interest is motivated toward enhancing the energy deposition rate over the classical Coulomb collision transfer rate. As an example, consider the single electron stopping power, dE/dx, given approximately by

$$\frac{dE}{dx} \simeq \frac{Z^2 \omega_p^2 e^2}{2\pi\beta^2 c^2} \tag{4.168}$$

For a plasma density $n_e \simeq 10^{15}/\text{cm}^3$ and $Z = 1$, $|dE/dx|$ is only 1.5×10^{-3} eV/cm for a 1-MeV electron. Collective effects on the other hand can increase this deposition rate by several orders of magnitude. In this section the important results pertaining to the heating of plasmas by linear intense beams through collective interactions are discussed and summarized.

Earlier in this chapter it was shown that an intense beam can propagate in the presence of a dense plasma by inducing a plasma current, approximately equal in magnitude to the beam current, but flowing in the opposite direction. A typical ordering of the resulting component longitudinal velocity distributions is shown in Fig. 4.11. There are two distinct collective processes which can heat the target plasma. The first type is that associated with the interaction between the plasma electrons (which carry the return current) and the plasma ions. The second important collective process is the two-stream (or cyclotron) interaction between the primary beam electrons and the plasma electrons (see Section 4.5.2).

The actual heating mechanism is due to the nonlinear saturation of unstable wave growth by nonlinear plasma processes. Such effects, including wave mixing and parametric instabilities, result in an energy transfer from the unstable wave to other plasma waves which do not interact with the beam, but couple instead to either the background plasma electrons or ions, and thereby heat the plasma. A high saturation level is necessary for efficient beam–plasma heating over short distances. In order to strongly

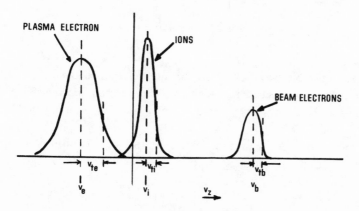

Figure 4.11. Typical ordering of the beam–plasma longitudinal velocity distributions.

decelerate the beam the wave electric potential must grow to a significant fraction of the beam kinetic energy.

4.6.1 Return Current Interaction

The plasma electrons which carry the induced return current attain a drift velocity through the plasma ions given by

$$v_d \simeq -v_b\left(\frac{n_b}{n_e}\right) \tag{4.169}$$

with v_d usually much larger than the electron thermal spread v_e. Microturbulence between the plasma electrons and the plasma ions leads to a decay of the return current in a characteristic magnetic diffusion time $\tau_B \simeq 4\pi\sigma r_b^2/c^2$ where the conductivity σ is given by

$$\sigma \simeq n_e e^2/m\nu \tag{4.170}$$

and ν is the "effective" collision frequency for the turbulence. In the very early stages of the interaction, the dominant energy transfer mechanism is the conventional Buneman instability between plasma electrons and ions. The frequency and maximum growth rate for this instability are given approximately by[30]

$$\omega_{k0} = \frac{1}{2}\left(\frac{m_e}{2m_i}\right)^{1/3}\omega_e \tag{4.171}$$

$$Im(\omega) = \frac{\sqrt{3}}{2}\left(\frac{m_e}{2m_i}\right)^{1/3}\omega_e \tag{4.172}$$

for $k_0 = \omega_e/|v_d|$.

The unstable wave grows at the rate determined by Eq. (4.172) until electron trapping occurs, in which case the electrons become spread symmetrically about the wave phase velocity over its trapping width $\Delta v_T \simeq (2eE_{\text{rms}}/mk_0)^{1/2}$.

An estimate for saturation of the Buneman instability is obtained by setting the trapping width equal to $|v_d| - v_e$, which yields[39]

$$\frac{E_{\text{rms}}^2}{8\pi n_e mv_d^2} = \frac{1}{8}\left(1 - \frac{v_e}{|v_d|}\right)^4 \tag{4.173}$$

for $v_e \ll |v_d|$ before trapping.

After trapping, however, $v_e \geq |v_d|$, and the instability passes into the ion acoustic stage.[40] Barring other processes, plasma heating via the ion acoustic instability will proceed until (a) the ions are trapped in the unstable wave, or (b) linear stability is achieved, or (c) energy loss due to wave convection leads to an equilibrium.

A detailed treatment of the energy balance of a return current heated plasma is given in Ref. 41. In terms of this analysis, the resistive energy loss W_R of the return current is given by the difference between the work done by the beam (to establish the return current) and the energy contained in the magnetic field due to both the beam current I_b and the plasma current I_p. If R and L denote the plasma resistance and the system inductance per unit length, then

$$W_R = \int_0^t dt\, RI_p^2 = LI_b\big[I_b + I_p(t)\big] - \tfrac{1}{2}L\big[I_b + I_p(t)\big]^2 \qquad (4.174)$$

where it has been assumed that the beam current is constant and $I_p(0) = -I_b$. Since currents in excess of the Alfven current I_A are not expected to propagate, a useful estimate of the heating time is to set

$$I_b + I_p(t_h) = I_A \qquad (4.175)$$

Since $I_p(t) \simeq I_b e^{-t/\tau_B}$, in approximate terms

$$t_h \simeq (I_A/I_b)\tau_B \qquad (4.176)$$

and the heating time is seen to depend on the effective collision frequency ν from Eq. (4.170). Numerical estimates indicate that the microturbulence yields initial values of ν as high as $(0.2\text{--}0.4)\omega_i$,[39] where ω_i is the ion plasma frequency, although ν is found to decline to less than $0.01\omega_i$ as the turbulence evolves.[42]

For beam currents in excess of the Alfven current the return current mechanism should be effective in heating relatively low-density plasmas $(n_e \lesssim 10^{14} \text{ cm}^{-3})$; however, for target plasmas much denser than the beam, the energy dissipation rate is much reduced because of the decrease in v_d with increasing n_e. The Buneman instability may not even develop, and the effective collision frequency is expected to be quite small.

4.6.2. Beam-Electron–Plasma-Electron Streaming Interaction

The second important type of microturbulence is the interaction between the primary beam electrons and the target plasma electrons. Depending on the presence and strength of an external longitudinal magnetic field, a number of interactions are possible, as discussed in Section 4.5.2.

For a cold beam, collisionless cold plasma interaction neglecting boundary effects, the efficiency for converting beam energy into plasma energy is determined primarily by the beam energy (γ_0), the beam-electron-plasma-electron density ratio (n_b/n_e), and the external magnetic field strength (B_z). An exact determination of the coupling efficiency requires the self-consistent solution of the two-dimensional, relativistic equations of motion; however, supported by extensive partial numerical simulations, the approximate parametric behavior of the coupling coefficient can be determined by assuming that the beam–plasma interaction is one-dimensional in space along the direction of wave propagation.[43]

In the presence of a strong magnetic field the most unstable waves are those associated with the lower branch of the two-stream instability which propagate parallel to the beam. The phase velocity of the most unstable wave ($k_\perp = 0$) is easily calculated to be

$$\beta_\omega = \frac{\omega_0}{k_0 c} \simeq \beta_0 \left[1 - \frac{1}{2\gamma_b} \left(\frac{n_b}{2n_e} \right)^{1/3} \right] \qquad (4.177)$$

where $\beta_0 = v_b/c$ and $k_0 \simeq \omega_e/v_b$.

Transforming to the rest frame of the wave, the beam energy is given by

$$\gamma_{b\omega} = \gamma_\omega \gamma_b (1 - \beta_\omega \beta_0) \qquad (4.178)$$

where $\gamma_\omega = (1 - \beta_\omega^2)^{-1/2}$. In the limit $(n_b/2n_e)^{1/3} < 2\gamma_b$

$$\gamma_\omega \simeq \gamma_b (1 + S)^{-1/2} \qquad (4.179)$$

where $S = \gamma_b \beta_0^2 (n_b/2n_e)^{1/3}$ is the so-called strength parameter.[44]

The dominant process that limits the wave amplitude is that of the beam electron trapping by the wave potential. Assuming that the wave instantaneously reaches its saturation amplitude at which time the beam electrons are trapped, then qualitatively the average beam velocity in the wave frame drops to zero and $\gamma_{b\omega} \to 1$. Transforming back to the lab frame yields

$$\gamma_b' \simeq \gamma_\omega \tag{4.180}$$

where γ_b' is the beam energy after the interaction. Hence, the energy conversion efficiency α can be written as

$$\alpha = \frac{\Delta E}{\gamma_b mc^2 n_b} = \frac{\gamma_b - \gamma_b'}{\gamma_b} = 1 - (1+S)^{-1/2} \tag{4.181}$$

which implies that the energy transfer efficiency is a function of the single parameter S. In the limit $S \ll 1$, $\alpha \approx \frac{1}{2}S$.

For $S \gtrsim 1$, on the other hand, the beam electrons are strongly relativistic, and it is necessary to carefully examine the particle orbits in the wave. If the beam electrons are uniformly distributed in phase at the instant of wave saturation, some electrons will be decelerated while others will be accelerated. The accelerated electrons, being relativistic, increase their effective mass but do not make a significant phase shift. This process leads to bunching of beam electrons in regions of phase space where they continue to extract energy from the wave, and results in a marked reduction in the energy transfer efficiency. A detailed consideration of the beam electron trajectories in the wave indicates that the approximate scaling is given by[44]

$$\alpha = S(S+1)^{-5/2} \tag{4.182}$$

Equation (4.182) indicates that the most efficient energy transfer occurs at $S = 2/3$; however, partial numerical simulation results indicate that the phase velocity of the wave slows as the electric field energy nears its maximum value. The net effect is that α peaks at $S \simeq 0.40 - 0.45$ and the maximum energy transfer saturates at 0.18 for larger S.[43]

The decrease in coupling efficiency caused by beam electron phase bunching can be overcome by allowing waves to propagate at an angle with respect to the beam. In the absence of a magnetic field the growth rate of the fastest-growing waves is given by Eq. (4.147). Assuming that the interaction is one dimensional (in space) along the direction of wave propagation ($\tan\theta = k_\perp / k_z$), Eq. (4.182) can be modified to take into account a finite $k_\perp$ as[43]

$$\alpha_\theta = S_\theta(1 + S_\theta)^{-5/2} \tag{4.183}$$

where

$$S_\theta = \beta_0^2 \frac{k_z^2}{k^2} \gamma_\parallel \left(\frac{n_b \gamma_\parallel}{2n_e \gamma_b} \right)^{1/3} \quad \text{and} \quad \gamma_\parallel = \left(1 - \beta_0^2 \frac{k_z^2}{k^2} \right)^{-1/2}$$

Note that the angular strength parameter S_θ decreases rapidly as θ increases, and that for $S > 2/3$, $\alpha_\theta > \alpha$ for waves with finite k such that $S_\theta \simeq 2/3$. In fact α_θ is sharply peaked at some θ_{max} since both the transverse ($S_\theta \ll 1$) and longitudinal ($S_\theta \simeq S$) interactions are inefficient. This observation leads to the oblique approximation[43]: the unstable spectrum is approximated by a single longitudinal wave number k_z and a single perpendicular wave number $k_\perp$ such that $S_{\theta max} \simeq 2/3$. The consequence of this approximation is that

$$\alpha = \begin{cases} S(1 + S)^{-5/2}, & S \lesssim 0.45 \\ 0.18, & S \gtrsim 0.45 \end{cases} \tag{4.184}$$

The coupling efficiency can be further increased by including the effects of a finite magnetic field due to the lower characteristic phase velocity of the oblique waves.[45]

In comparison with two-dimensional numerical simulations, the one-dimensional oblique approximation appears to predict the two-dimensional dynamics of the deeply trapped electrons which account for most of the beam energy loss; however, it overestimates the magnitude of the electric field energy. These results indicate that the waves are damped by strong nonlinear processes, the most likely being nonlinear Landau damping. Significant bulk heating of electrons is observed, roughly symmetric with respect to the beam. In addition, high-energy tails are formed perpendicular to the beam and in the direction of the beam.[43, 45]

It should be mentioned that several experimental investigations of plasma heating performed prior to the theoretical development outlined here have failed to obtain the predicted high conversion efficiencies.[46-52] This discrepancy is believed to be due, for example, to beam scatter in the anode foil, plasma density gradients, etc. Hence verification of the theory must await the outcome of future heating experiments which have been designed within the assumptions of the theory.

4.7. Summary

When an intense charged particle beam is injected into a dense plasma redistribution of the plasma electrons effectively neutralizes the beam space charge, thereby reducing the radial electric field to zero. As a result the beam will constrict, or pinch, owing to the azimuthal self-magnetic field. For beam currents in excess of the Alfven limit the beam self-magnetic field is sufficient to reverse the direction of beam electron trajectories at the outer edge of the beam. However, depending upon the plasma conductivity, the induction electric field at the beam head (associated with the rising beam current) will drive a plasma current opposite to the beam current. This counterstreaming plasma return current can effectively cause the net current, and hence the azimuthal magnetic field in the system, to vanish, thereby allowing the propagation of primary beam currents in excess of the Alfven limit.

Current neutralization eventually decays owing to the finite plasma conductivity in a characteristic magnetic diffusion time. During this period a steady state exists in which no net self-fields act on the beam particles, and the beam transport is limited by the various instabilities associated with the motion of the beam through the plasma. Such unstable oscillations are usually subdivided into two classes: macroscopic oscillations with characteristic scale lengths of the order of the beam radius or larger; and microscopic oscillations of length scale smaller than the beam radius.

Examples of the macroscopic instabilities are the hose (kink) and sausage distortions which can cause considerable deformation of the equilibrium beam configuration. An intense beam propagating through a plasma can be hydromagnetically unstable to transverse perturbations because of the finite plasma conductivity. When the beam axis undergoes a small displacement, the magnetic field axis lags behind for times on the order of the magnetic diffusion time, and a restoring force tends to return the beam to its original position. For finite conductivity this interaction can be so phased that these transverse (hose) perturbations can grow indefinitely. Similarly, radially symmetric perturbations can also be destabilized by finite plasma conductivity leading to growth of the sausage instability.

In addition to the macroscopic instabilities, several microscopic instabilities can develop when an intense beam propagates through a plasma, depending on the frequency and parameter ranges of interest. For the case of an electron beam propagating in an ionic plasma whose ion density equals the beam electron density (an external magnetic field is assumed to prevent beam constrictions) the electron–ion streaming instability (Buneman mode) can develop.

If the plasma density greatly exceeds the beam density, the beam may be considered as a perturbation which can excite natural plasma oscillations. The instabilities divide naturally into three regions depending on the relative magnitude of the electron plasma frequency (plasma electron density) and the electron cyclotron frequency (external magnetic field strength). In the dense-plasma–weak-field limit the cyclotron instability degenerates into the two-stream instability and the growth rate is peaked at large angles. When the cyclotron frequency becomes comparable to the plasma frequency, the upper hybrid oscillation mode is driven unstable at large angles by the beam cyclotron resonance, while the two-stream mode becomes longitudinal. In the rarefied-plasma–strong-field limit four distinct instabilities are possible: (i) two-stream instability, (ii) hybrid two-stream instability, (iii) electron cyclotron instability, and (iv) upper hybrid instability, in order of increasing wave propagation angle.

Analysis of the above-mentioned instabilities is possible within the framework of the electrostatic approximation. In addition, current neutralized intense electron beams can also be unstable to transverse electromagnetic modes. In particular, development of the Weibel instability can tear the beam into filaments.

The microturbulence associated with the development of the various microscopic instabilities enhances the beam–plasma energy deposition rate over that associated with classical Coulomb collisions, leading directly to the possibility of heating a dense plasma to thermonuclear temperatures by intense electron beams. There are two distinct collective processes which can heat the target plasma. The first is that associated with the interaction between the plasma electrons (which carry the return current) and the plasma ions. In the early stage of the interaction, the dominant energy transfer mechanism is the conventional Buneman electron–ion two-stream instability. At the end of this stage, the instability passes into the ion acoustic stage. Plasma heating via the ion acoustic instability proceeds until (i) the ions are trapped in the unstable wave, (ii) linear stability is achieved, or (iii) energy loss due to wave convection leads to an equilibrium. For beam currents in excess of the Alfven limit, the return current mechanism should effectively heat relatively low-density plasmas; however, for target plasmas much denser than the beam, the energy dissipation rate is reduced because the plasma electron drift velocity decreases with increasing density ratio. The Buneman instability may not develop and the effective collision frequency is expected to be small.

The second important type of microturbulence is the interaction between the primary beam electrons and the target plasma electrons. As previously discussed, a number of interactions are possible, depending on the presence and strength of an external longitudinal magnetic field. For a cold beam, collisionless cold plasma interaction, the efficiency for converting beam energy into plasma energy is essentially determined by a single quantity, the so-called strength parameter. The approximate behavior of the coupling coefficient can be determined by assuming that the beam–plasma interaction is one-dimensional in space along the direction of wave propagation (one-dimensional oblique approximation).

In comparison with two-dimensional numerical simulations the one-dimensional oblique approximation is found to predict the two-dimensional dynamics of the deeply trapped electrons which account for most of the beam energy loss; however, it overestimates the wave electric energy, indicating strong wave damping due most probably to nonlinear Landau damping. The two-dimensional simulations indicate substantial bulk electron heating with the possibility of significant energy transfer.

Problems

1. Derive the Alfven current limit, Eq. (4.1).

2. Derive the Green's functions, Eqs. (4.22) and (4.23).

3. (a) Derive the perturbation fields [analogous to Eqs. (4.31)–(4.36)] for the case of a beam of radius r_b in an infinite homogeneous plasma; i.e., use the Green's function Eq. (4.22). The beam current density is specified as

$$j_{bz} = \begin{cases} - envg(u), & 0 \leq r < r_b \\ 0, & r > r_b \end{cases}$$

where

$$g(u) = \begin{cases} 1, & u > 0 \\ 0, & u < 0 \end{cases}$$

(b) In the high-conductivity limit compare the B_θ expression obtained from part (a) with Eq. (4.34).

(c) Plot the qualitative behavior of the plasma current density, the net current density, and the azimuthal magnetic field as a function of distance from the beam front.

(d) Assuming $r_b = 5$ cm and $\sigma = 10^{11}$ sec^{-1}, calculate the length of the current neutralization region.

4. (a) Obtain an expression for $F(g)$, Eq. (4.99), valid in the limit of large conducting wall radius, $R \to \infty$.

(b) Obtain the hose growth rate for (a) above. What is the form of the low-frequency decay constant?

5. Using the hose instability dispersion relation, Eq. (4.103), with $\omega'_\beta = 0$, show that the peak growth rate in k_z is given by

$$\text{Max}[-Im(k_z)] \approx 0.3\omega_\beta/\beta_0 c$$

6. Suppose that $f(k_{\beta 0})$ is the distribution of betatron wave numbers with normalization

$$1 = \int_0^\infty dk_{\beta 0} f(k_{\beta 0})$$

In this case Eq. (4.129) becomes

$$-ik_u = \alpha \int_0^\infty dk_{\beta 0} f(k_{\beta 0}) \left(\frac{k_z^2 - 2k_{\beta 0}^2}{4k_{\beta 0}^2 - k_z^2} \right)$$

(a) Show that the above equation can be expressed as

$$-ik_u = \frac{1}{2}\alpha \left[-1 + \frac{1}{4}\pi i |k_z| f\left(\frac{|k_z|}{2} \right) + P \int^\infty dk_{\beta 0} \frac{f(k_{\beta 0})\left(\frac{1}{2}k_z\right)^2}{k_{\beta 0}^2 - \left(\frac{1}{2}k_z\right)^2} \right]$$

(b) Suppose that $f(k_{\beta 0})$ has a maximum at $\overline{k_{\beta 0}}$ with half-width $2\Delta k_{\beta 0}$. Show that the peak growth rate occurs for $\frac{1}{2}k_z \approx \overline{k_{\beta 0}} - \Delta k_{\beta 0}$, and has magnitude

$$\text{Re}(-ik_u)_{\max} \approx \frac{1}{2}\alpha \left(-1 + \frac{\overline{k_{\beta 0}}}{2\Delta k_{\beta 0}} \right)$$

(c) What is the frequency spread required for sausage mode stabilization?

7. Using Eq. (4.133) show that the current density necessary to excite oscillations for a particular value of ω is given by (Buneman mode)

$$j = \frac{m v_{bz}^3}{4 \pi e} \frac{m_i}{m_e} k_z^2$$

8. Show that for $\rho_{00} v_{bz}/R > \min(\omega_e, \Omega_e)$ the two-stream instability in an overdense plasma is possible only on the upper branch, ω, while for $\rho_{00} v_{bz}/R < \min(\omega_e, \Omega_e)$, instability can develop on both branches, $\omega_{\pm}$.

9. If $f(x, p, t)$ is the beam distribution in the wave frame of the two-stream instability interaction, the energy lost by the beam electrons in the lab frame averaged over a wavelength is given by

$$\frac{\Delta E}{n_b \gamma m c^2} = \frac{\gamma_w}{\gamma_0} \int_{-\lambda/2}^{\lambda/2} (dx/\lambda) \int_{-\infty}^{\infty} dp\, f(x, p, t) \{ \gamma_R - \gamma(x, p, t)$$

$$+ \beta_w [p_R - p(x, t)] \}$$

Show that use of the nonrelativistic "rigid-rotor" model for the beam distribution in which the beam is described initially by $f(x, p, t=0) = \delta(p - p_R)$, and after one-half revolution in phase space is described by $f(x, p, t=t_s) = \delta(p + p_R)$, yields for the coupling coefficient

$$\alpha = \tfrac{1}{2} S (1+S)^{-1} \approx \tfrac{1}{2} S \qquad \text{for } S \ll 1$$

5

Propagation of Intense Beams through Neutral Gas

5.1. Introduction

When an intense relativistic electron beam is injected into neutral gas the beam transport may be characterized as either unneutralized beam propagation (Chapter 3), or transport through a plasma (Chapter 4), depending upon how rapidly the beam can ionize the gas thereby creating a space charge and/or current-neutralizing background. Because of the several parameters which determine the gas ionization rate, a quantitative analysis of beam transport through neutral gas in the general case is very difficult.

In descriptive terms it is helpful to characterize the neutral gas transport by the ratio of the beam current to the space charge limit, I/I_l, and the ratio of the beam current to the Alfvén limit, I/I_A. Writing the space charge limiting current as $I_l(t) = I_{l0}[1 - f_e(t)]^{-1}$, where f_e is the space charge neutralization fraction, if the peak injected beam current I_0 does not exceed the limit I_{l0}, the electrostatic potential can never exceed the beam kinetic energy, and propagation is possible regardless of the gas pressure; however, if $I_0 > I_{l0}$, the beam behavior depends critically on the ionization rate.[1]

Assuming that the beam is space charge neutralized, $f_e \simeq 1$, efficient propagation can occur if the beam current does not exceed the Alfven current limit, $I_A = \beta\gamma mc^3/e$. However, if $I_0 > I_A$ the self-magnetic field energy of the propagating beam current would exceed the particle kinetic energy.[2] In this case propagation of the total beam current can only occur if there is some degree of current neutralization. Since current neutralization is largely determined by the plasma conductivity, and the conductivity depends in turn on the plasma electron–neutral collision frequency, current neutralization is also a sensitive function of the background gas pressure.

The behavior of propagating electron beams can be qualitatively analyzed in terms of the radial force equation

$$F_r = \gamma m \ddot{r} = -eE_r\left[1 - f_e - \beta^2(1 - f_m)\right] \qquad (5.1)$$

where f_e and f_m denote the degrees of beam space charge and magnetic neutralization. Depending upon the beam parameters, the background gas ionization, and the plasma conductivity, a number of situations are possible.

Finally, for the propagation of an intense beam over substantial distances ($\gtrsim 10$ m) through dense neutral gas, it is necessary to consider the energy lost by the beam in the creation and transport through the plasma channel. Also important is the possible development of various beam–plasma instabilities. In the following sections the topics briefly mentioned in this section are treated in more detail, including a summary of the important gas ionization processes. To facilitate the presentation, the beam transport is characterized according to the ratio of the beam current I_b, to the space charge limit I_l, or the Alfvén limit, I_A, and is ordered according to the background neutral gas pressure.

5.2. Beam-Induced Neutral Gas Ionization Processes

Because of the dominant role of background conductivity in determining the beam transport characteristics, this section is devoted to summarizing the important gas ionization processes.

5.2.1. Electron Impact Ionization

Neutral gas atoms become ionized as the result of direct collisions with primary beam electrons. In this case the increase in the background ion density can be written as

$$\frac{\partial n_i(t)}{\partial t} = \frac{n_b(t)}{\tau_e} \qquad (5.2)$$

where τ_e is the characteristic electron impact ionization time. For the case of hydrogen gas a useful approximation for τ_e is[3]

$$\tau_e \approx 5.0 [P(\text{Torr})]^{-1} \text{ nsec} \qquad (5.3)$$

over the beam electron kinetic energy interval $\varepsilon_e = 0.3 - 4$ MeV.

The ionization of a neutral gas atom produces a positive ion and a secondary electron and both types of secondary particles can contribute to further gas ionization under appropriate conditions.

5.2.2. Electron Avalanche

The phenomenon of electron avalanche, i.e., the further ionization of the background gas by secondary electrons can be modeled as

$$\frac{\partial n_e(t)}{\partial t} = \frac{n_b(t)}{\tau_e} + \frac{n_e(t)}{t_e} - \frac{n_e(t)}{t_s} \qquad (5.4)$$

$$\frac{\partial n_i(t)}{\partial t} = \frac{n_b(t)}{\tau_e} + \frac{n_e(t)}{t_e} \qquad (5.5)$$

where t_e is the effective electron avalanche time and t_s is the effective secondary escape time. For electron avalanche to be an important effect, it is necessary to have $t_e < t_s$. A useful estimate for t_e is $t_e \sim t_F/(8 \ln 10)$, where t_F, the "formative time," depends on the magnitude of the accelerating field E and on the type and pressure P of the background gas.[3] Based on the data E/P must be of the order of $10 - 10^2$ V/cm Torr for effective electron avalanching.[4] Before the beam becomes charge neutralized $E \sim 10^6$ V/cm, and the secondary electrons are quickly lost radially to the guide walls in a time of the order $t_s \sim R/c$. Hence, electron avalanching effects can usually be neglected unless $f_e \approx 1$.

On the other hand, if the beam space charge is neutralized, the large radial electrostatic fields do not exist and t_s can increase by several orders of magnitude. Further, the inductive electric field at the beam head ($E_z \propto I_0/t_r$) is typically of the order $\leq 10^3$ V/cm. Thus, for (hydrogen) pressures ≥ 1 Torr, it is expected that electron avalanche can lead to rapid breakdown of the background gas once charge neutralization has been achieved.

5.2.3. Ion Ionization

Because of the large accelerating fields which can exist in an intense electron beam environment, it is not unreasonable to expect that ionization of the neutral gas could be enhanced by energetic background ions. Such a process can be modeled as

$$\frac{\partial n_i(t)}{\partial t} = \frac{n_i(t)}{t_i} \tag{5.6}$$

where t_i is the effective ion avalanche time.

An important competing reaction is charge exchange. A detailed consideration of the energetics of these reactions indicates that (for the case of hydrogen) if $\varepsilon_i > 50$ keV then ionization is favored over charge exchange; if $\varepsilon_i \leqslant 50$ keV, then charge exchange dominates. For ion energies in the range 50 keV $< \varepsilon_i < 1$ MeV a useful approximation for t_i is given by[3]

$$t_i \simeq 0.33 [P(\text{Torr})]^{-1} \text{nsec} \tag{5.7}$$

which is much less than τ_e. Hence, under appropriate circumstances it is expected that ion ionization can play a dominant role in the charge neutralization process.

5.3. Neutral Gas Transport for $I_b/I_l < 1$

5.3.1. Low-Pressure Limit

For $I_b/I_l < 1$ a deep stationary potential well (virtual cathode) does not form, and the beam electrons are not slowed appreciably in the axial direction. If the background gas pressure is low enough that the beam charge neutralization time τ_N exceeds the beam pulse time t_b, then the gas ionization is volumetric being due primarily to beam electron impact ionization. For the case of hydrogen, this criterion is stated approximately as [see Eq. (5.3)]

$$P_N(\text{Torr}) < \frac{5}{t_b(\text{nsec})} \tag{5.8}$$

In this pressure limit the beam particle trajectories may be analyzed in terms

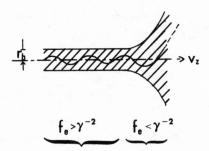

Figure 5.1. Expansion of an unneutralized charged particle beam.

$$f_e > \gamma^{-2} \qquad f_e < \gamma^{-2}$$

of the radial force equation

$$F_r = \gamma m \ddot{r} = -eE_r + ev_b B_\theta = -eE_r\left(1 - f_e - \beta^2\right) \tag{5.9}$$

For regions in which $0 \leqslant f_e \leqslant \gamma^{-2}$ the beam electrons experience a net repulsive radial force and the beam expands. Integration of Eq. (5.9) leads to the envelope equation for the trajectory of an electron at the beam edge

$$\ddot{r}_b = \frac{\omega_b^2}{2\gamma_b r_b}\left(1 - f_e - \beta^2\right) \tag{5.10}$$

The solution of Eq. (5.10) is given by

$$\frac{t\omega_b}{2\gamma_b^{1/2}}\left(1 - f_e - \beta^2\right)^{1/2} = r_b D\left[(\ln r_b)^{1/2}\right] \tag{5.11}$$

where $D(x)$ represents Dawson's integral defined by

$$D(x) = e^{-x^2}\int_0^x e(t^2)\,dt \tag{5.12}$$

For $f_e = \gamma_e^{-2}$, $\ddot{r}_b = 0$. For regions in which $f_e > \gamma_e^{-2}$, the beam electrons will undergo betatron oscillations in the radial direction maintaining an envelope radius on the order of the beam radius, as indicated in Fig. 5.1.

5.3.2. Intermediate-Pressure Regime

For somewhat higher pressures such that $\tau_N \lesssim t_r$, where t_r is the beam rise time, $f_e \approx 1$ early in the pulse and gas ionization due to electron avalanche is expected to be important. In this case the secondary electron density may be modeled as

$$
\begin{aligned}
n_e &\simeq 0, & t < \tau_N \\
\frac{\partial}{\partial t}\left[n_e(t) + n_b(t)\right] &\simeq \frac{n_b(t)}{\tau_e} + \frac{n_e(t)}{t_e}, & t > \tau_N
\end{aligned}
\tag{5.13}
$$

which attempts to account for radial loss of secondary electrons until $t = \tau_N$, and the effects of stationary ions for $t > \tau_N$. Assuming that $\tau_N \approx t_r$ and that n_b is a constant for $t > \tau_N$, integration of Eq. (5.13) yields

$$
n_e(t) = \frac{n_b t_e}{\tau_e}\left[e^{(t - \tau_N)/t_e} - 1\right]
\tag{5.14}
$$

For a weakly ionized gas, the conductivity is given by

$$
\sigma \simeq e^2 n_e / m\nu_c
\tag{5.15}
$$

where ν_c is the momentum transfer collision probability. [Typically ν_c $(\sec^{-1}) \approx 10^9 P$ (Torr).] Hence, the exponentiation of the secondary electron density due to electron avalanche ionization also rapidly increases the conductivity of the background gas and leads to the formation of a reverse plasma current according to $j_p = \sigma E_z$, where E_z is the axial electric field with both electrostatic and inductive components.

A more detailed examination of charge and current neutralization for a beam with $I_b \ll I_l$ injected into neutral gas was performed in Ref. 5. Satisfactory agreement between theory and experiment was obtained for gas pressures $P \lesssim 1$ Torr. Discrepancies observed at higher pressures have been attributed to collective processes, in particular, the resistive hose instability.[6]

5.3.3. High-Energy Beam Transport in the High-Pressure Regime

For a high-energy particle beam injected into a very dense neutral gas background electrostatic breakdown and the formation of a plasma channel occur very rapidly. Since the remainder of the pulse is subjected to only magnetic self-pinch forces, the possibility arises for self-pinched beam transport over substantial distances; however, this simple picture must be modified to account for collisions between beam and background particles and beam energy loss processes, both of which increase the beam transverse energy and can lead to beam expansion.

Assuming that the macroscopic effects of these processes occur over time scales long compared with a characteristic betatron period, the change in the perpendicular kinetic energy per unit length is given by[7]

$$\Delta U = \Delta E_{\perp \text{coll}} + \Delta E_{\perp \dot{\gamma}} - \Delta W \qquad (5.16)$$

where $\Delta E_{\perp \text{coll}}$ and $\Delta E_{\perp \dot{\gamma}}$ represent the increase in perpendicular energy due to collisions and beam energy loss, and ΔW accounts for the work done by the expanding beam against the confining azimuthal magnetic field. Assuming an isotropic perpendicular velocity distribution the internal energy U is given by

$$U = 2\pi \int_0^\infty dr\, rp \qquad (5.17)$$

where the plasma pressure p is given by $p = n_e \gamma m \langle v_\perp^2 \rangle / 2$. The condition for a radial force equilibrium in which the plasma pressure is balanced by the self-magnetic field of the beam is

$$\frac{1}{8\pi} \frac{1}{r^2} \frac{\partial}{\partial r} \left(r^2 B_\theta^2 \right) + \frac{\partial p}{\partial r} = 0 \qquad (5.18)$$

or, in terms of the beam current density

$$\frac{j_b B_\theta}{e} + \frac{\partial p}{\partial r} = 0 \qquad (5.19)$$

After solving Eq. (5.19) for p and substituting the resulting expression into Eq. (5.17) an integration by parts yields

$$U = \frac{\pi}{c} \int_0^\infty j_b B_\theta r^2\, dr \qquad (5.20)$$

The work done by the beam in expanding against the magnetic field is given by

$$\Delta W = \int_0^\infty (F_r \cdot \Delta r) n_b 2\pi r\, dr \qquad (5.21)$$

where $F_r = -(e/c)v_z B_\theta$ is the force on an electron and $\Delta r = v_r dt$ is the expansion increment. If the beam is assumed to expand in self-similar fashion, i.e., without change in the radial current profile, then the expansion velocity is given by

$$v_r = \frac{r}{a} \frac{da}{dt} \tag{5.22}$$

where a is the radial scale length. Evaluating Eq. (5.21) yields

$$\Delta W = \left(\frac{2\pi}{c} \right) \frac{da}{a} \int_0^\infty dr \, r^2 j_b B_\theta$$

$$= 2U \frac{da}{a} \tag{5.23}$$

The contributions to the perpendicular kinetic energy due to collisions and beam energy loss processes can be evaluated in the following manner. As the result of a collision, the change in the perpendicular kinetic energy of a beam electron is given by

$$\Delta \varepsilon_\perp \simeq \Delta \left(\frac{p_\perp^2}{2\gamma m} \right) = \frac{\boldsymbol{p}_\perp \cdot \Delta \boldsymbol{p}}{\gamma m} + \frac{(\Delta p)^2}{2\gamma m} - \frac{p_\perp^2}{2\gamma m} \frac{\Delta \gamma}{\gamma} \tag{5.24}$$

When averaged over many collisions the first term on the right-hand side of Eq. (5.24) vanishes. Further, the effects of all beam energy loss processes can be incorporated by generalizing the last term on the right-hand side, i.e.,

$$\Delta E_{\perp \dot{\gamma}} = -U \frac{\Delta \gamma}{\gamma} \tag{5.25}$$

Defining the quantity $\varepsilon_{\perp \, \text{coll}}$ as

$$\varepsilon_{\perp \, \text{coll}} = \sum \frac{(\Delta p_\perp)^2}{2\gamma m} \tag{5.26}$$

(representing the result of many collisions) $\Delta E_{\perp \, \text{coll}}$ is evaluated as

$$\Delta E_{\perp \, \text{coll}} = \int_0^\infty 2\pi r \, dr \, n_e \varepsilon_{\perp \, \text{coll}} \tag{5.27}$$

Substitution of Eqs. (5.23), (5.25), and (5.27) into Eq. (5.16) yields

$$\frac{\Delta U}{U} + \frac{\Delta(a^2)}{a^2} + \frac{\Delta\gamma}{\gamma} = \frac{\Delta E_{\perp \, \text{coll}}}{U} \tag{5.28}$$

which has the formal solution

$$\frac{U\gamma a^2}{U_0 \gamma_0 a_0^2} = \exp \int_0^z dz \, \frac{1}{U} \left(\frac{\Delta E_{\perp \, \text{coll}}}{\Delta z} \right) \tag{5.29}$$

For the simple case of a beam with a uniform density profile, i.e.,

$$j_b = \begin{cases} I_b/\pi r_b^2, & 0 \leqslant r \leqslant r_b \\ 0, & r > r_b \end{cases} \tag{5.30}$$

with $r_b = r_b(z)$, Eq. (5.29) becomes

$$r_b = r_{b0} \left(\frac{\gamma_0}{\gamma} \right)^{1/2} \exp \left[\int_0^z dz \, \frac{c^2}{I_b^2} \left(\frac{\Delta E_{\perp \, \text{coll}}}{\Delta z} \right) \right] \tag{5.31}$$

The quantity $(\Delta E_{\perp \, \text{coll}}/\Delta z)$ represents the perpendicular kinetic energy gained per unit length as the result of many small angle deflections. At high relativistic energies a convenient approximate expression is

$$\left(\frac{\Delta E_{\perp \, \text{coll}}}{\Delta z} \right) \approx \frac{1}{2} \frac{E}{\lambda_R} \left[\frac{21}{E(\text{MeV})} \right]^2 \tag{5.32}$$

where λ_R, the radiation length, is defined by[7]

$$\lambda_R = \left[4 n_e Z(Z+1) \frac{e^2}{\hbar c} \left(\frac{e^2}{mc^2} \right)^2 \ln \left(\frac{137}{Z^{1/3}} \right) \right]^{-1} \tag{5.33}$$

Evaluating $(\Delta E_{\perp \, \text{coll}}/\Delta z)$ using Eq. (5.33) and substituting the resulting expression into Eq. (5.31) yields

$$r_b = r_{b0} \left(\frac{\gamma_0}{\gamma} \right)^{1/2} \exp \left(2 \int_0^z dz \left\langle \frac{1}{\lambda_R} \right\rangle \frac{C_N}{W} \right) \tag{5.34}$$

where $W = I_b E$ is the beam power in watts, $C_N = 3.7 \times 10^{12}$ is a constant, and $\langle 1/\lambda_R \rangle$ is the value of $1/\lambda_R$ averaged over $n_e(r)$. Equation (5.34) is essentially the famous result derived by Nordsieck in 1959 (apart from the factor of 2 in the exponential).[8] It is readily apparent that very high beam power (10^{12}–10^{13} W) is required for beam transport over distances of several λ_R without catastrophic beam expansion in the high-energy, pinched beam propagation mode.

5.4. Neutral Gas Transport for $I_l \lesssim I_b \lesssim I_A$

5.4.1. Low-Pressure Regime

When the injected electron beam current exceeds the space-charge-limiting current for the case of injection into a low-pressure neutral gas in the absence of external magnetic fields, then beam propagation away from the anode region is effectively prevented by the formation of a deep quasistationary electrostatic potential well (a virtual cathode) due to the beam space charge. Although numerical simulations of this injection stage indicate that the magnitude of the potential minimum and its position oscillate in time, the average values are approximately given by analytic expressions for the well depth and penetration length at the time of the first electron reflection. Typically $|e\phi_0| = \alpha \xi_e$, where $1 \leq \alpha \leq 1.5$.[9]

The shape of the deep potential well may be determined if the beam density distribution is known. If there is no external magnetic field then the electron density will rapidly drop to zero for axial distances from the anode greater than the location of the potential minimum due essentially to space charge repulsion. (The quasistatic current flow patterns consist of an injected current, a current reflected by the virtual cathode back to the anode, and a radial current from the virtual cathode to the drift tube walls as indicated in Fig. 5.2.) Hence, it is reasonable to assume that the beam electron density distribution can be approximated by a disk of uniform charge density of radius b and depth d, i.e.,

$$n_b(r, z) = \begin{cases} n_{b0}, & 0 \leq r \leq b, \quad 0 \leq z \leq d \\ 0, & \text{otherwise} \end{cases} \tag{5.35}$$

Carrying out the integration over the Green's function (Section 3.2.4) yields

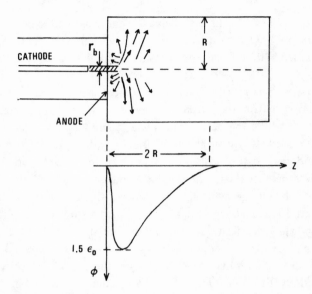

Figure 5.2. Qualitative current flow patterns for an electron beam with $I_b/I_l > 1$ injected into a low-pressure neutral gas.

(for $z > d$, $r = 0$, $L \gg R$, and $L \gg z$)[9]

$$\phi(0,z) \approx (8\pi b e n_{b0} R) \sum_{n=1}^{\infty} \frac{J_1(\lambda_n b/R)}{\lambda_n^3 [J_1(\lambda_n)]^2} [\cosh(\lambda_n d/R) - 1]$$

$$\times \exp(-\lambda_n z/R) \tag{5.36}$$

Typically, the first term is dominant in the summation and $\phi(0,z)$ decays like $\exp(-\lambda_1 z/R)$ provided $d \lesssim R$.

The deep potential well (which prevents beam propagation) will persist until ionization of the background gas produces sufficient charge neutralization to reduce the depth of the potential well, $|e\phi_0|$ below the injected electron beam kinetic energy. (Almost complete space charge neutralization is required). Unlike the case of $I_b < I_l$, the ion avalanche process can now play a significant role; since the charge neutralization time due only to electron impact ionization is typically tens of nanoseconds, background ions may be accelerated and trapped in the deep potential well. As discussed in Section 5.2.3, ions (protons) with kinetic energy in excess of 50 keV are much more effective in ionizing the gas than the beam electrons and can substantially reduce the time to achieve charge neutralization. For typical parameters 50 keV corresponds to the potential well depth at an axial position L from the anode approximately equal to a drift tube diameter $(2R)$.[9]

When the beam space charge is almost neutralized a nonadiabatic transition occurs from the virtual cathode state to a beam propagating state. The speed at which the beam front (i.e., the moving rise-time structure of the beam) moves away from the anode region is determined primarily by the background gas pressure with three distinct regimes being apparent.

During the virtual cathode stage a potential well of length L forms and decays over the charge neutralization time τ_N; hence, the beam front could be considered to translate self-consistently with a front velocity of the order $\beta_f c \approx L/\tau_N$. However, if the background gas pressure is so low that $\beta_f c < \beta_{i0} c$, where $\beta_{i0} c$ corresponds to the velocity of the fastest ions created during the deep well stage, then it may be expected that the beam front velocity will be determined by these fast ions. In essence, the fast ions can extend the equilibrium well length such that the front propagates self-consistently with speed $\beta_{i0} c$.

In the opposite extreme, if the gas pressure is high enough that the charge neutralization time τ_N is less than the time required for the finite rise-time beam current to exceed the space charge limit, then the virtual cathode never forms and the beam never stops. This "runaway" pressure p_R is roughly defined as the pressure for which $\tau_N = (I_l/I_b)t_r$. In this case a useful estimate for the beam front velocity can be obtained by requiring that the beam power consumed in establishing the injected beam current equals the power associated with the translating beam. Since the magnetic field energy per unit length of a uniform beam is given by

$$\xi_B = (I_0/2c)^2[1 + 4\ln(R/r_b)] \tag{5.37}$$

the power balance requires[9]

$$VI_b - (\gamma_f - 1)mc^2 I_b/e = (I_b/2c)^2[1 + 4\ln(R/r_b)]\beta_f c \tag{5.38}$$

where V is the effective diode voltage, $\gamma_f = (1 - \beta_f^2)^{-1/2}$, and it has been assumed that all injected electrons are slowed to the velocity $\beta_f c$ at the beam front. After the transition stage, the beam propagates in a quasiequilibrium manner in which the beam dynamics are coupled with the ionization processes.

5.4.2. Intermediate-Pressure Regime

The beam runaway pressure essentially denotes the boundary of the intermediate-pressure regime in which current neutralization effects can become important; however, it should be recognized that a lack of current neutralization does not prevent beam propagation since $I_b < I_A$.

The experimentally observed net current behavior may be qualitatively explained[10] via a simple model in which the plasma current varies according to $j_p = \sigma E_z$, where σ is the (scalar) conductivity, and E_z is the inductive electric field generated by the time-changing azimuthal magnetic field associated with the net current.[11] Assuming that E_z over the beam cross section is approximately given by the inductive electric field on axis, then

$$j_p = \sigma E_z \approx \sigma E_z(0) = -\frac{\sigma}{c}\frac{\partial}{\partial t}\int_0^R B_\theta(r')\,dr' \tag{5.39}$$

where

$$B_\theta(r') = \frac{2\pi(j_p + j_b)}{c}\begin{cases} r', & r' < r_b \\ r_b^2/r, & r' > r_b \end{cases} \tag{5.40}$$

In writing Eq. (5.40) it has been assumed that both the beam and plasma currents have the same (uniform) radial profile. Substituting Eq. (5.40) into Eq. (5.39) and performing the integration yields

$$j_p = -\frac{\sigma\pi r_b^2}{c^2}\left[1 + 2\ln(R/r_b)\right]\frac{\partial}{\partial t}(j_p + j_b) \tag{5.41}$$

If the conductivity is a complicated function of time then solving for j_p would necessitate the solution of a difficult nonlinear problem. For simplicity it will be assumed that the conductivity is zero until the onset of charge neutralization and exponentiation of the secondary electron density by electron avalanche ionization. After this gas "breakdown" time, t_B, σ is assumed to be a constant. Representing the time-dependent current pulse in terms of a current-rise region, a constant-current region, and a decay region, i.e.,

$$j_b = j_{b0}\begin{cases} t/t_1, & 0 \leqslant t \leqslant t_1 \\ 1, & t_1 \leqslant t \leqslant t_2 \\ (t_2 - t)/t_2, & t_2 \leqslant t \leqslant t_3 \end{cases} \tag{5.42}$$

the solution of Eq. (5.41) yields

$$
j_p = \begin{cases}
0, & 0 \leqslant t \leqslant t_B \\
a_1 k\sigma \left[1 - e^{-(t-t_B)/k\sigma} \right], & t_B < t < t_1 \\
a_2 k\sigma e^{-(t-t_1)/k\sigma}, & t_1 < t < t_2 \\
k\sigma \left[a_3 - a_4 \left(1 - e^{-(t-t_3)/k\sigma} \right) \right], & t_2 < t < t_3 \\
a_5 k\sigma e^{-(t-t_3)k\sigma}, & t > t_3
\end{cases}
\tag{5.43}
$$

where $k = (\pi r_b^2/c^2)\,[1 + 2\ln(R/r_b)]$. The constants a_i are related by requiring that j_b remain continuous from region to region.

An examination of Eq. (5.43) indicates that in order to achieve high-current neutralization two conditions must be fulfilled: (1) Breakdown must occur early in the pulse; and (2) the breakdown conductivity σ_B must be high.[10]

A more exact theory of current neutralization which considers in detail the atomic physics processes occurring in the gas, and simultaneously computes the gas conductivity and solves the Maxwell equations, has been shown to give excellent agreement with experiment for pressures in excess of ~ 1 Torr.[12] However, several experimental discrepancies have been observed at somewhat lower pressures, including anomalously high plasma density, significant gas ionization continuing after the beam pulse (indicating a hot plasma component), a net current in excess of the injected beam current, and strong microwave emission.[13] These observations are consistent with, and have been attributed to, the presence of two-stream microturbulence.[14]

The dispersion relation for a cold relativistic beam interacting with plasma [Eq. (4.134)] modified to account for collisions between plasma electrons and neutral gas molecules can be written as

$$
0 = 1 - \frac{\omega_e^2}{\omega(\omega + i\nu_m)} - \frac{\omega_b^2}{\gamma_b^3} \frac{\left[\gamma_b^2 (k_\perp/k)^2 + (k_z/k)^2 \right]}{(\omega - \mathbf{v}_b \cdot \mathbf{k})^2}
\tag{5.44}
$$

where ν_m is the plasma-electron–gas-molecule collision frequency (assumed to scale linearly with pressure).[14] The collisionless and collisional growth rates are given by

$$
Im(\omega)_{NC} = \frac{\sqrt{3}}{2} \left[\omega_b^2 \omega_e \frac{\left(\gamma_b^2 k_\perp^2/k^2 + k_z^2/k^2 \right)}{\gamma_b^3} \right]^{1/3}
\tag{5.45}
$$

$$Im(\omega)_C = \frac{\omega_b}{\gamma^{3/2}}\left[\omega_e \frac{\left(\gamma_b^2 k_\perp^2/k^2 + k_z^2/k^2\right)}{2\nu_m}\right]^{1/2} \tag{5.46}$$

with the condition $Im(\omega)_{NC} \approx \mathbf{k} \cdot \Delta\mathbf{v}$ marking the transition to the kinetic regime.

Stabilization of the two-stream mode occurs when the beam velocity spread becomes comparable to the collisional, cold-beam growth rates. The approximate criterion for this phase mix damping is $Im(\omega)_C \lesssim \mathbf{k} \cdot \Delta\mathbf{v}$, with equality indicating the stability boundary.[15] To estimate the parametric dependence of the two-stream stability criterion assume that $k_\perp \simeq r_b^{-1}$ and $k_z \simeq \omega_e/v_b$ for maximum growth. Since transverse velocity spread ($\Delta v_\perp$) is due to betatron oscillations in the self-magnetic field a measure of the spread is $\Delta v_\perp/v_b \simeq (I_b/2I_A)^{1/2}$. Finally, for a monoenergetic beam $\Delta v_\parallel/c \simeq \Delta v_\perp^2/(2c)^2$. Hence, the approximate stability criterion becomes[14]

$$\frac{\omega_b}{\gamma_b^{3/2}}\left[\frac{\omega_e^2\left(\gamma_b^2/r_b^2 + \omega_e^2/v_b^2\right)}{2\nu_m\left(1/r_b^2 + \omega_e^2/v_b^2\right)}\right]^{1/2} = c\left[\frac{1}{r_b}\left(\frac{I_b}{2I_A}\right)^{1/2} + \frac{\omega_e}{v_b}\frac{I_b}{4I_A}\right] \tag{5.47}$$

For fixed pressure the qualitative features of Eq. (5.47) indicate that stability should result for sufficiently high or low plasma density. In both regimes the growth rate varies as $n_e^{1/4}$. On the other hand, the phase mixing term approaches a constant for low plasma density, but varies as $n_e^{1/2}$ in the high-density regime. It is also apparent that sufficiently high pressure completely suppresses the mode, while the density required for stability rises very rapidly as the pressure is decreased. This behavior of the stability boundary is indicated in Fig. 5.3 for a given set of beam parameters. Also indicated are the kinetic, collisional, and collisionless mode regimes. Note

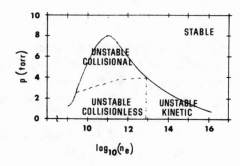

Figure 5.3. Two-stream instability boundaries in electron density–pressure space for a few MeV electron beams.

that for intermediate pressures the mode should be unstable for a range of plasma densities; however, classical beam ionization processes (both direct and avalanche) produce densities larger than this range and the mode is suppressed. For gas pressure below ~ 1 Torr air, the two-stream mode results in a hot plasma electron component that ionizes the gas on the microsecond time scale resulting in the anomalously high plasma densities experimentally observed. In addition, the net momentum transfer from beam electrons to plasma electrons drives a plasma current in the same direction as the beam resulting in "current multiplication."[13, 14]

5.4.3. High-Pressure Regime

At high gas pressure the plasma conductivity is too low to sustain current neutralization throughout the pulse. Although the beam should propagate since $I_b < I_A$, the results of experiments performed in this parameter regime indicated severe beam attenuation that could not be attributed to an increase in transverse temperature due to Coulomb scattering by the gas nuclei. A radial breakdown mechanism has been proposed to explain the loss process, but the experimental results were insufficient to clearly establish the physical mechanism of the proposed radial breakdown.[16] A more probable alternative is that the beam degradation was due to the onset of the resistive hose instability (Section 4.4.1). The essential condition for hose stability is that the plasma conductivity increase sufficiently rapidly so that the characteristic magnetic diffusion time is of the same order of magnitude as the beam pulse length early in the beam pulse. In the high-pressure regime the avalanching conductivity is reduced by collisions, thereby increasing the rate of hose growth. In the pulse body, a good qualitative estimate of maximum hose growth is predicted by a simple model to be[14]

$$Y \propto \left(\frac{\xi}{\xi_0} \right)^{0.69/(d\lambda/d\xi)} \tag{5.48}$$

where λ is the characteristic magnetic dipole diffusion length and $\xi = ct - z$ is the distance from the beam front. Detailed numerical simulations of the hose instability using the EMPULSE code are in good agreement with the experimental observations over a wide range of pressures (1–200 Torr).[6, 14]

The general features of beam transport in this current regime are illustrated in Figure 5.4, including net current, beam transport efficiency, and the various stability regimes.

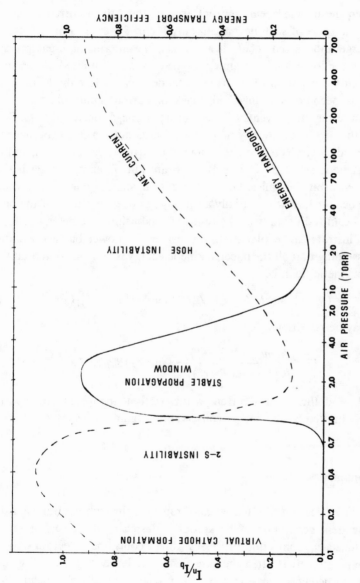

Figure 5.4. Pressure-dependent beam transport characteristics for a few MeV electron beam with $I_l < I_b < I_A$.

5.5. Neutral Gas Transport for $I_b > I_A$

When the injected beam current exceeds the Alfven current limit, the efficient transport of such beams can only occur if there is a high degree of both charge and current neutralization, i.e., the intermediate gas pressure regime. In the low-pressure regime propagation is first limited by the space charge fields (formation of a virtual cathode). Even after the beam space charge has become neutralized, the lack of current neutralization in the low-pressure regime prevents efficient beam propagation for $I_b > I_A$. Similarly, in the high-pressure regime propagation is not permitted because the plasma conductivity is too low to sustain the current neutralization, and the electromagnetic self-energy exceeds the beam kinetic energy, regardless of the detailed orbit dynamics. Unless there is substantial current neutralization, the beam can lose essentially all its energy over distances of the order of a few beam radii simply as the result of establishing the beam self-fields.

To illustrate this explanation, reconsider the power balance equation (5.38), assuming that all the injected kinetic energy is consumed in establishing the magnetic field, i.e.,

$$(\gamma_0 - 1)mc^2 I_b/e = (I_b/2c)^2 [1 + 4\ln(R/r_b)]\beta_0 c \qquad (5.49)$$

which may be rewritten as

$$I_b = \frac{4(mc^3/e)(\gamma_0 - 1)}{\beta_0 [1 + 4\ln(R/r_b)]} \sim (mc^3/e)\beta_0\gamma_0 = I_A \qquad (5.50)$$

If $I_b \gtrsim I_A$ and there is no current neutralization, then beam transport is effectively limited by the energy conservation argument.

5.6. Summary

The novel feature of intense beam transport through neutral gas is the time-dependent generation of background plasma as the result of several beam-induced gas ionization processes (direct collisional, electron and ion avalanching, etc.). It is often the case that a single beam pulse will exhibit propagation behavior varying from that characteristic of unneutralized beam transport to that characteristic of transport through a dense plasma depending on distance from the propagating beam front. Because of the general nature of the problem a quantitative description of neutral gas transport phenomena is very difficult.

To proceed in this analysis, it is helpful to examine the transport according to beam current and background gas pressure regimes. The important current parameters are the space-charge-limiting current I_l and the Alfven current limit, I_A. In general terms, if the beam current exceeds I_A, propagation is only possible when there is both charge and magnetic neutralization; if $I_l \lesssim I_b \lesssim I_A$, current neutralization is not necessary for propagation; and for $I_b < I_l$, propagation is always possible.

The background gas pressure can be ordered according to (i) a low-pressure regime for which the charge neutralization time is of the order of, or greater than, the beam pulse duration; (ii) an intermediate-pressure regime for which the charge neutralization time is of the order of the beam rise time (or less), thereby leading to rapid electron avalanching and high plasma conductivity; and (iii) a high-pressure regime in which the high plasma-electron–neutral-gas-molecule collision frequency substantially reduces the plasma conductivity. With this ordering, the effects associated with current neutralization appear in the intermediate pressure regime. (It should be noted that such an ordering is somewhat arbitrary because regime boundaries can vary with the beam parameters.)

With such an organization the general features of beam transport in neutral gas are presented in Table 5.1. It should be realized, however, that

Table 5.1. Summary of Beam Transport Characteristics as a Function of Beam Current and Background Pressure

	Low pressure $f_e \sim f_m \sim 0$ ($\lesssim 0.5$ Torr)		Intermediate pressure $f_e \sim f_m \sim 1$ ($1-10$ Torr)	High pressure $f_e \sim 1, f_m < 1$ ($\gtrsim 10$ Torr)
$I < I_l$	Propagation with radial expansion	Two-stream instability	Stable beam propagation:	Pinched beam Propagation: Nordsieck Expansion
$I_l < I < I_A$	Virtual cathode formation		Beam expansion controlled by transverse kinetic energy	Hose Instability
$I > I_A$				Magnetic power balance limit

this summary provides only general guidelines. In particular, the possible development of beam plasma instabilities, namely, the two-stream and resistive hose instabilities, can markedly alter this simple picture. Once beam propagation has been established the subsequent beam behavior can be qualitatively analyzed in terms of the radial force equation including the effects of both charge and current neutralization. Finally for pinched beam transport over substantial distances (>10 m) in dense neutral gas, it is necessary to include the effects of beam scattering and energy loss in the creation of, and transport through, the resulting plasma channel.

Problems

5.1. Consider a relativistic electron beam of 1 cm radius with a current pulse that rises to a peak of 10 kA in 20 nsec. Over what pressure ranges would electron avalanching be an important conductivity generation process?

5.2. Plot the Nordsieck beam expansion ratio (r_b/r_{b0}) as a function of distance for a 50-MV, 20-kA electron beam assuming that $\gamma = \gamma_0$.

5.3. Using Eqs. (5.42) and (5.43) plot the ratio of j_p/j_b for the case $t_1 = 10$ nsec, $t_2 = 20$ nsec, $t_3 = 30$ nsec, for several values of σ and τ_B. What conditions must be fulfilled to achieve a high current neutralization?

5.4. Show that an exchange of momentum between the beam electrons and plasma electrons can lead to a current multiplication factor of order ($\gamma - 1$).

6

High-Power Sources of Coherent Radiation

The recent application of intense relativistic electron beam technology to the generation of coherent electromagnetic radiation has already produced unprecedented power levels over a wavelength band ranging from centimeters to a few hundred microns. The great interest in these new concepts stems from the fact that the available power from conventional microwave tube technology decreases rapidly as the wavelength drops below ~ 10 cm. Although substantial efforts will be required to develop practical devices, there are now several groups at the university, laboratory, and industrial levels actively engaged in research toward these goals.

In this chapter our interest will be concentrated on three distinct device concepts: (1) the relativistic magnetron, (2) the electron cyclotron maser, and (3) the free electron laser. These devices not only span the frequency range of interest, but are representative in terms of the technological problems that must be considered. The relativistic magnetron operates by converting electrostatic potential energy into microwave energy via the interaction of a rotating space charge cloud with a reentrant slow wave structure. The typical wavelength range is from one to ten centimeters. In the electron cyclotron maser (or gyrotron) transverse electron kinetic energy is coupled to a selected waveguide mode via an instability that depends on the energy dependence of the relativistic electron cyclotron frequency. Efficient operation at millimeter and submillimeter wavelengths is anticipated. In a free electron laser a low-frequency pump wave undergoes stimulated backscattering from a relativistic electron beam. Since the back-scattered radiation undergoes a double Doppler upshift from the pump frequency, operation in the submillimeter and infrared, possibly even extending to optical wavelengths, is anticipated.

213

6.1. The Relativistic Microwave Magnetron

Historically the magnetron was the first crossed-field electron tube to generate high-power coherent microwave radiation at decimeter and centimeter wavelengths. Cross sections of several typical microwave magnetrons are shown schematically in Fig. 6.1. A voltage is applied between the coaxial cathode and anode in the presence of an axial magnetic field. Electrons emitted from the cathode cylinder are caused to $\mathbf{E} \times \mathbf{B}$ drift azimuthally in the interaction space. The anode contains a resonant slow wave structure which can support several possible modes of oscillation corresponding to integral numbers of wavelengths around the structure. Those waves which have phase velocities nearly equal to the electron drift velocity can interact strongly with the electrons. The forces due to the rf fields tend to produce azimuthal electron bunching and particle migration either toward the anode or back to the cathode, depending on the phase of the interaction. These resultant rotating electron "spokes" supply additional energy to the rf which, in turn, produces further electron bunching. In this fashion a diode current is produced, but the electrons which reach the anode have very little kinetic energy. Hence, the essential feature of magnetron operation is that the electron bunches drift in synchronism with the rf waves supported by the anode structure. A considerable amount of the electrical potential energy associated with the radial dc electric field can then be converted into rf energy, provided that most electrons are well bunched and reach the anode after several cycles. For high magnetron efficiency, the dc position of the drifting electron cloud (i.e., without rf interaction) must be relatively close to the cathode. In this case the electron kinetic energy remains small, and a large fraction of the potential energy can be converted into rf energy.

Although this qualitative understanding of magnetron tube operation has existed for several years,[1] a detailed quantitative description of the important space charge dynamics still does not exist, primarily because of the strongly nonlinear, two-dimensional nature of the wave–particle interaction. It should also be understood that even the problem of the smooth bore magnetron (anode without resonator vanes) has not yet been satisfactorily resolved. Despite this lack of theoretical understanding, experimental persistence has led to the development of an impressive array of microwave magnetron tubes. As an example, a 10-cm S-band magnetron with a power of 3 MW at an efficiency of 60% had been built by the end of World War II.

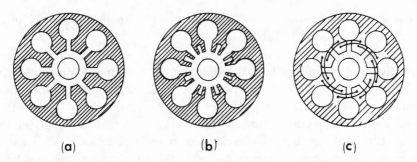

Figure 6.1. Examples of various microwave resonator systems: (a) unstrapped, (b) rising sun, and (c) strapped.

The various devices and methods resulting from the large wartime effort are described in the monograph *Microwave Magnetrons*.[1]

In conventional magnetrons[1,2] the cathode is a thermionic emitter. Also, since the voltage between the anode and cathode does not exceed a few tens of kilovolts, typical currents drawn from the cathode do not exceed a few hundred amperes. Hence, the maximum rf power output is generally limited to a few megawatts, although the efficiencies are usually quite high ($\gtrsim 50\%$). In recent years, however, application of the high-voltage pulsed power technology to the magnetron tube has yielded unprecedented rf powers in the range of hundreds of megawatts to several gigawatts, although the efficiencies have been limited to 10%–30%.[3-7] In this case the applied voltage is of the order of one megavolt and the electron current (~ 100 kA) is provided by a cold cathode plasma emitter. The operating parameters of a

Table 6.1. Comparison of Conventional and Relativistic
Pulsed Magnetrons

Parameter	Conventional	Relativistic
Voltage	$\lesssim 100$ kV	~ 1 MV
Cathode	Thermionic and secondary emission	Field emission
Current	~ 100 A	$\gtrsim 10$ kA
Pulse duration	$\gtrsim 1$ μsec	$\lesssim 100$ nsec
Rise time	$\lesssim 200$ kV/μsec	~ 100 kV/nsec
Power	$\lesssim 10$ MW	$\gtrsim 1$ GW
Efficiency	$\sim 50\%$	$\lesssim 30\%$

conventional high-power pulsed magnetron are compared with those of a relativistic magnetron in Table 6.1.

In this section the general features of magnetron design and operation are discussed in terms of several specific topics, including (1) preoscillation phenomena, (2) anode circuits, and (3) electron dynamics (spoke formation) and the qualitative operating regimes. Special attention is devoted to the problem of the relativistic magnetron. In particular, it is important to emphasize that the high efficiencies of conventional tubes resulted from intense engineering optimization studies; the relativistic magnetron is still in its infancy, and higher efficiencies are undoubtedly possible.

6.1.1. Preoscillation Phenomena

The initial startup phase of a magnetron tube consists of a magnetically insulated rotating beam. Since there is essentially no rf interaction this state can be simply approximated by the smooth bore magnetron diode without anode resonator vanes. For the relativistic magnetron diode we note that the anode–cathode gap must be small enough to produce significant explosive electron emission. From Section 2.2.3 the required field strength at the cathode is of the order of a few hundred kilovolts per centimeter. Hence, for an anode of radius R and cathode of radius r_c, a conservative criterion for the magnitude of the applied voltage ϕ_0 is that

$$\frac{|\phi_0|}{r_c \ln(R/r_c)} \gtrsim 500 \text{ kV/cm} \tag{6.1}$$

If it is assumed that there is no time variation in any of the field quantities, then the conservation laws of energy and canonical angular momentum are given by

$$\gamma = 1 + e(\phi_0 + \phi)/mc^2 \tag{6.2}$$

$$\gamma r v_\theta(r) = eB(r^2 - r_c^2)/2mc \tag{6.3}$$

Since $\gamma = (1 - v_\theta^2/c^2)^{-1/2}$, Eq. (6.3) can be written as

$$\left(\frac{eB}{2mc^2}\right)\frac{(r^2 - r_c^2)}{r} = (\gamma^2 - 1)^{1/2} \tag{6.4}$$

For the critical case when the electron cloud just reaches the anode ($\phi = 0$ at $r = R$), Eq. (6.4) gives the critical magnetic field for magnetic insulation (see Section 2.4.3)

$$B_c = \left(\frac{mc^2}{e} \right) \left(\frac{2R}{R^2 - r_c^2} \right) \left[\frac{2e\phi_0}{mc^2} + \left(\frac{e\phi_0}{mc^2} \right)^2 \right]^{1/2}$$ (6.5)

Equation (6.5) is often termed the "Hull cutoff."[8] According to Eq. (6.4) if $B > B_c$, then a steady state equilibrium solution exists for the electron space charge. Since the actual form of the equilibrium depends on various assumptions about the initial conditions the equilibrium is not unique and has been the subject of much debate. The two classes of solution that have received the most attention are (a) the Brillouin or parapotential flow,[9-12] in which the electrons move parallel to the electrode surfaces with drift velocities given by $v_\theta = E_r c / B$, and (b) the double-stream model,[13-17] in which the electrons make cycloidal orbits which begin and end on the cathode surface. (The parapotential flow model is discussed in detail in Section 2.3.2, while the double-stream model is treated in Section 2.4.3.) If the rise time of the voltage pulse is slow compared with the electron cyclotron period (adiabatic approximation), then the system might be expected to settle into Brillouin flow. In fact, recent numerical simulations[18] indicate that the preoscillation phase does exhibit many of the characteristics of relativistic Brillouin flow. Guided by these results we will adapt the model of Section 2.3.2 to the case of cylindrical geometry.

It is easily verified that the principal assumption of Brillouin flow, i.e., force balance along each equipotential, trivially satisfies Eq. (6.4). (See Problem 6.1.) The radius of the space charge boundary can be derived by the matching conditions for the potential and the electric field at the boundary. Setting $\phi = -\phi_b$ at $r = r_b$, Eq. (6.4) yields

$$\frac{eB}{2mc^2} r_b \left(1 - \frac{r_c^2}{r_b^2} \right) = \left(\gamma_b^2 - 1 \right)^{1/2}$$ (6.6)

In the region outside the space charge the solution of Laplace's equation gives

$$\phi = -\frac{\phi_b}{\ln(R/r_b)} \ln(R/r)$$ (6.7)

Matching the electric field calculated from both Eqs. (6.4) and (6.7) indicates that

$$
\left(\frac{eB}{2mc^2}\right)^2 \left(1 + \frac{r_c^2}{r_b^2}\right) r_b \left(1 - \frac{r_c^2}{r_b^2}\right) = \frac{\gamma_b(\gamma_0 - \gamma_b)}{r_b \ln(R/r_b)} \tag{6.8}
$$

Equations (6.5) and (6.8) then determine the radius of the space charge boundary. For radii $r < r_b$ the space charge distribution, calculated from Poisson's equation, is given by

$$
n(r) = (4\pi e\gamma)^{-1} \left(\frac{eB}{2mc^2}\right)^2 \left[1 + 3\frac{r_c^4}{r^4} - \frac{r^2}{\gamma^2}\left(\frac{eB}{2mc^2}\right)^2 \left(1 - \frac{r_c^4}{r^4}\right)^2\right] \tag{6.9}
$$

where γ is determined from Eq. (6.4).

It is now well known that even though ϕ_0 and B are chosen such that the diode should be strongly cut off, inevitably there is some flow of electron current to the anode. Although numerous investigations have been made of the effect of various parameters on this phenomenon (e.g., electron emission velocities, magnetic field tilt, cathode eccentricity, fringing fields, etc.) none of these effects can account for the violation of the cutoff formula. It is now generally accepted that the only viable explanation is a breakdown of the steady state assumption itself, i.e., time-varying electromagnetic oscillations (instabilities) develop resulting in a disruption of the steady state flow.

The results of recent relativistic magnetron diode experiments[3] have tended to confirm earlier theoretical pictures of the physical mechanism of this disruption. Fluctuations are not observed until $B > B_c$ when Brillouin flow can be established. From Eq. (6.3) it is apparent that the flow is strongly sheared, which suggests that free energy is available to drive the diocotron instability (Section 3.7.2). As a result of geometrical system resonances the microwave emission from the unstable rotating electron cloud is enhanced at the resonant frequencies of the system.

Annular sheared electron flows are known to be unstable to the diocotron effect in which a wave on one surface interacts with a wave on the opposite surface. In the case of the magnetron diode it might be thought that only one surface exists since the beam is confined on one side by the cathode. As pointed out by Buneman,[19] however, if a plasma covers the cathode surface the situation is equivalent to an ion–vacuum interface superimposed on the electron background, and diocotron modes are again possible. Bergeron[20] has numerically evaluated the dispersion for this type of system. His results indicate an instability that is qualitatively similar to the conventional diocotron, but whose characteristic frequency is somewhat lower owing to the higher inertia of the ion surface. This result is consistent with the experimental findings of Orzechowski and Bekefi,[3] although a conclusive comparison of measured instability wavelength with the theory could not be performed.

6.1.2. Anode Circuits

While the sheared rotating electron cloud can be unstable in the smooth bore magnetron, the electromagnetic (em) modes of the smooth annular gap always have phase velocities which are greater than the speed of light. Hence, a strong interaction between the em waves and the electron space charge is not possible; the emission spectrum is broad and the efficiency is poor. On the other hand, if periodically spaced cavities are introduced into the anode structure, then a certain set of em modes of the interaction space can have phase velocities less than c. Phase synchronism between any one of these modes and the sheared electron space charge can result in a strong wave–particle interaction and the efficient generation of microwave power.

For conventional magnetrons three types of resonant anode structures have received considerable attention. In order of historical development these are (1) the unstrapped system, (2) the strapped system, and (3) the alternating unstrapped, or rising-sun system, examples of which were previously shown in Fig. 6.1. In the unstrapped system identical resonators are coupled only through the em fields in the interaction space. In the strapped systems special coupling links are provided between pairs of identical neighboring resonators. In rising-sun systems, alternate resonators are different. Although the unstrapped system has been superseded in conventional magnetron design by the other systems, these more complicated structures have thus far been avoided in relativistic magnetrons in order to better answer more basic questions associated with the higher voltages and currents.

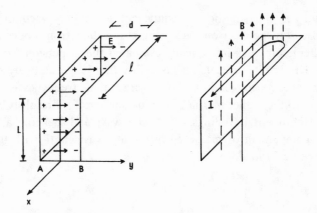

Figure 6.2. Schematic diagram of the rectangular-slot side resonator magnetron.

Regardless of the particular type of resonant anode structure, its primary function is to provide an rf field of the desired frequency and configuration to interact with the electron space charge. In addition, a good resonant system should ensure stable operation against small changes in frequency and against discrete frequency jumps. Although somewhat obsolete, we will discuss the unstrapped resonant system in some detail for two reasons. First, a rather complete theoretical description is available, and second, the important concepts essential for understanding the more complicated systems can be seen in their simplest form in the unstrapped system.

To rigorously compute the frequencies of the natural modes of oscillation it is necessary to solve Maxwell's equations in vacuum for the anode block resonators alone, and then for the interaction space alone (without electron space charge). Setting the rf admittance of one equal to the rf admittance of the other over their common interface then leads to a transcendental equation for the desired natural frequencies (Ref. 1, Chap. 2, p. 49). To illustrate this procedure consider the case of the rectangular-slot side resonator magnetron of Fig. 6.2.

The rectangular side resonator has a boundary (dashed line) across which the admittance is to be measured. The most important modes are those with which the electrons can strongly couple. They require strong electric field lines going from anode segment to anode segment with little axial variation. Hence, it will be assumed from the outset that the electric field is transverse, with no axial variation. One-quarter of a period later, the electric field and charges will have disappeared, to be replaced by currents flowing around the inside of the cavity producing an axial magnetic field, which is also assumed to have no axial variation.

The voltage across the gap $A-B$, defined by $\int_A^B \mathbf{E} \cdot d\mathbf{l}$, clearly depends on the path of integration; however, this is not a significant point if the path used to calculate the side resonator admittance does not greatly differ from the path used to calculate the admittance of the interaction space. There are several definitions of the admittance. That most frequently used is

$$Y = \frac{I}{V} \frac{V^*}{V^*} = \frac{2P^*}{V^2}$$

where $P^* = \frac{1}{2} VI^*$ is the complex conjugate of the complex power. (V and I denote peak rather than rms values of voltage and current.) Since the complex power can be written in terms of the Poynting vector as

$$P^* = \frac{1}{2} \int_S (\mathbf{E}^* \times \mathbf{B}) \cdot \hat{n} \, dS \tag{6.10}$$

then the admittance can be written as

$$Y = \frac{L \int_A^B (\mathbf{E}^* \times \mathbf{B}) \cdot \hat{n} \, dl}{|\int_A^B \mathbf{E} \cdot d\mathbf{l}|^2} \tag{6.11}$$

where the path of integration is the same for both integrals. Although this path has no particular significance the resonant frequencies to be determined will depend to some extent on the assumed electric field. It has been shown that the resonant frequencies computed on the basis of Eq. (6.11) are somewhat less sensitive to the assumed field distribution than are those computed on the basis of I/V, for example.[21]

Assuming a harmonic time dependence ($\sim e^{i\omega t}$),

$$E_x = \frac{c}{i\omega} \frac{\partial B_z}{\partial y} \tag{6.12}$$

$$E_y = -\frac{c}{i\omega} \frac{\partial B_z}{\partial x} \tag{6.13}$$

The boundary conditions for $\mathbf{E}$ are that (1) $E_y = 0$ at $x = l$, (2) $E_x = 0$ at

$y = \pm d/2$, (3) E_y must have the constant value E_0 at $x = 0$, and (4) that all fields must be continuous. Application of the second condition implies that the solutions of the wave equation must have the form

$$E_x^n \sim -\frac{2\pi n}{d} \sin\left(\frac{2\pi n y}{d}\right) e^{\pm k_x x}$$

$$E_y^n \sim \mp k_x \cos\left(\frac{2\pi n y}{d}\right) e^{\pm k_x x}$$

$$B_z^n \sim \frac{i\omega}{c} \cos\left(\frac{2\pi n y}{d}\right) e^{\pm k_x x} \tag{6.14}$$

where

$$k_x = \left[\left(\frac{2\pi n}{d}\right)^2 - \frac{\omega^2}{c^2}\right]^{1/2}$$

and n is any positive integer. Application of the condition that $E = E_0$ for all y at $x = 0$ restricts the solution to $n = 0$, so that $E_x = 0$. At $x = l$, $E_y = 0$; hence,

$$E_y\big|_{x=l} = -\frac{i\omega}{c}\left(\alpha e^{i\omega l/c} - \beta_e^{-i\omega l/c}\right) = 0 \tag{6.15}$$

$$E_y\big|_{x=0} = E_0 = -\frac{i\omega}{c}(\alpha - \beta) \tag{6.16}$$

Solving for the constants α and β yields

$$E_y = E_0 \frac{\sin[\omega(l-x)/c]}{\sin(\omega l/c)} \tag{6.17}$$

From which it follows that

$$B_z = -iE_0 \frac{\cos[\omega(l-x)/c]}{\sin(\omega l/c)} \tag{6.18}$$

Since both E_y and B_z are independent of y, the integrations of Eq. (6.11) are trivial and the admittance is found to be

$$Y_r = -\frac{iL}{d}\cot(\omega l/c) \tag{6.19}$$

The admittance of the interaction space can be determined by similar methods. In this case it is convenient to use the cylindrical coordinates, r, θ, and z. Assuming as before that $E_z = 0$, and that $\mathbf{E}$ has no axial variation, it follows that

$$B_z = B_z(r, \theta) e^{i\omega t} \tag{6.20}$$

$$B_r = B_\theta = 0$$

$$E_r = \frac{c}{i\omega r} \frac{\partial B_z}{\partial \theta} \tag{6.21}$$

$$E_\theta = -\frac{c}{i\omega} \frac{\partial B_z}{\partial r} \tag{6.22}$$

As before it is assumed that E_θ is constant across each gap, and further, that the field at any gap differs from the field at an adjacent gap by only a constant phase factor. E_θ must vanish at the cathode and the anode segments. Explicitly,

$$E_\theta(r_c, \theta) = 0$$

$$E_\theta(R, \theta) = \begin{cases} E_0 e^{i2\pi n/Nj}, & (2\pi j/N - \psi) < \theta < (2\pi j/N + \psi) \\ 0, & \text{otherwise} \end{cases} \tag{6.23}$$

In Eq. (6.23) j denotes the gap number ($0 \leq j \leq N-1$), where N is the number of anode segments, n is an integer, and 2Ψ is the angle subtended by a side resonator cavity.

The fundamental set of solutions for the wave equation are the functions $J_m(kr)e^{im\theta}$ and $N_m(kr)e^{im\theta}$, where $k = \omega/c$. Application of the boundary conditions determines the field solutions as given by

$$E_\theta(r, \theta) = E_0\left(\frac{N\psi}{\pi}\right) \sum_{m=-\infty}^{\infty} \left(\frac{\sin m\psi}{m\psi}\right) \frac{Z_m'(kr)}{Z_m'(kR)} e^{im\theta} \tag{6.24}$$

$$E_r(r, \theta) = -iE_0\left(\frac{N\psi}{\pi kr}\right) \sum_{m=-\infty}^{\infty} m\left(\frac{\sin m\psi}{m\psi}\right) \frac{Z_m(kr)}{Z_m'(kR)} e^{im\theta} \tag{6.25}$$

$$B_z(r, \theta) = -iE_0\left(\frac{N\psi}{\pi}\right) \sum_{m=-\infty}^{\infty} \left(\frac{\sin m\psi}{m\psi}\right) \frac{Z_m(kr)}{Z_m'(kR)} e^{im\theta} \tag{6.26}$$

where

$$Z_m(kr) = J_m(kr) - \frac{J'_m(kr_c)}{N'_m(kr_c)} N_m(kr) \qquad (6.27)$$

Finally, substituting Eqs. (6.24) and (6.26) into the admittance integrals yields

$$Y_i = i \frac{NL}{2\pi R} \sum_{m=-\infty}^{\infty} \frac{\sin m\psi}{m\psi} \frac{Z_m(kr)}{Z'_m(kR)} \qquad (6.28)$$

As indicated previously the system resonances are found by setting the admittance looking out of a side resonator (the negative of the admittance looking in) equal to the admittance looking into the interaction space, i.e., $Y_i + Y_r = 0$, where Y_r is the admittance of the side resonator. In general this involves the solution of a transcendental equation [examine Eqs. (6.19) and (6.28), for example], which is usually obtained by numerical means.

In Fig. 6.3 schematic representations of both admittances are graphed as a function of frequency for an $N = 8$ magnetron. The resonances are those values of ω at which intersection occurs. There is an infinite number of resonances, but only the lowest few $n \leqslant N/2$ are of interest. The resonances are usually designated as n_p, where n denotes the principal reso-

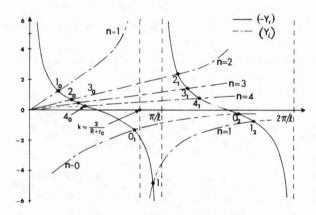

Figure 6.3. Graphical representation of the rf admittances as a function of frequency. The solid line is the admittance of the resonator, while the dashed lines are the admittances of the interaction space. The points of intersection give the oscillation frequencies.

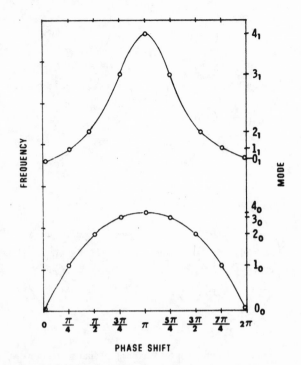

Figure 6.4. Brillouin dispersion diagram for the modes of an $N = 8$ magnetron. The slope of a straight line drawn from the origin gives the phase velocity of a mode in segments per second.

nance (i.e., a resonance of the lowest-frequency root of the resonance equation), while p denotes the resonances of the pth higher order group. The Brillouin dispersion diagram (Fig. 6.4) is a particularly useful method of presenting this information. Note that each resonance is a doublet (degeneracy of order 2) except those for which $n = 0$ or $N/2$. This result is a consequence of the fact that while the same wavelength is obtained for either $-n$ or n (except for the cases $n = 0$ or $N/2$), the structure of the em field is different, and the two fields are linearly independent.

For the lowest resonance group ($p = 0$) the system is observed to be highly dispersive; as n increases the phase velocity (given by the slope of a straight line drawn from the origin) decreases. The higher-p resonance groups are similar in character to the first group, except that the frequencies are much higher. They are ordinarily of little interest in magnetron devices, and are not usually studied in detail, although some attempts have been

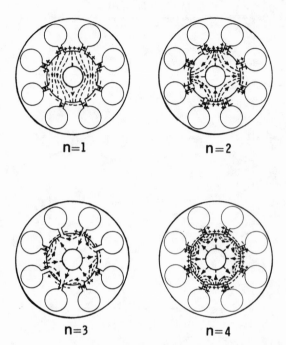

Figure 6.5. Charge and field distributions for the four principal modes of an $N = 8$ magnetron.

made to operate magnetrons in the 0_1 mode because it is well separated and nondegenerate.

The phase differences between the resonators for the $N = 8$ magnetron for $n = 1, 2, 3, 4$ are $\pi/4$, $\pi/2$, $3\pi/4$, and π. The charge and electric field configurations for these modes are presented in Fig. 6.5. The degenerate forms of each mode correspond to a rotation of the standing wave pattern so that the positions of the nodes and antinodes are interchanged. Most conventional magnetrons are designed to operate in the $n = N/2$ mode, for which the rf fields in adjacent resonators are 180° out-of-phase with one another; it is also the so-called "π mode."

It is usually the case that actual magnetrons are not perfectly symmetric, and the lack of symmetry tends to split the degenerate modes into two nondegenerate modes with slightly different resonant frequencies. This splitting usually leads to lack of uniformity in tubes operating in these modes.

The symmetric unstrapped resonant system is usually satisfactory only for $N \lesssim 8$, because the modes crowd together for high values of n and N. This effect restricts their operation to comparatively long wavelengths (~ 10 cm). The strapped and rising-sun systems were developed to improve the mode separation for higher-N systems, thus permitting efficient operation at shorter wavelengths.

6.1.3. Interaction of the Electron Space Charge and the rf Fields

Although microwave magnetrons have now been manufactured for over 40 years, our understanding of the interaction of the electron space charge with the electromagnetic field of the resonator is still essentially qualitative. As a result, there has been some difficulty in designing magnetron tubes of increased power at shorter wavelengths without sacrificing mode stability. In this section we present this qualitative understanding of magnetron operation, and supplement where possible with the results of recent numerical simulation studies of the newer relativistic magnetrons.

As the starting point for the discussion we return to the Brillouin steady state of Section 6.1.1. The motion of individual electrons consists of a small orbit, fast rotation at the cyclotron frequency, superposed on the slower rotation of the electron cloud about the cathode, given essentially by the ratio of E_r / B. In the steady state a logarithmic potential exists between the anode and the electron cloud. Now assume that a small rf field of one of the modes of the resonator structure from Section 6.1.2 is superposed onto the dc field as shown in Fig. 6.6. As the electrons encounter the rf perturbations their velocity must change depending on the phase of the interaction. Consider an electron at point A at the instant for which the fields are as shown. The rf field at this point tends to accelerate the electron toward the anode. As the electron speeds up, the $\mathbf{v} \times \mathbf{B}$ force increases the curvature of the electron path and the electron strikes the cathode with considerable energy. An electron at point B, however, is in a decelerating rf electric field. As the electron velocity is reduced the curvature also decreases. If the rf oscillation frequency is correct, this electron will always be in a decelerating field as it passes before the anode segments. The result is that the electron will eventually strike the anode following the indicated path. Although the electron "gains" the energy associated with the applied dc potential, it gives up most of this energy to the rf field as a result of the rf retardation.

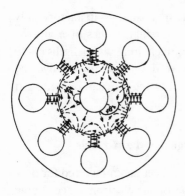

Figure 6.6. Schematic representation of electron orbits in an oscillating magnetron.

Now, considering the behavior of the space charge as a whole, following the above discussion the rotating Brillouin electron cloud tends to be distorted in the presence of the rf into a smaller cylinder with spokelike ridges which extend to the anode (Fig. 6.7). As a rotating space charge spoke passes in front of an anode segment, a positive charge is induced on its surface. The induced charge then flows around the back of the adjacent cavities to the two adjacent anode segments, constituting a displacement current, while the space charge spoke rotates to a position in front of the next anode segment. In addition to this rf displacement current, the electrons which strike the anode constitute a conduction current; however, since this current is 90° out of phase with the displacement current it does not contribute to the oscillation buildup.

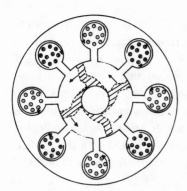

Figure 6.7. Spoke formation in an oscillating magnetron.

The concepts discussed above lead to a simple relationship between the operating voltage, the magnetic field, the rf frequency, and the magnetron geometry that has been well supported by experiment. For efficient magnetron operation the applied potential ϕ_0 and the static magnetic field strength must be such that the rotation of the electron space charge about the cathode maintains the proper phase relationship with the rf fields in the resonators. For the nth mode of operation the phase difference between the resonators is given by $2\pi n/N$, while the distance between resonator segments for an electron at radius r is $2\pi r/N$. For rf oscillations at frequency $f = \omega/2\pi$, the time required for the phase of the rf fields to vary by the amount $2\pi n/N$ is $\tau = 2\pi n/(\omega N)$. Hence, phase synchronism will be achieved if the electron rotates with the (synchronous) velocity

$$v_\theta = 2\pi r/(N\tau) = r\omega/n \qquad (6.29)$$

Substituting this expression into Eq. (6.3) for the relativistic Brillouin cloud yields

$$B = \frac{2\gamma mc}{e} \frac{r^2}{r^2 - r_c^2} \frac{\omega}{n} \qquad (6.30)$$

When Eq. (6.30) is evaluated for the case of the space charge cloud extending to the anode, the result is termed the Buneman–Hartree threshold.[19]

$$B_{\text{BH}} = \frac{2mc}{e} \left(1 + \frac{e\phi_0}{mc^2}\right) \frac{R^2}{R^2 - r_c^2} \left(\frac{\omega}{n}\right) \qquad (6.31)$$

The ratio of B_{BH}/B_c from Eqs. (6.5) and (6.31) is

$$\frac{B_{\text{BH}}}{B_c} = \frac{\left[1 + e\phi_0/mc^2\right](\omega R/nc)}{\left[2e\phi_0/mc^2 + \left(e\phi_0/mc^2\right)^2\right]^{1/2}} \qquad (6.32)$$

This ratio is a function of the applied voltage and the phase velocity only, and is independent of the anode–cathode gap.

The Hull cutoff curve and Buneman–Hartree curves are shown in Fig. 6.8. To the left of the cutoff parabola the magnetron is not cutoff and no oscillations occur as anode current is drawn. To the right of the cutoff curve

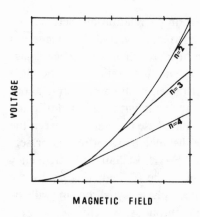

MAGNETIC   FIELD

Figure 6.8. Buneman–Hartree threshold voltages and the Hull parabola presented as a function of magnetic field strength.

no current flows unless oscillations exist to drive spoke formation. Note that as n decreases, the electron must travel a greater distance around the cathode in one period and a larger value of E/B is required. This is apparent from the $n = 2$ and $n = 3$ lines which were calculated assuming the same frequency as for the $n = N/2 = 4$ mode.

The Hull cutoff and the Buneman–Hartree threshold provide important scaling relationships for magnetron designs. For example, in the nonrelativistic limit Eq. (6.31) becomes

$$\phi_0 = \frac{B_{\mathrm{BH}} \pi R^2}{2 \lambda n} \left(1 - \frac{r_c^2}{R^2} \right) \qquad (6.33)$$

For a constant voltage ϕ_0 and a constant ratio of R/r_c, Eq. (6.33) implies that $\lambda \sim R^2 B_{\mathrm{BH}}$. Since the rf wavelength scales as the circumference of the anode block, $\lambda \sim R$, then $B_{\mathrm{BH}} \sim 1/\lambda$. Hence, at a constant voltage ϕ_0, a change in the physical dimensions by a scale factor (s) will result in a wavelength change by a factor s, the operating magnetic field will change by s^{-1}, and the operating characteristics of the new scaled magnetron should be similar to those of the original tube. This simple analysis also illustrates the two major difficulties in making very-short-wavelength magnetrons: The physical size of the magnetron becomes very small, and a large magnetic field is necessary.

By putting some of the remarks of this section into a more quantitative format it is possible to estimate the performance of a magnetron.[22] Assuming that the electric potential is independent of z in the interaction space, then in the nonrelativistic limit the single-particle equations of motion are simply

$$\ddot{x} = -\frac{e}{m}\left(E_x + \frac{\dot{y}B}{c} \right) \tag{6.34}$$

$$\ddot{y} = -\frac{e}{m}\left(E_y + \frac{\dot{x}B}{c} \right) \tag{6.35}$$

where B denotes the strength of a uniform magnetic field applied parallel to the z axis. Multiplying Eq. (6.35) by i and adding Eq. (6.34) yields

$$\ddot{\xi} = i\Omega_0 \dot{\xi} + \frac{2e}{m}\frac{\partial \phi}{\partial \xi^*} \tag{6.36}$$

where the complex coordinate $\xi = x + iy = re^{i\theta}$, $\xi^* = x - iy = re^{-i\theta}$, $\Omega_0 = eB/mc$, and $\mathbf{E} = -\nabla\phi$. It is assumed that the electric potential has both dc and rf contributions, i.e., $\phi = \phi_{dc} + \phi_{rf}$, where

$$\phi_{dc} = -\phi_0 \left(\frac{R^2 - r^2}{R^2 - r_c^2} \right) \tag{6.37}$$

$$\phi_{rf} = -\frac{2\phi_1}{\pi}\left(\frac{R^n}{R^{2n} - r_c^{2n}} \right)\left(r^n - \frac{r_c^{2n}}{r^n} \right)\sin n\theta \cos \omega t \tag{6.38}$$

[Note that both Eqs. (6.37) and (6.38) are approximate solutions of Laplace's equation, neglecting dc and rf components of electron space charge, and assuming $p = 0$ for the rf component.]

In the absence of oscillations ($\phi_1 = 0$), the solution of Eq. (6.36) is given by

$$\xi = S_+ e^{i\omega_+ t} + S_- e^{i\omega_- t} \tag{6.39}$$

where S_+ and S_- are constants set by the initial conditions, and

$$\omega_{\pm} = \frac{\Omega_0}{2}\left\{ 1 \pm \left[1 - \left(\frac{2\Omega}{\Omega_0} \right)^2 \right]^{1/2} \right\} \tag{6.40}$$

where $\Omega^2 = 2e\phi_0/m(R^2 - r_c^2)$. Note that the motion given by Eq. (6.40) generally corresponds to the superposition of two circular motions; one has smaller radius S_+ and rotates at the cyclotron frequency Ω_0, while the other corresponds to a slower rotation about the cathode at the $E \times B$ drift velocity.

Now assuming that the rf field is added as a perturbation ($\phi_1 \ll \phi_0$), then in first approximation the electron motion can again be described by Eq. (6.39), although S_+ and S_- are no longer constant but vary slowly with time.[22] Substituting Eq. (6.39) into Eq. (6.36) yields an equation for the perturbed coefficients given by

$$(2i\omega_+ - i\Omega_0)\dot{S}_+ e^{i\omega_+ t} + (2i\omega_- - i\Omega_0)\dot{S}_- e^{i\omega_- t} = \frac{e}{m} e^{i\theta}\left(\frac{\partial}{\partial r} + \frac{i}{r}\frac{\partial}{\partial\theta}\right)\phi_{\text{rf}}$$

$$(6.41)$$

If it is now assumed that resonance occurs between the angular velocity ω_- and the rf component of the em field traveling in the same direction ($\omega = n\omega_-$), then $dS_+/dt = 0$. Writing $S_- = Se^{i\delta}$, where δ denotes the phase angle of the electron to the traveling field, then the real and imaginary parts of Eq. (6.41) yield

$$\frac{dS}{dt} = -K\left(\frac{R^n}{R^{2n} - r_c^{2n}}\right)\left(S^{n-1} - \frac{r_c^{2n}}{S^{n+1}}\right)\cos n\delta \qquad (6.42)$$

$$\frac{d\delta}{dt} = K\left(\frac{R^n}{R^{2n} - r_c^{2n}}\right)\left(S^{n-1} + \frac{r_c^{2n}}{S^{n+1}}\right)\frac{\sin n\delta}{S} \qquad (6.43)$$

where $K = -e\phi_1 n/m\pi(\omega_+ - \omega_-)$.

The envelope of the electron flow can be determined from Eqs. (6.42) and (6.43) by integrating and then eliminating t. The result is

$$\left(S^n - \frac{r_c^{2n}}{S^n}\right)\sin n\delta = \left(S_0^n - \frac{r_c^{2n}}{S_0^n}\right)\sin n\delta_0 = \text{const} \qquad (6.44)$$

which is shown in Fig. 6.9. Those electrons which originate at the cathode at phase angle $\cos n\delta < 0$ reach the anode and tend to converge at $n\delta = \pi, 3\pi, \ldots$, which are the phase angles of the strongest retarding field in the traveling rf wave. In contrast, electrons originating at phase angles $\cos n\delta > 0$

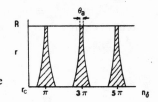

Figure 6.9. Locus of electron orbits in the interaction space of an oscillating magnetron.

tend to return to the cathode (back-bombardment). In this manner rotating electron spokes tend to form at the anode.

Once the electron paths have been determined, it is relatively easy to compute the dc anode current by calculating the width θ_a of the electron spoke, the radial velocity at the anode, and the electron density, and then substituting into the formula

$$I_a = -2\pi\rho RL\left(\frac{\theta_a}{2\pi}\right)\left(\frac{dS}{dt}\right)_a \tag{6.45}$$

where ρ is the electron charge density in the spoke.

Thus far the calculation has completely neglected the effects of the electron space charge. It is now assumed that the cathode is covered with a Brillouin layer; the potential ϕ_b at the boundary of the layer r_b is given by Eq. (6.4), i.e., (nonrelativistically)

$$\phi_b = -\phi_0 + \frac{eB^2}{8mc^2}r_b^2\left(1 - \frac{r_c^2}{r_b^2}\right)^2 \tag{6.46}$$

Outside the Brillouin cloud, the dc potential is again assumed to vary quadratically according to

$$\phi_{dc} = \frac{\phi_b\left(r^2 - r_b^2\right)}{\left(R^2 - r_b^2\right)} \tag{6.47}$$

In this case $\Omega^2 = -2e\phi_b/m(R^2 - r_b^2)$ in Eq. (6.40), and assuming synchronism at the edge of the layer yields

$$\phi_0 = \frac{eB^2}{8mc^2}r_b^2\left(1 - \frac{r_c^2}{r_b^2}\right) - \frac{\pi}{n\lambda}\left(B - \frac{2\pi mc^2}{en\lambda}\right)\left(R^2 - r_b^2\right) \tag{6.48}$$

If Z_0 is the load impedance then the efficiency of converting the potential energy into rf energy is given by

$$\eta = \frac{\phi_1^2/2Z_0}{\phi_0 I_a} \qquad (6.49)$$

The electronic efficiency η can be obtained by calculating the energy loss when the electron strikes the anode. The energy that is available for conversion into rf energy at the boundary of the synchronous Brillouin layer is

$$\varepsilon_b = -e\phi_b + \tfrac{1}{2}m(\omega_- r_b)^2 \qquad (6.50)$$

while the total energy of the electrons which strike the anode is

$$\varepsilon_a = \frac{1}{2}m(\omega_- R)^2 + \frac{1}{2}m\left(\frac{dS}{dt}\right)_a^2 \qquad (6.51)$$

where $(dS/dt)_a$ is calculated from Eq. (6.42), assuming $\delta = \pi$. Ignoring any losses due to cathode back-bombardment the electronic efficiency is given by

$$\eta = \frac{\varepsilon_b - \varepsilon_a}{e\phi_0} = \eta_0 - \frac{\beta\phi_1^2}{\phi_0^2} \qquad (6.52)$$

where

$$\eta_0 = \frac{-e\phi_b + \tfrac{1}{2}m\omega_-^2\left(r_b^2 - R^2\right)}{e\phi_0}$$

$$\beta = \tfrac{1}{2}m\phi_0\left[\frac{en}{m\pi R(\omega_+ - \omega_-)}\right] \qquad (6.53)$$

Combining Eqs. (6.49) and (6.52) yields an expression for ϕ_1, given by

$$\frac{\phi_1}{\phi_0} = \frac{\left(1 + \beta\gamma^2\eta_0\right)^{1/2} - 1}{\beta\gamma} \qquad (6.54)$$

where $\gamma = 4Z_0 I_a/\phi_1$. The anode current can be computed from Eq. (6.45)

where

$$\rho = \frac{4\pi}{r}\frac{d}{dr}\left(r\frac{d\phi_0}{dr}\right) = -\frac{m}{2\pi e}\omega_+\omega_-$$ (6.55)

$$\left(\frac{dS}{dt}\right)_a = -\frac{e\phi_1 n}{\pi mR(\omega_+ - \omega_-)}$$

$$\theta_a = 2\sin^{-1}\left\{\left(\frac{S_0}{S}\right)^n\left[\frac{1-(r_c/S_0)^{2n}}{1-(r_c/S)^{2n}}\right]\right\}$$ (6.56)

With these results the expression for γ becomes

$$\gamma = \frac{8}{\pi}\left(\frac{LC}{\gamma}\right)\frac{\omega_+}{\omega_+ - \omega_-}\sin^{-1}\left\{\left(\frac{r_b}{R}\right)^n\left[\frac{1-(r_c/r_b)^{2n}}{1-(r_c/R)^{2n}}\right]\right\}$$ (6.57)

Assuming a value for r_b, then ϕ_0 can be computed from (6.48), ϕ_1 from Eq. (6.54), η from Eq. (6.52), and I_a from Eq. (6.45). A magnetron performance chart computed on the basis of these arguments is presented in Fig. 6.10, assuming π mode operation ($n = N/2$). It has become customary to plot the voltage along the ordinate and the anode current along the abscissa. On this graph the lines of constant magnetic field strength appear as approximately parallel lines with positive slope. High-power operation corresponds to the upper-right-hand corner, while high-efficiency operation corresponds to the upper-left-hand corner. Compared with typical mag-

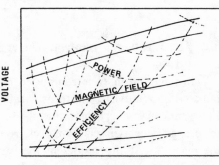

Figure 6.10. Qualitative magnetron performance chart.

netron performance charts, the chart of Fig. 6.10 indicates all the correct tendencies with the exception that very high efficiencies at very low anode currents are not observed experimentally. This phenomenon is due to the neglect of leakage current from the cathode in the theory.

As a final remark, note that from Eq. (6.3) the rotation rate at the edge of the Brillouin cloud is given (nonrelativistically) by

$$\omega_- = \frac{\Omega_0}{2}\left[1 - \left(\frac{r_c}{r_b}\right)^2\right] \tag{6.58}$$

Since ω_- varies with r_b, it is not always equal to the angular velocity of the traveling rf field. This gives rise to the important magnetron effect termed "frequency pushing."

In comparison with the qualitative cold fluid theory results presented above, recent particle simulation code results[18] do indicate the existence of a Brillouin-like sheath around the cathode and the subsequent development of space charge "spokes" that move in synchronism with the operating rf mode of the magnetron. However, in the quiescent phase the simulations indicate that pure laminar flow does not exist, as evidenced by spreads in both the radial and axial momentum distributions. Moreover, in the fully oscillating state the electrons that lie above the synchronous layer are found to be widely scattered in phase space due to the strong wave–particle interaction. In addition, back-bombardment of the cathode is found to occur over most of the cathode surface, rather than only where the electrons are out of phase with the rf. Hence, a cold fluid analysis of the fully oscillating state of the magnetron is inadequate; a worthwhile goal is to develop a semiempirical model of magnetron operation that supplements the cold fluid theory by modeling the results of the simulations in a consistent manner.

6.2. The Electron Cyclotron Maser (ECM)

In the electron cyclotron maser (or gyrotron) electromagnetic energy is radiated by relativistic electrons gyrating about an external magnetic field. Coherent emission is achieved as the result of orbital phase bunching due to the energy-dependent electron cyclotron frequency. Since the free energy for the maser instability resides in the rotational energy of the electrons, the instability exists only if the perpendicular energy exceeds a threshold value.

Table 6.2. Power Levels and Efficiencies from Cyclotron Masers
Driven by cw Sources

Wavelength (mm)	B Field (kG)	Voltage (kV)	Current (A)	Power (kW)	Efficiency (%)
2.78	40.5	27	1.4	12	31
1.91	28.9	18	1.4	2.4	9.5
0.92	60.6	27	0.9	1.5	6.2

Conversion of streaming energy to rotational energy, which can be achieved by a spatially varying magnetic field, is the classical analog of pumping a molecular energy level to produce a population inversion. Since the ECM mechanism does not rely on the physical structure of a waveguide or cavity, efficient operation at millimeter and submillimeter wavelengths appears to be possible.

The linear mechanism of the ECM was first recognized in 1958 by Twiss,[23] and later, independently by Gaponov[24] (classically) and Schneider[25] (quantum mechanically). While the experiments of Pantell[26] were perhaps the first work involving the ECM, the first definitve confirmation of the ECM mechanism was provided by the experiments of Hirshfield and Wachtel.[27] The efficiencies observed in the initial experiments were typically limited to only a few percent.

More recent efforts have generally followed one of two paths: (1) development in the Soviet Union of practical ECM tubes using conventional electron beams produced by thermionic cathodes for CW and long-pulse operation (Table 6.2); and (2) basic laboratory studies in the U.S. centered about the short-pulse intense relativistic electron beam technology (Table 6.3). Excellent reviews of these efforts are now available in the literature.[28-30] In this section the physical mechanism of the electron

Table 6.3. Power Levels and Efficiencies from Cyclotron Masers Driven
by Intense Relativistic Electron Beams

Wavelength (cm)	B Field (kG)	Voltage (MV)	Current (kA)	Power (MW)	Efficiency (%)
4	9	3.3	80	1000	0.4
2	15	2.6	40	350	0.3
0.8	5.5	0.6	15	8	0.1

cyclotron maser is first described, followed by the development and discussion of the linear dispersion relation. Finally, the nonlinear evolution of the instability is briefly examined in order to estimate the limitations to efficiency and power.

6.2.1. Physical Mechanism of the Electron Cyclotron Maser

The classical picture[31, 32] of the ECM mechanism can be understood by considering the orbit dynamics of electrons in a uniform magnetic field subject to a perturbing electric field of frequency ω_0 (Fig. 6.11). The electrons are initially assumed to be distributed uniformly on a sample ring with Larmor radius $r_L = v_\perp /(\Omega/\gamma_\perp)$, where $\Omega = eB/mc$ and $\gamma_\perp = [1 - (v_\perp /c)^2]^{-1/2}$. The rate of change of electron energy is given by $\dot\varepsilon = - ev_x E_x$, or

$$\dot\gamma_\perp = - \frac{e}{mc} \frac{(\gamma_\perp^2 - 1)^{1/2}}{\gamma_\perp} \cos\left[\left(\frac{\Omega}{\gamma_\perp}\right)t + \phi\right] \qquad (6.59)$$

where ϕ is the initial phase of rotation with respect to the electric field. First assume that the frequency of the electric field is equal to the relativistic cyclotron frequency. With the initial choice of field direction as shown in Fig. 6.11a ($t = 0$) electrons below the x axis ($-\pi < \phi < 0$) will lose energy and tend to spiral inward. Since $\gamma_\perp$ decreases, the relativistic cyclotron frequency of these particles increases and their phase tends to slip ahead of

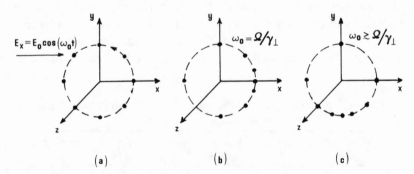

Figure 6.11. Test particle orbits in velocity space illustrating the electron cyclotron mechanism. (a) Initial particle positions; (b) bunched particles after several cycles with $\omega_0 = \Omega/\gamma_\perp$; (c) bunched particles after several cycles with $\omega_0 > \Omega/\gamma_\perp$.

the wave. Similarly, the electrons initially above the x axis ($0 < \phi < \pi$) tend to gain energy, and their phase falls behind the wave. After an integral number of wave periods the particles will have become bunched uniformly about the x axis (Fig. 6.11b). Since equal numbers will be gaining and losing energy, there will be neither net emission nor absorption. If, however, $\omega_0 \gtrsim \Omega/\gamma_\perp$, then the particles will, on the average, traverse a coordinate space angle less than 2π in a wave period $2\pi/\omega_0$. In this case the bunch tends to slip behind the wave as a whole (Fig. 6.11c). Since the net kinetic energy of the ensemble now decreases as a result of further interaction, emission exceeds absorption, and the wave amplitude increases.

The simple physical picture given above also indicates two important nonlinear saturation mechanisms. First, since the instability depends on the electron transverse energy, sufficient depletion of the free energy source must result in saturation of the instability. Depending on the initial parameters, however, the particles can become "phase trapped" and saturation may occur before all the free energy is depleted. Note that if ω_0 remains greater than $\Omega/\gamma_\perp$ the particles continue to slip behind the wave. On the other hand, if the particles lose sufficient energy such that $\omega_0 < \Omega/\gamma_\perp$, then the phase slippage will reverse and the bunch will tend to oscillate about the x axis. In either case the electron bunch will eventually be located above the x axis after some integral number of wave periods, and will begin to extract energy from the wave.

6.2.2. Linear Theory of the ECM Mechanism

From the previous discussion it is apparent that the cyclotron maser effect depends on the detailed momentum space distribution of the gyrating electrons. Such an instability cannot be analyzed within the framework of a macroscopic fluid model, but must be described with a kinetic approach in which the electron distribution function $f(\mathbf{x}, \mathbf{p}, t)$ and the average electric and magnetic fields, $\mathbf{E}(\mathbf{x}, t)$ and $\mathbf{B}(\mathbf{x}, t)$, evolve self-consistently according to the Vlasov–Maxwell equations (see Section 1.4):

$$\left[\frac{\partial}{\partial t} + \mathbf{v} \cdot \frac{\partial}{\partial \mathbf{x}} - e\left(\mathbf{E} + \frac{\mathbf{v} \times \mathbf{B}}{c} \right) \cdot \frac{\partial}{\partial \mathbf{p}} \right] f(\mathbf{x}, \mathbf{p}, t) = 0 \qquad (6.60)$$

$$\nabla \times \mathbf{E} = -\frac{1}{c} \frac{\partial \mathbf{B}}{\partial t} \qquad (6.61)$$

$$\nabla \times \mathbf{B} = -\frac{4\pi e}{c} \int d^3 p \ \mathbf{v} f(\mathbf{x}, \mathbf{p}, t) + \frac{1}{c} \frac{\partial \mathbf{E}}{\partial t} \tag{6.62}$$

$$\nabla \cdot \mathbf{E} = -4\pi e \int d^3 p \ f(\mathbf{x}, \mathbf{p}, t) \tag{6.63}$$

$$\nabla \cdot \mathbf{B} = 0 \tag{6.64}$$

To simplify the analysis we assume a planar model[31] (Fig. 6.12) in which a thin sheet beam is assumed to propagate parallel to a uniform applied magnetic field, $B_0 \hat{e}_z$, between two conducting planes located at $x = \pm a$. All quantities are assumed independent of the y coordinate, and it is further assumed that the z component of the unperturbed macroscopic beam velocity is constant. The following linear theory follows that of Mondelli.[33]

The Vlasov equilibrium is obtained by setting $\partial/\partial t = 0$ in Eqs. (6.60)–(6.64), and looking for solutions of the resultant stationary equations. If the beam space charge density is sufficiently small, the equilibrium beam self-fields may be neglected, and the equilibrium relativistic Vlasov equation becomes

$$\left[u_\perp (\cos\theta) \frac{\partial}{\partial x} + \Omega \frac{\partial}{\partial \theta} \right] f_0 = 0 \tag{6.65}$$

where $\Omega = eB/mc$, and $\mathbf{u} = \gamma \mathbf{v} = (u_\perp \cos\theta, u_\perp \sin\theta, u_z)$. It is easily verified that solutions of the form $f_0 = f_0(u_\perp, u_\parallel, \xi)$ satisfy Eq. (6.65), where $\xi = x - (\gamma u_\perp)/\Omega \sin\theta$ is the constant of the motion that corresponds to the guiding center coordinate.

From Eqs. (6.61)–(6.64) the beam current can excite electromagnetic fields in the parallel plate waveguide. Restricting the analysis to TE modes,

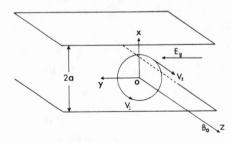

Figure 6.12. The electron cyclotron maser configuration in planar geometry.

the wave equation is given by

$$\left[\frac{d^2}{dx^2} + \left(\frac{\omega^2}{c^2} - k_z^2\right)\right]\delta A_y = -\frac{4\pi}{c}\delta j_y \qquad (6.66)$$

where

$$\delta E_y = -\frac{1}{c}\frac{\partial \delta A_y}{\partial t} = \frac{i\omega}{c}A(x)e^{i(k_z z - \omega t)} \qquad (6.67)$$

$$\delta B_x = -\frac{\partial \delta A_y}{\partial z} = -ik_z A(x)e^{i(k_z z - \omega t)} \qquad (6.68)$$

$$\delta B_z = \frac{\partial \delta A_y}{\partial x} = \frac{dA}{dx}e^{i(k_z z - \omega t)} \qquad (6.69)$$

The perturbed current density, δj_y, that gives rise to the TE wave through Eq. (6.66), must be found by integrating the linearized Vlasov equation. Since the beam density is assumed to be very low, the zero-order beam self-fields can be ignored, and Eq. (6.60) yields

$$\frac{df_1}{dt} = \frac{e}{m}\left(\delta \mathbf{E} + \frac{1}{\gamma c}\mathbf{u}\times\delta\mathbf{B}\right)\cdot\frac{\partial f_0}{\partial \mathbf{u}} \qquad (6.70)$$

where f_1 denotes the perturbation of the equilibrium electron distribution function. Equation (6.70) can be explicitly written as

$$\frac{df_1}{dt} = \frac{e}{mc}\left[\left(i\omega - \frac{ik_z u_z}{\gamma}\right)A(x)(\sin\theta)\frac{\partial f_0}{\partial u_\perp}\right.$$
$$\left. + \frac{ik_z u_\perp}{\gamma}(\sin\theta)A(x)\frac{\partial f_0}{\partial u_z}\right]e^{i(k_z t - \omega t)} \qquad (6.71)$$

where it has been recognized that $\partial f_0/\partial\xi = 0$.

The integration of Eq. (6.71) must be performed along the characteristics of the zero-order orbits according to

$$z'(t') = z + v_z(t' - t)$$

$$\theta'(t') = \theta + \Omega(t' - t)/\gamma$$

$$x'(t') = x + (u_\perp/\Omega)(\sin\theta' - \sin\theta)$$

$$u'_\perp(t') = u_\perp$$

$$u'_z(t') = u_z$$

Transforming the integration from t' to θ' yields

$$f_1 = \frac{i}{B_0} \int^\theta d\theta' \left[(\gamma\omega - k_z u_z)(\sin\theta') \frac{\partial f}{\partial u_\perp} + k_z u_\perp (\sin\theta') \frac{\partial f_0}{\partial u_z} \right] A(x')$$

$$\times e^{i(k_z z - \omega t)} e^{i(k_z u_z - \gamma\omega)(\theta' - \theta)/\Omega} \tag{6.72}$$

Since $x' = x + (u_\perp/\Omega)(\sin\theta' - \sin\theta)$, in the limit of small Larmor radius $A(x') \simeq A(x)$, which can be carried through the integral. Performing the remaining integration yields

$$f_1 = \frac{ieA(x)}{2mc} e^{i(k_z z - \omega t)} \left[(\gamma\omega - k_z u_z) \frac{\partial f_0}{\partial u_\perp} + k_z u_\perp \frac{\partial f_0}{\partial u_z} \right]$$

$$\times \left[\frac{e^{i\theta}}{(\gamma\omega - k_z u_z - \Omega)} - \frac{e^{-i\theta}}{(\gamma\omega - k_z u_z + \Omega)} \right] \tag{6.73}$$

The perturbed current density can be calculated as

$$\delta j_y = -e \int v_y f_1 \, dp^3 = -(e/\gamma) \int u_\perp (\sin\theta) f_1 u_\perp \, du_\perp \, du_z \, d\theta \tag{6.74}$$

Assuming that the equilibrium distribution function is of the form

$$f_0 = n_0 h(x) g(u_\perp, u_z) \tag{6.75}$$

where $g(u_\perp, u_z)$ is appropriately normalized, the integration over θ of Eq. (6.74) yields

$$\delta j_y = \frac{\omega_b^2}{8\gamma c} A(x) h(x) G(\omega, k_z) e^{i(k_z z - \omega t)} \tag{6.76}$$

where $\omega_b^2 = 4\pi e^2 n_0/m$, and

$$G(\omega, k_z) = \int u_\perp^2 \, du_\perp \, du_z \left[(\gamma\omega - k_z u_z) \frac{\partial g}{\partial u_\perp} + k_z u_\perp \frac{\partial g}{\partial u_z} \right]$$

$$\times \left[\frac{1}{(\gamma\omega - k_z u_z - \Omega)} + \frac{1}{(\gamma\omega - k_z u_z + \Omega)} \right] \tag{6.77}$$

Evaluation of Eq. (6.77) for a cold beam distribution function given by

$$g(u_\perp, u_z) = \frac{\delta(u_\perp - u_{\perp 0})\delta(u_z - u_{z0})}{2\pi u_{\perp 0}} \tag{6.78}$$

which could result from passing a cold beam through a spatially varying magnetic field, yields

$$G(\omega, k_z) = -\frac{1}{\pi} \left\{ \left[\frac{1}{(\omega - k_z v_{z0} - \Omega/\gamma)} + \frac{1}{(\omega - k_z v_{z0} + \Omega/\gamma)} \right] (\omega - k_z v_{z0}) \right.$$

$$+ \frac{v_{\perp 0}^2}{2} \left(\frac{\omega^2}{c^2} - k_z^2 \right)$$

$$\left. \times \left[\frac{1}{(\omega - k_z v_{z0} - \Omega/\gamma)^2} + \frac{1}{(\omega - k_z v_{z0} + \Omega/\gamma)^2} \right] \right\} \tag{6.79}$$

For a very thin beam with half-width $\Delta \ll a$, and with a centroid at $x = x_0$, then $h(x) \approx 2\Delta\delta(x - x_0)$ and

$$\delta j_y \approx \frac{\omega_b^2 \Delta}{4\gamma c} A(x_0)\delta(x - x_0)e^{i(k_z z - \omega t)}G(\omega, k_z) \tag{6.80}$$

The wave equation, Eq. (6.66), then becomes

$$\left(\frac{d^2}{dx^2} + \lambda^2 \right) A(x) = -\frac{\pi \omega_b^2 \Delta}{\gamma c^2} A(x_0)\delta(x_0)G(\omega, k_z) \tag{6.81}$$

where $\lambda^2 = [(\omega^2/c^2) - k_z^2]$. The appropriate boundary conditions for $A(x)$ are that $A(\pm a) = 0$, and $A(x)$ is continuous at $x = x_0$. Equation (6.81) implies that at $x = x_0$, dA/dx must satisfy the jump condition given by

$$\left. \frac{dA}{dx} \right|_{x_0 + \varepsilon} - \left. \frac{dA}{dx} \right|_{x_0 - \varepsilon} = -2\Delta\lambda^2 A(x_0) - \frac{\pi \omega_b^2 \Delta}{\gamma c^2} A(x_0)G(\omega, k_z) \tag{6.82}$$

in the limit that $\varepsilon \to 0$. Carrying out these calculations yields the dispersion relation given by

$$\frac{\lambda}{2}\left[\frac{\sin 2\lambda a}{\sin \lambda(x_0 + a)\sin \lambda(x_0 - a)}\right] - \Delta\lambda^2 \equiv H(\lambda)$$

$$= -\frac{\omega_b^2 \Delta}{2\gamma c^2}\left\{\left[\frac{1}{(\omega - k_z v_{z0} - \Omega/\gamma)} + \frac{1}{(\omega - k_z v_{z0} + \Omega/\gamma)}\right](\omega - k_z v_{z0})\right.$$

$$\left. - \frac{1}{2}\lambda^2 v_{\perp 0}^2\left[\frac{1}{(\omega - k_z v_{z0} - \Omega/\gamma)^2} + \frac{1}{(\omega - k_z v_{z0} + \Omega/\gamma)^2}\right]\right\}$$

$$\text{(6.83)}$$

If $\lambda^2 < 0$, the function $H(\lambda)$ is monotonically increasing and waveguide modes are cut off; only beam perturbations are possible. On the other hand, if $\lambda^2 > 0$ then $H(\lambda)$ has resonances at the zeros of $\sin \lambda(x_0 + a)\sin \lambda(x_0 - a)$, and propagating waveguide modes are possible. It is the coupling between positive energy waveguide modes and negative energy beam cyclotron modes that is responsible for the electron cyclotron maser instability.

In the limit of vanishing beam space charge ($\Delta \to 0$ while holding ω_b constant), it is easily verified that Eq. (6.83) yields the usual waveguide modes, where $\lambda^2 = k_n^2$ and $k_n = n\pi/2a$. Hence, the only way that unstable waveguide modes can occur is if $(\omega - k_z v_{z0} - \Omega/\gamma)$ is close to zero; i.e., frequencies and wave numbers in the vicinity of the intersection of the beam cyclotron mode dispersion relation, $\omega_0 - k_z v_{z0} = \Omega/\gamma$, and the waveguide dispersion relation, $\omega_0^2 = k_{z0}^2 c^2 + k_n^2 c^2$. These frequencies are given by (Fig. 6.13)

$$\omega_{0\pm} = \frac{\Omega}{\gamma}\gamma_z^2\left\{1 \pm \beta_z\left[1 - \left(\frac{\Omega_0 \gamma}{\gamma_z \Omega}\right)^2\right]^{1/2}\right\} \tag{6.84}$$

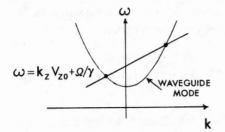

$$\omega = k_z V_{z0} + \Omega/\gamma$$

WAVEGUIDE MODE

Figure 6.13. Intersection of the vacuum waveguide dispersion relation with the Doppler-shifted cyclotron frequency.

where $\beta_z = v_{z0}/c$, $\gamma_z^2 = (1 - \beta_z^2)^{-1}$, and $\Omega_0 = k_n c$ is the waveguide cutoff frequency.

To find the growth rate of the instabilities we expand the dispersion relation, Eq. (6.83), about the fast-cyclotron mode as

$$\omega = k_z v_{z0} + \Omega/\gamma + \delta = \omega_0 + \delta \tag{6.85}$$

where $\delta \ll \omega_0$. Substituting Eq. (6.85) into Eq. (6.83) and keeping terms of order δ^3 yields

$$H(\lambda)\delta^2 + H'(\lambda)\frac{\delta^3\omega_0}{\lambda c^2} + \frac{\omega_b^2\Delta}{2\gamma c^2}\left[\frac{\Omega\delta}{\gamma} - \frac{\lambda^2 v_{\perp 0}^2}{2}\right] = 0 \tag{6.86}$$

In the limit of small beam space charge $\lambda \simeq k_n$, and $H(k_n) = 0$. Assuming that $x_0 = 0$ (the beam is centered in the waveguide), then

$$H'(\lambda) \approx H'(k_n) = \frac{2ak_n}{1 - (-1)^n} \tag{6.87}$$

and for n odd, Eq. (6.86) becomes

$$\delta^3 = -\frac{\omega_b^2\Delta}{2a\omega_0\gamma}\left(-\frac{\Omega\delta}{\gamma} + \frac{k_n^2 v_{\perp 0}^2}{2}\right) \tag{6.88}$$

If $\delta \ll \gamma k_n^2 v_{\perp 0}^2 / 2\Omega$, then the growth rate is given by

$$\delta_i \approx \frac{\sqrt{3}}{2}\left(\frac{\omega_b^2\Delta k_n^2 v_{\perp 0}^2}{4a\gamma\omega_0}\right)^{1/3} \tag{6.89}$$

In the limit of small perpendicular energy it is necessary to retain all the terms of Eq. (6.88). In this case it is easy to show that there are no complex roots (and no instability) unless the perpendicular energy exceeds the threshold criterion given by

$$\left(\frac{\gamma k_n v_{\perp \mathrm{cr}}}{\Omega}\right)^2 = \left(\frac{2}{3}\right)^{3/2}\frac{\omega_b}{(\omega_0\Omega)^{1/2}}\left(\frac{\Delta}{a}\right)^{1/2} \tag{6.90}$$

At $v_\perp = v_{\mathrm{cr}}$ the stabilizing and destabilizing terms of Eq. (6.83) just balance, and there is no free energy available for driving the instability.

6.2.3. Nonlinear Saturation Mechanisms for the ECM Instability

A comprehensive study of the self-consistent nonlinear evolution of the ECM instability, performed by Sprangle and Drobot,[32] indicates that two mechanisms are primarily responsible for saturation of the unstable wave. They are (1) depletion of the rotational free energy, and (2) phase trapping of the particles by the excited wave. The limiting saturation mechanism depends on the initial beam parameters.

6.2.3.1. Free Energy Depletion. If $v_\perp \gtrsim v_{\perp cr}$ from Eq. (6.90), then the maximum free energy per particle available to the wave is

$$\varepsilon_f = (\gamma_\perp - \gamma_{\perp cr}) m_0 c^2 \qquad (6.91)$$

where $\gamma_\perp = (1 - v_\perp^2 / c^2)^{-1/2}$, and $\gamma_\perp \gtrsim \gamma_{\perp cr}$. Hence, if the particles were to lose all of the free energy given by Eq. (6.91), then the conversion efficiency would be

$$\eta = \frac{(\gamma_\perp - \gamma_{\perp cr})}{(\gamma_\perp - 1)} \qquad (6.92)$$

6.2.3.2. Phase Trapping. If $\gamma_\perp \gg \gamma_{\perp cr}$, Eq. (6.92) is not a good approximation for the conversion efficiency because the particles can become phase trapped in the wave before the excess free energy is depleted. Recalling from Section 6.2.1 that the wave frequency must be slightly greater than the relativistic cyclotron frequency, $\Omega / \gamma_\perp$, for an instability to develop, then in the beam frame

$$\omega_0 - \Omega / \gamma_\perp = \delta \gtrsim 0 \qquad (6.93)$$

initially, where δ is the frequency shift as given by the dispersion relation, Eq. (6.88). As the wave grows and extracts energy from the particles the average γ decreases until $\omega_0 - \Omega / \langle \gamma_\perp \rangle = -\delta$. At saturation $\langle \gamma_\perp \rangle$ is minimum and the average change in the perpendicular energy of the particles is given by

$$\langle \Delta \gamma_\perp \rangle_s = \gamma_\perp - \langle \gamma_\perp \rangle_s = 2\gamma_\perp (\delta / \omega_0) \qquad (6.94)$$

Hence, in this case the conversion efficiency is given by

$$\eta = 2(\delta / \omega_0) \gamma_\perp / (\gamma_\perp - 1) \qquad (6.95)$$

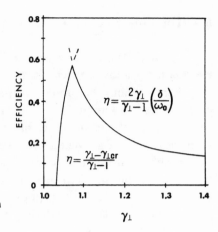

Figure 6.14. The efficiency of the cyclotron maser as a function of energy.

Since the growth rate is a function of δ, the efficiency cannot be made arbitrarily high by simply increasing the frequency shift.

From Eqs. (6.92) and (6.95), it is apparent that the efficiency is maximum when $\gamma_\perp = \gamma_{\perp\,cr}(1 - 2\delta/\omega_0)^{-1}$, and is given by (Fig. 6.14)

$$\eta_{max} \approx \frac{2\gamma_{\perp\,cr}\delta/\omega_0}{\gamma_{\perp\,cr} - (1 - 2\delta/\omega_0)} \qquad (6.96)$$

These simple estimates are in qualitative agreement with the results of detailed numerical simulations although Eq. (6.96) overestimates the observed efficiencies, typically by a factor of ~ 2.

While ECM experiments using intense relativistic electron beams have generally produced efficiencies of less than 2%,[34-37] these results are in good agreement with the theory; they are due to a combination of excessive beam temperature introduced by the method of producing the population inversion, and the fact that $\gamma_\perp$ was not optimized for maximum efficiency.[32] Based on the theory outlined here it should be possible to use the ECM mechanism for the efficient production (tens of per cent) of short-wavelength radiation, particularly in the submillimeter range when higher cyclotron harmonics are considered. Moreover, it may be possible to dramatically increase the efficiency by mistuning the relativistic cyclotron frequency in order to postpone electron phase trapping.[32]

6.3. The Free Electron Laser (FEL)

The free electron laser is a device which operates via the stimulated backscatter of a low-frequency wave from the "free" particles of a relativistic electron beam. The primary reason for interest in the FEL is that the backscattered radiation from relativistic beams can undergo a dramatic frequency upshift (of the order of γ^2), and is readily tunable over a wide frequency range. The eventual development of high-power FELs operating in the submillimeter and infrared spectral regions is expected to find widespread applications in several areas, including plasma heating, photochemistry, molecular spectroscopy, radar, and communications. Recent reviews of stimulated scattering research are now available in the literature.[38-40]

The stimulated scattering of photons by a group of electrons was first discussed by Kapitza and Dirac,[41] while the first analysis of high-frequency radiation from an electron beam in the presence of a modulating magnetic field was made by Motz.[42] In 1960 Phillips reported the first experiment involving stimulated emission with a nonrelativistic electron beam-"wiggler" configuration.[43] His device, the Ubitron, was capable of generating kilowatts of millimeter radiation. The first proposal for generating very short-wavelength radiation by stimulated scattering from a relativistic electron beam was due to Pantell et al.[44]

The two principal types of scattering processes which occur in FEL experiments are wave–particle (Compton) scattering, and wave–wave (Raman) scattering.[45-48] If the incident pump wavelength in the electron beam frame is much greater than the Debye wavelength, then the wave–wave process dominates and scattering occurs from plasma oscillations. On the other hand, wave–particle scattering dominates when the pump wavelength is comparable to or smaller than the Debye wavelength. In this case the scattering is off single "shielded" particles.

Experiments demonstrating wave–particle stimulated scattering with an output in the infrared have been performed at Stanford University using a low-current, high-energy electron beam from a linear accelerator.[49, 50] Observed efficiencies were less than 0.01%; attempts to improve the efficiency have focused on the use of storage rings to continuously recirculate the beam through the wave generation region.

The first relativistic beam experiments demonstrating stimulated scattering in the Raman regime were performed by Granatstein et al.,[51, 52] and Marshall et al.[53] Utilizing intense relativistic electron beam generators, superradiant FEL oscillators were demonstrated by producing megawatt power levels in short interaction regions (~ 30 cm) at wavelengths ranging from 2 mm to 400 μm, and with efficiencies as high as 0.1%. More recently, McDermott et al.[54] reported the realization of a collective Raman FEL for the first time. The experiment was designed so as to permit several passes of feedback by employing a quasioptical cavity. Laser output of 1 MW at 400 μm and line narrowing to $\Delta\omega/\omega \approx 2\%$ were observed, compared to $\Delta\omega/\omega \approx$ 10% for the earlier superradiant oscillator studies.

In the remainder of this section, the physical mechanism of the FEL is first discussed, followed by a presentation of the classical linear theory for the Raman scattering regime. Finally, the nonlinear mechanisms, which limit the ultimate power and efficiency, are briefly examined.

6.3.1. Physical Mechanism of the Free Electron Laser

6.3.1.1. Single-Electron Limit (Compton Scattering). The basic FEL operating principles can be easily understood by considering the motion of a single free electron as it passes through the magnetic field as shown in Fig. 6.15. It is assumed that the static magnetic field is linearly polarized, i.e., aligned in a single plane, according to

$$\mathbf{B}(z) = B_0 \hat{x} \sin k_0 z \tag{6.97}$$

where $k_0 = 2\pi/\lambda_0$, and λ_0 is the period of the field. With Eq. (6.97) the relativistic equation of motion for the electron becomes

$$\dot{v}_y = -\left(v_z \Omega_0/\gamma\right)\sin k_0 z \tag{6.98}$$

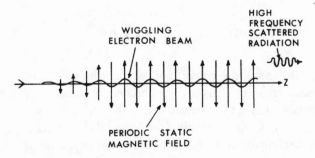

Figure 6.15. Schematic of the free electron laser interaction.

$$\dot{v}_z = \left(v_y \Omega_0 / \gamma \right) \sin k_0 z \qquad (6.99)$$

where $\Omega_0 = eB_0/mc$. Since the situation of interest is essentially uniform motion along the z axis with only small perturbations, we assume that $v_z \approx v_0 \approx c \gg v_x, v_y$. In this case the approximate solution of Eqs. (6.98) and (6.99) is

$$v_y \approx \frac{v_0 \Omega_0}{\gamma \omega_0} \cos \omega_0 t \qquad (6.100)$$

$$v_z \approx v_0 - \frac{v_0 \Omega_0^2}{4\gamma^2 \omega_0^2} \cos 2\omega_0 t \qquad (6.101)$$

where $\omega_0 = k_0 c$. Integrating Eqs. (6.100) and (6.101) yields the approximate particle trajectory given by

$$y = \frac{v_0 \Omega_0}{\gamma \omega_0^2} \sin \omega_0 t \qquad (6.102)$$

$$z = v_0 t - \frac{v_0 \Omega_0^2}{8\gamma^2 \omega_0^3} \sin 2\omega_0 t \qquad (6.103)$$

Hence, the periodic magnetic field has resulted in a sinusoidal "wiggling" of the electron trajectory. Moreover, in a reference frame traveling at speed v_0, the electron describes a figure-8 orbit in the $y-z$ plane.

Both the transverse and longitudinal accelerations cause the electron to emit electromagnetic energy in accordance with the classical radiation law[55]

$$\frac{dP}{d\Omega} = \frac{e^2}{4\pi c} \frac{\left| \hat{n} \times \left[(\hat{n} - \boldsymbol{\beta}) \times \dot{\boldsymbol{\beta}} \right] \right|^2}{(1 - \hat{n} \cdot \boldsymbol{\beta})^5} \qquad (6.104)$$

where $dP/d\Omega$ denotes the power radiated per unit solid angle, $\hat{n}$ is a unit vector directed from the charge to the observation point, and $\boldsymbol{\beta} = \mathbf{v}/c$. Since we are only considering relativistic particles, however, the radiation from the longitudinal component is negligible (of order $1/\gamma^2$) compared to that from the transverse component. Consequently, we may approximate the radiation intensity by that due to the transverse contribution alone. In this case Eq.

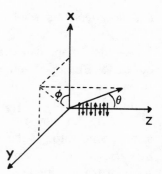

Figure 6.16. Polar angles (θ, ϕ) for radiation emitted by an accelerated electron traversing the periodic magnetic field.

(6.104) becomes

$$\frac{dP}{d\Omega} \simeq \frac{2}{\pi} \frac{e^2 \dot{v}_y^2}{c^3} \frac{\gamma^6}{\left(1 + \gamma^2 \theta^2\right)^3} \left[1 - \frac{4\gamma^2 \theta^2 \cos^2 \phi}{\left(1 + \gamma^2 \theta^2\right)^2} \right] \qquad (6.105)$$

where θ and ϕ are the polar angles as depicted in Fig. 6.16.

Since the motion of the radiating electron is periodic, when Eq. (6.105) is integrated over the electron's path through the magnetic field it is found that the frequency spectrum of the radiation is discrete, containing frequencies that are integral multiples of the fundamental, i.e.,

$$\omega_R = \frac{n\omega_0}{\left[1 - (v_0/c)\cos\theta\right]} \qquad (6.106)$$

for $\phi = 0$. Along the z axis ($\theta = 0$) the fundamental is dominant. In this case $\omega_R = \omega_0(1 - v_0/c)^{-1}$. In the limit of $v_0 \to c$, the frequency of the radiation becomes approximately

$$\omega_R \approx 2\gamma^2 \omega_0 \qquad (6.107)$$

Hence, the radiation frequency appears as a double Doppler upshift of the magnet frequency. In fact, viewed from the electron rest frame, the magnet period λ_0 is Lorentz contracted to λ_0/γ, and $\omega_0 \to \gamma\omega_0$. In other words, the static magnetic field appears as an electromagnetic wave. Hence, the spontaneous emission process in the FEL can be considered as the Compton scattering of virtual photons from free electrons. Similarly, if the static magnetic field were replaced by an actual electromagnetic wave of frequency

ω_0, the interaction would result in back-scattered radiation at frequency $(1 + \beta_0)^2 \gamma_0^2 \omega_0 \approx 4 \gamma_0^2 \omega_0.$[55]

While the spontaneous emission just described is an important prerequisite for FEL operation, the actual amplification mechanism is due to the stimulated emission that occurs as the result of the interaction of the upshifted radiation with the free electrons. Returning to the simple picture of Fig. 6.15, it is now assumed that em radiation of frequency $\omega_R \approx 2\gamma^2 \omega_0$ is also passing through the same cavity. Depending on the relative phase between the particle's oscillations in the magnetic field and the varying fields of the radiation, either stimulated emission or absorption can occur.

Since the wavelength of the radiation pulse is much less than either the cavity dimensions or the pulse length, it can be accurately represented by a plane wave traveling at c. Further, since the electron also travels at $\beta_0 c \approx c$, the electron and the radiation remain essentially overlapped during the transit through the magnet. Although the radiation fields alone do not affect the electron trajectory, the interaction of the transverse radiation electric field with the transverse oscillations of the electron induced by the static wiggler magnetic field causes the electron to either gain or lose energy according to

$$\dot{\gamma} = -\frac{e}{m} v_y E_y \qquad (6.108)$$

Assuming that both v_y and E_y are at their instantaneous maximum amplitudes, then according to Fig. 6.17, if

$$\beta_0 = 1 - \frac{\omega_0}{\omega_R} \qquad (6.109)$$

then exactly one-half wavelength of light ($\lambda_R = 2\pi c / \omega_R$) will pass over the electron as the electron traverses one-half of a magnet wavelength ($\lambda_0 = 2\pi c / \omega_0$). The resonance condition, Eq. (6.109), corresponds to Eq. (6.107).

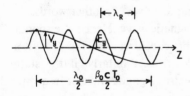

Figure 6.17. The mechanism for stimulated emission. (a) Both the transverse particle velocity and the transverse electric field of the radiation are at the instantaneous maximum amplitudes and are in phase. (b) For resonance, one-half wavelength of light must pass over the electron as it traverses one-half of a magnet wavelength.

Hence, the work done on the electron via Eq. (6.108) by the radiation stored in the cavity as the result of spontaneous emission retains the same sign over many wavelengths. If $\dot{\gamma} > 0$, then radiation energy is absorbed; if $\dot{\gamma} < 0$ then the radiation fields grow, corresponding to stimulated emission.

From this simple analysis it is apparent that the electron's relative position within a radiation wavelength will determine whether it consistently gains or loses energy as it travels through the wiggler magnet. The correlation between electron position and energy change in a FEL results in a bunching of electrons in phase space. Depending on the details of the interaction the bunched electrons can undergo coherent deceleration so that the efficiency of energy extraction can be greatly enhanced.

6.3.1.2. Intense Beam Limit (Collective Raman Scattering). For high current beams the simple single-particle analysis presented in Section 6.3.1.1 becomes inappropriate because the beam can respond to the fields in the FEL cavity as a coherent plasma entity. In this case the strong collective fields can be exploited by using stimulated Raman scattering (SRS) to transform beam energy into high-frequency radiation.

Consider a monoenergetic beam with energy $mc^2(\gamma - 1)$ and velocity $v_0 = c(\gamma^2 - 1)^{1/2}/\gamma$ traveling through a rippled magnetic field of wavelength λ_0. In a reference frame comoving with the beam at velocity v_0, the static magnetic field appears as an electromagnetic pump wave at frequency $\omega_p' = \gamma(2\pi/\lambda_0)v_0$. The pump wave results in the development of a body "idler" space charge or cyclotron wave, i.e., $\omega_i' = \omega_b'$, where $\omega_b' = (4\pi n e^2/\gamma m)^{1/2}$ or $\omega_i' = \Omega = eB/mc$, where B is the strength of a guiding magnetic field. Conservation of momentum and energy yield the three-wave phase-matching conditions described by the Manley–Rowe equations

$$\omega_p' = \omega_i' + \omega_R'$$

$$k_p' = k_i' + k_R' \tag{6.110}$$

where ω_R', k_R' denote the frequency and wave number of the back-scattered electromagnetic wave. The corresponding laboratory frame frequencies (unprimed) can be found from the primed frequencies via the relativistic transformation equations:

$$\omega_p' = \gamma(\omega_p + k_p v_0) \tag{6.111}$$

$$\omega_R' = \gamma(\omega_R - k_R v_0) \tag{6.112}$$

Combining Eqs. (6.110)–(6.112) yields

$$(\omega_R - k_R v_0) = (\omega_p + k_p v_0) - \omega_i'/\gamma \tag{6.113}$$

For the case of the static wiggler field, $\omega_p = 0$ and $k_p c = \omega_0$. Since $\omega_R k_R = c$, Eq. (6.113) becomes

$$\omega_R = \frac{\omega_0}{1 - \beta_0} - \frac{\omega_i'/\gamma}{1 - \beta_0} \tag{6.114}$$

If ω_b and Ω are small compared to ω_0, then Eq. (6.114) gives the same result as Eq. (6.107) for the single-particle case.

When the static field is replaced by an actual electromagnetic field of frequency ω_0, then $\omega_p k_p = c$ and Eq. (6.113) yields

$$\omega_R = \left(\frac{1 + \beta_0}{1 - \beta_0} \right) \omega_0 - \frac{\omega_i'/\gamma}{1 - \beta_0} \tag{6.115}$$

If $\omega_i \ll \omega_0$ and $\beta_0 \approx 1$, then

$$\omega_R \approx 4\gamma^2 \omega_0 \tag{6.116}$$

A rearrangement of Eq. (6.110) indicates that

$$\frac{\omega_R'}{k_R'} = \frac{\omega_p' - \omega_i'}{k_p' - k_i'} \tag{6.117}$$

and

$$\frac{\omega_i'}{k_i'} = \frac{\omega_p' - \omega_R'}{k_p' - k_R'} \tag{6.118}$$

According to Eq. (6.117) the phase velocity of the pump wave beating with the idler wave is exactly equal to the phase velocity of the scattered wave. Hence, the beat disturbance moves synchronously with the scattered radiation. Moreover, from Eq. (6.118) the beating of the pump and scattered wave moves in phase with the plasma wave.

From elementary wave theory the amplitude of the beat disturbance of two waves is equal to the product of the individual wave amplitudes times a phase factor. Assume initially that the amplitudes of both the idler and scattered waves are small, but that the pump amplitude is large. While the beat disturbance of the pump and idler waves is small, it will eventually cause substantial growth of the scattered radiation because of the phase synchronism. Idler wave growth will also occur for similar reasons. As each wave grows, however, the interaction itself becomes stronger, i.e., the system is unstable, and the backscattered radiation grows exponentially.

It has been tacitly assumed that the idler wave phase velocity in the beam frame is greater than the beam thermal spread, i.e.,

$$v_i' = \frac{\omega_i'}{k_i'} \gg v_{th} \tag{6.119}$$

If $\omega_i' = \omega_b$, then Eq. (6.119) implies that

$$k_D' \gg k_p' + k_R' \tag{6.120}$$

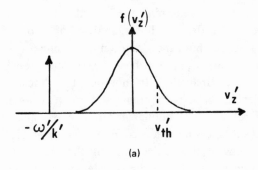

(a)

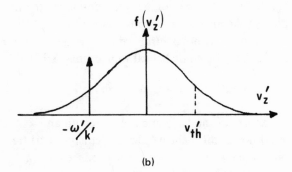

(b)

Figure 6.18. A comparison of the electron distribution and the phase velocity of the idler wave for (a) stimulated Raman scattering, (b) stimulated Compton scattering.

where $k'_D = \omega_b/v'_{th}$ is the Debye wave number. Thus, for stimulated Raman scattering the Debye length of the electron ensemble ($\lambda'_D = 2\pi/k'_D$) must be smaller than the wavelengths of either the pump or scattered waves. In order to satisfy this requirement the beam must be dense and cold. If the electron distribution were tenuous or warm, then the phase velocity of the space charge wave would overlap the electron distribution as shown in Fig. 6.18b, and only electrons in the vicinity of $v'_z = \omega'/k'$ would participate in the stimulated scattering process, i.e., stimulated Compton scattering as discussed in Section 6.3.1.1.

6.3.2. Linear Theory of the Free Electron Laser in the Raman Scattering Limit

There has been a considerable amount of research in both the linear and nonlinear regimes of free electron lasers. Sprangle et al.[56] studying the scattering of an electromagnetic wave by a magnetized relativistic electron beam, were the first to determine the growth rate for the weak pump scattering process. The detailed linear analysis of Kroll and McMullin[57] yielded growth rates for both the Compton and Raman scattering regimes. The high-gain (Raman) regime has been further analyzed by Hasegawa,[58] and by Bernstein and Hirshfield.[59] In subsequent nonlinear analyses, growth rates and efficiencies have been obtained for all high-gain processes by Sprangle and Drobot.[60] The nonlinear problem has also been examined using numerical particle simulation codes by Kwan et al.,[61] and Lin and Dawson[62] have showed that efficiencies of the order of 10% might be possible by appropriately varying the wavelength of the wiggler field and the pump amplitude. In this section we will summarize the linear theory of the FEL, following the analysis of Kwan et al.[61, 63]

Consider a completely nonneutralized relativistic electron beam propagating along the z axis of a static helical wiggler magnetic field described by

$$\mathbf{B}_0 = B_0\left(\cos k_0 z \hat{e}_x + \sin k_0 z \hat{e}_y\right) \qquad (6.121)$$

where B_0 is a measure of the magnetic field strength and $k_0 = 2\pi/\lambda_0$ is the wave number. Although this helical field model is not curl free, it is a good approximation to the field near the axis produced by a helical current distribution, and is sufficient for our purposes. In most situations of practical interest for Raman scattering the beam must be guided through the interaction region by a solenoidal magnetic field. While this field configuration also permits the cyclotron mode to be used as an idler wave, the guide field couples the motion of the beam particles from one perpendicular direction to the other and the system is no longer isotropic. Since the wave phenomena become rather complex in this instance we drop the guide field from the subsequent analysis.

Only spatial variations along the z axis will be considered for the electron beam, pump, and scattered fields. Choosing the Coulomb (or transverse) gauge, $\nabla \cdot \mathbf{A} = 0$, Maxwell's equations yield the wave equation

$$\left(\frac{\partial^2}{\partial z^2} - \frac{1}{c^2} \frac{\partial^2}{\partial t^2} \right) \mathbf{A}_\perp = - \frac{4\pi}{c} \mathbf{j}_\perp \tag{6.122}$$

which relates the transverse component of the vector potential to the transverse component of the current density.

The motion of the electron beam in the electromagnetic fields and the static wiggler field is governed by the equation of motion

$$\dot{\mathbf{p}} = - e \left[\mathbf{E} + \frac{1}{c} (\mathbf{v} \times \mathbf{B}) \right] \tag{6.123}$$

where $\mathbf{B}$ is the total magnetic field. [The longitudinal component of the stress tensor may be added to the right-hand side of Eq. (6.123) to provide an approximate description of thermal effects.[61,63]] In the following it is assumed that the beam is sufficiently tenuous, so that the beam self-fields may be neglected in the analysis; however, the perturbed fields resulting from longitudinal particle bunching will be retained.

We adopt a linearization of the form $\mathbf{A}_\perp = \mathbf{A}_0 + \mathbf{A}'_\perp$, $\mathbf{v} = \mathbf{v}_0 + \mathbf{v}'$, etc., where $\mathbf{B} = \nabla \times \mathbf{A}$. From Eq. (6.123) the solution of the zero-order equation of motion of the beam through the helical field is simply

$$\mathbf{v}_0 = v_0 \hat{e}_z - \frac{\Omega}{\gamma k_0} (\cos k_0 z \hat{e}_x + \sin k_0 z \hat{e}_y) \tag{6.124}$$

where $\Omega = eB_0 / mc$.

The linearization of Eq. (6.123) yields

$$\frac{d}{dt} \mathbf{p}' = - e \left[\mathbf{E}' + \frac{1}{c} (\mathbf{v}' \times \mathbf{B}_0 + \mathbf{v}_0 \times \mathbf{B}') \right] \tag{6.125}$$

while the linearization of Eq. (6.122) yields

$$\left(\frac{\partial^2}{\partial z^2} - \frac{1}{c^2} \frac{\partial^2}{\partial t^2} \right) \mathbf{A}'_\perp = -\frac{4\pi}{c} \mathbf{j}'_\perp \qquad (6.126)$$

Since $\mathbf{j}_\perp = -en\mathbf{v}_\perp$, the perturbed current density is given by

$$\mathbf{j}'_\perp = -e(n_0 \mathbf{v}'_\perp + n' \mathbf{v}_{\perp 0}) \qquad (6.127)$$

where n_0 and n' are the zeroth- and first-order electron densities.

Since the system is assumed uniform in the transverse directions, it follows that the transverse canonical momentum of a particle is a constant of the motion, i.e.,

$$\mathbf{p}_\perp = \frac{e}{c} \mathbf{A}_\perp \qquad (6.128)$$

Substituting Eqs. (6.128) and (6.127) into Eq. (6.126) yields

$$\left(\frac{\partial^2}{\partial t^2} - \frac{1}{c^2} \frac{\partial^2}{\partial t^2} - \frac{\omega_{b0}^2}{\gamma_0 c^2} \right) \mathbf{A}'_\perp = \frac{4\pi e n'}{c} \mathbf{v}_{\perp 0} \qquad (6.129)$$

where $\omega_{b0}^2 = 4\pi n_0 c^2/m$.

Careful consideration of Eq. (6.125) indicates that the transverse components simply reexpress Eq. (6.128); using Eqs. (6.121), (6.124), and (6.128) the longitudinal component can be explicitly written as

$$\gamma_0^3 m \left(\frac{\partial}{\partial t} + v_0 \frac{\partial}{\partial z} \right) v'_z = -eE'_z - \frac{e}{c} \left[\frac{\Omega}{\gamma} \left(A'_x \sin k_0 z - A'_y \cos k_0 z \right) \right.$$

$$\left. - \frac{\Omega}{\gamma_0 k_0} \left(\frac{\partial A'_x}{\partial z} \cos k_0 z + \frac{\partial A'_y}{\partial z} \sin k_0 z \right) \right]$$

$$(6.130)$$

where it has been recognized that $\dot{p}'_z \approx \gamma_0^3 m \dot{v}'_z$.

Eqs. (6.130) and (6.129), together with the linearized continuity equation

$$\frac{\partial}{\partial t} n' + v_0 \frac{\partial n'}{\partial z} + n_0 \frac{\partial v'_z}{\partial t} = 0 \qquad (6.131)$$

and Poisson's equation

$$\frac{\partial}{\partial z} E_z' = -4\pi e n'$$ (6.132)

form a closed set of equations that can be used to determine the dispersion relation. Assuming solutions of the form

$$\xi = \sum_k \xi_k e^{i(kz - \omega t)}$$ (6.133)

where

$$\xi_k = \frac{e^{i\omega t}}{2\pi} \int \xi e^{-ikz} \, dt$$ (6.134)

then Eqs. (6.129) and (6.131) yield

$$A_{xk}' = -\frac{2\pi e \Omega c}{\gamma_0 k_0} \left(n_{k+k_0}' + n_{k-k_0}' \right) \left(\omega^2 - k^2 c^2 - \frac{\omega_{b0}^2}{\gamma} \right)^{-1}$$ (6.135)

$$A_{yk}' = \frac{2\pi i e \Omega c}{\gamma_0 k_0} \left(n_{k+k_0}' - n_{k-k_0}' \right) \left(\omega^2 - k^2 c^2 - \frac{\omega_{b0}^2}{\gamma} \right)^{-1}$$ (6.136)

Similarly, Eqs. (6.131) and (6.132) give

$$E_{z,k+k_0}' \left[\omega - (k + k_0) v_0 \right] = 4\pi i e n_0 v_{z,k+k_0}'$$ (6.137)

Substituting Eq. (6.133) into Eq. (6.130) and using Eqs. (6.135)–(6.137) to eliminate the perturbed variables finally yields the dispersion relation given by

$$\left\{ \left[\omega - (k + k_0) v_0 \right]^2 - \frac{\omega_{b0}^2}{\gamma_0^3} \right\} \left(\omega^2 - k^2 c^2 - \frac{\omega_{b0}^2}{\gamma} \right) = \frac{\omega_{b0}^2 \Omega^2 (k + k_0)^2}{2\gamma_0^5 k_0^2}$$

(6.138)

Equation (6.138) is the appropriate expression for Raman scattering in the limit of small Debye length. The right-hand side of Eq. (6.138) can be considered as a coupling between the emitted wave and the plasma wave on the electron beam. The dispersion diagram of the normal modes of the

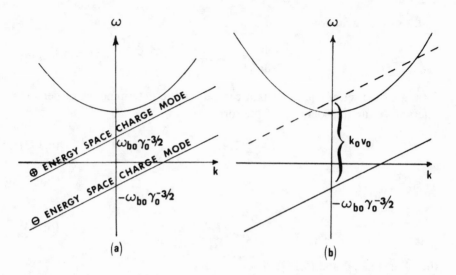

Figure 6.19. Dispersion diagram of the system normal modes (a) in the absence of a static wiggler magnet; (b) in the presence of a static periodic magnetic field of wave number k_0.

system, neglecting the mutual effects between the beam and electromagnetic modes, is presented in Fig. 6.19. The hyperbola is the dispersion curve of the electromagnetic mode with

$$\omega^2 - k^2 c^2 = \omega_{b0}^2 / \gamma \qquad (6.139)$$

The other two lines represent the positive- and negative-energy beam space charge modes with ($k_0 = 0$)

$$\omega = k v_0 \pm \omega_{b0} \gamma_0^{-3/2} \qquad (6.140)$$

An instability will only occur if the negative-energy space charge mode interacts with the positive-energy electromagnetic mode. For the case of Fig. 6.19a, there is no interaction and no instability. However, the wiggler magnetic field modifies Eq. (6.140) as

$$\omega = (k + k_0) v_0 \pm \omega_{b0} \gamma_0^{-3/2} \qquad (6.141)$$

so that there now exists the possibility of coupling and instability, as shown in Fig. 6.19b.[64]

If the coupling term is much less than unity, then the dispersion relation may be expanded around the natural frequencies according to

$$\omega = \omega_1 + \delta = \omega_2 + \delta + \mu \qquad (6.142)$$

where $\omega_1 = (k^2c^2 + \omega_{b0}^2/\gamma)^{1/2}$ and $\omega_2 = (k + k_0)v_0 - \omega_{b0}/\gamma_0^{3/2}$. Substituting Eq. (6.142) into Eq. (6.138) yields

$$\omega = \omega_1 + \frac{1}{2}\left\{ -\mu \pm i\left[\frac{\omega_{b0}\Omega^2(k + k_0)^2}{2\gamma_0^{7/2}k_0^2\omega_1} - \mu^2 \right]^{1/2} \right\} \qquad (6.143)$$

Eq. (6.143) indicates that for a given frequency mismatch, there is no instability unless the strength of the wiggler field exceeds a threshold value given by

$$\Omega_{cr} = \frac{\gamma_0^{7/4}\mu k_0}{(k + k_0)}\left(\frac{2\omega_1}{\omega_{b0}} \right)^{1/2} \qquad (6.144)$$

If $\Omega > \Omega_{cr}$, then the growth rate is given by[56,57,60,61]

$$\omega_i = \tfrac{1}{2}\mu\left[(\Omega/\Omega_{cr})^2 - 1 \right]^{1/2} \qquad (6.145)$$

which corresponds to the cold beam, weak pump regime of Sprangle et al.[38] Evidently a strong pump wave is required for a reasonably large growth rate in a laboratory device. Growth rates for the case of a large coupling term, the strong pump limit, available in the literature,[56,60] indicate a $(\Omega)^{2/3}$ scaling.

6.3.3. Nonlinear Saturation Mechanism for the Free Electron Laser

Of primary importance is the fraction of electron beam energy that can be transferred into high-frequency radiation. Although a detailed theoretical answer can only be determined through complex particle–wave computer simulations,[61-63] Sprangle and Drobot have deduced several semiquantitative upper limits for the energy transport efficiency in various operational regimes.[60]

In the cold beam, weak pump, Raman regime, the dominant saturation mechanism appears to be electron trapping in the electrostatic idler wave.[61]

As the interaction proceeds and the beam loses energy the space charge wave will eventually reach an amplitude at which it can trap most of the electrons. Hence, at the trapping time the electron beam velocity, on the average, is approximately equal to the phase velocity of the space charge wave. Assuming that all the energy lost by the beam electrons appears in the scattered electromagnetic wave, then the efficiency η is given approximately by

$$\eta \approx \frac{(\gamma_0 - \gamma_{ph})}{(\gamma_0 - 1)} \tag{6.146}$$

where $\gamma_{ph} = [1 - (v_{ph}/c)^2]^{-1/2}$. From Eq. (6.141)

$$v_{ph} \approx v_0 - \omega_{b0}\gamma_0^{-3/2} \tag{6.147}$$

Hence, if the beam is strongly relativistic, then

$$\gamma_{ph} \approx \gamma_0 \left(1 - \frac{\omega_{b0}}{2\gamma_0^{3/2}k_0c} \right) \tag{6.148}$$

and Eq. (6.146) for the efficiency becomes[60, 61]

$$\eta \approx \frac{\omega_{b0}}{2\gamma_0^{3/2}k_0c} \tag{6.149}$$

Note that in this weak pump limit the FEL efficiency is independent of pump amplitude.

In the cold beam, strong pump, Raman regime, Sprangle and Drobot show that the phase velocity of the most unstable wave is given approximately by[60]

$$v_{ph} \simeq v_0 - 2^{-1/3} \left(\frac{\omega_{b0}\Omega}{\gamma_0^{3/2}k_0^2c^2} \right)^{2/3} \frac{c}{4\gamma_0^2} \tag{6.150}$$

In this case Eq. (6.146) yields

$$\eta \approx \frac{1}{8\gamma_0^2} \left(\frac{2\omega_{b0}\Omega}{k_0^2c^2} \right)^{2/3} \tag{6.151}$$

Since both the growth rate and the efficiency vary as the two-thirds power of the pump strength in the strong pump limit, this appears to be the most attractive regime for generating high frequency radiation. However, in both the weak and strong pump limits the efficiency is a decreasing function of the scattered wave frequency.

It should be noted that the cold beam assumption implicitly used in deriving the efficiency expressions is only valid if the beam thermal velocity is much less than the difference between the axial beam velocity and the phase velocity of the space charge wave, i.e.,

$$v_{th} \ll v_0 - v_{ph} \qquad (6.152)$$

Since the thermal energy spread $W_{th} \approx \gamma_0^3 mc v_{th}$ for strongly relativistic electron beams, the allowed thermal energy spread for both regimes is given approximately by

$$W_{th}/W_0 \approx \eta \qquad (6.153)$$

The simple single-pass efficiency estimates given by Eqs. (6.149) and (6.151) indicate that with a high-quality beam intrinsic efficiencies of 1%–10% might be possible at infrared and submillimeter wavelengths, dropping to 0.1%–1% at optical wavelengths. Possible (but as yet untried) techniques for increasing these estimates include the use of depressed collectors to recover the unused electron kinetic energy, and axially varying the parameters of the interaction region to delay the onset of particle–wave trapping.[62, 65]

6.4. Summary

In this chapter we have considered in some detail three potential sources of high-power coherent radiation that are based on high-voltage, relativistic electron beam technology. In a relativistic magnetron a high voltage is applied across a coaxial anode–cathode gap. Electrons field-emitted from the cathode are caused to azimuthally drift in the interaction space as the result of an externally applied axial magnetic field. The anode contains a resonant slow wave structure which can support electromagnetic waves. Those waves which have phase velocities nearly equal to the electron drift velocity interact strongly with the space charge and lead to the formation of electron "spokes" that rotate in synchronism with the rf waves. Although electrons migrate from cathode to anode through the spoke structures, they strike the anode with very little energy; in effect, the electrical potential energy associated with the radial electric field is converted directly into rf energy. Since the resonant wavelengths are determined by the physical construction of the anode slow wave structure, the characteristic operating regime for a relativistic magnetron is 1–10 cm. Gigawatt power levels have already been achieved at $\lesssim 30\%$ efficiency.

In the case of the electron cyclotron maser electromagnetic radiation is produced as the result of relativistic electrons gyrating about an external magnetic field. Coherent emission results from orbital phase bunching due to the energy-dependent relativistic electron cyclotron frequency. Since the free energy for the instability resides in the rotational motion of the electrons, conversion of electron streaming energy to rotational energy can be considered as the classical analog of molecular pumping to produce an energy level population inversion. As this mechanism does not rely on the physical structure of a waveguide or cavity, it may be possible to achieve efficient operation at millimeter and submillimeter wavelengths. Although the observed efficiency levels to date have been rather low ($\lesssim 1\%$), they are consistent with the theoretical estimates. Hence, optimized experiments are expected to produce short-wavelength radiation at much higher efficiency.

The free electron laser operates via the stimulated backscatter of a lower-frequency pump wave from the "free particles" of a relativistic electron beam. Since the frequency of the backscattered radiation is doubly Doppler upshifted from the pump frequency, extremely short-wavelength radiation, extending from submillimeter to perhaps even optical wavelengths, may be possible. For cold, high-current electron beams the dominant scattering process is stimulated Raman scattering of the pump wave from an idler beam mode. Utilizing this mechanism superradiant FEL oscillators have produced megawatt power levels with efficiencies as high as 0.1%. In addition a collective Raman FEL experiment utilizing a quasioptical cavity has demonstrated line narrowing of the backscattered radiation to $\sim 2\%$, compared to line widths of $\sim 10\%$ for the superradiant oscillator experiments.

Problems

6.1. (a) Show that Eq. (6.4) follows directly from the parapotential flow assumption
$\mathbf{E} + 1/c(\mathbf{v} \times \mathbf{B}) = 0$.

(b) Why would you expect this result?

6.2. Show that the electron space charge density of the Brillouin flow in a nonrelativistic magnetron diode is given by

$$n(r) = \frac{B^2}{8\pi mc^2}\left[1 + \left(\frac{r_c}{r}\right)^4\right]$$

Note that $\omega_p \sim \omega_c$.

6.3. Show that the admittance of a cylindrical side resonator of radius a and height L is given by

$$Y = \frac{iL}{2\pi a}\left[\frac{J_0(ka)}{J_0'(ka)} + 2\sum_{m=1}^{\infty}\left(\frac{\sin m\psi}{m\psi}\right)^2 \frac{J_m(ka)}{J_m'(ka)}\right]$$

where $k = \omega/c$.

6.4. Note that from Eq. (6.3) the rotation rate at the edge of the Brillouin cloud is given (nonrelativistically) by

$$\omega_- = \frac{\Omega_0}{2}\left[1 - \left(\frac{r_c}{r_b}\right)^2\right]$$

Since ω_- varies with r_b, it is not always equal to the angular velocity of the traveling rf field. This gives rise to the important magnetron effect termed "frequency pushing," in which the space charge spoke induces a leading out-of-phase component of rf current on the segments. Following the analysis presented in the text obtain the magnetron performance chart, including the frequency pushing characteristics.

6.5. Derive Eq. (6.79), assuming the cold beam distribution function of (6.78).

6.6. If the beam is given transverse energy by passing through a foil, then the distribution function is given by

$$g(u_\perp, u_z) = \frac{\delta(u_z - u_{z0})}{2\pi u_{z0}^2}\exp\left(\frac{u_\perp^2}{2u_{\perp 0}^2}\right)$$

In this case there is no longer a population inversion in the perpendicular direction, but there is still a temperature anisotropy. Although there can be unstable TM modes, show that the temperature anisotropy cannot produce TE mode instability.[31]

6.7. Show the intermediate steps leading to Eqs. (6.130) and (6.138).

6.8. Show that the saturated electric field amplitude at the time of electron trapping in the weak pump Raman regime is given by[60]

$$|\mathbf{E}|_{\text{sat}} = \frac{m_0 c^2}{e} k_0 \gamma_0^{1/2} \left(\frac{\omega_{b0}}{\gamma_0^{1/2} k_0 c} \right)^{3/2}$$

7

Collective Ion Acceleration with Linear Intense Relativistic Electron Beams

7.1. Introduction

Because of technological limitations on rf power sources and dielectric breakdown strength, the accelerating fields in conventional particle accelerators are generally restricted to 10^4–10^5 V/cm. As a result, acceleration concepts which employ the very large self-fields of an intense relativistic electron beam have received considerable attention in recent years. In these new schemes, the accelerating fields are not imposed externally, but arise from the collective action of a larger number of particles (electrons) on a smaller number of particles (positive ions). Since the accelerating fields are not limited by electrical breakdown, collective methods may eventually lead to the compact, economical acceleration of intense currents of light or heavy ions to hundreds of MeV per nucleon. Potential areas of application of such beams include controlled thermonuclear research, electronuclear breeding, basic nuclear physics, material studies, and radiation therapy.

Much of the original interest in the use of intense electron beams for collective ion acceleration began with the proposals of Veksler.[1] In 1956 several possible acceleration mechanisms were described including (a) an inverse Čerenkov drag, (b) a coherent impact of a small ion bunch with a larger collection of electrons, and (c) an external electromagnetic wave coupled to a quasineutral ion bunch. Although there were apparently no extensive investigations of these early ideas, Veksler and his coworkers continued their collective acceleration efforts introducing the electron ring accelerator (ERA) concept in the late 1960s.[2] In the ERA approach a localized space charge potential well is produced by a high-density ring of electrons. A smaller ion cluster is trapped in the potential well and accelerated by controlling the motion of the electron ring. Although considerable progress has been made in ERA development (see the reviews of Refs. 3–8), as the electron density increases, the ring becomes susceptible to instabilities, including the negative-mass instability, the ion–electron instability, and the resistive-wall instability. Estimates for the growth rates of these instabilities appear to limit the peak achievable accelerating fields to 5×10^5 V/cm.[2]

Probably the first successful demonstration of collective acceleration was achieved by Plyutto and coworkers during the 1960s.[9–14] The essential feature of these experiments was the extraction of an electron beam and an ion bunch from a relatively low-voltage plasma-filled diode. Although the collective nature of the acceleration process was recognized, no definitive theories of the process have been advanced. This work has been reviewed recently and will not be considered further.[14]

During the same time period the development of the pulse power technology by J. C. Martin and coworkers led to the production of high-current pulsed electron beams. Rabinovich had reviewed several possible approaches for collective acceleration[15] using these high-current beams in linear geometry when in 1968 Graybill and Uglum unexpectedly discovered collectively accelerated ions upon injection of an intense electron beam into a low-pressure neutral gas.[16] The Graybill and Uglum results were subsequently verified and extended,[17,18] with groups at several laboratories obtaining accelerating fields of the order of 1 MV/cm.[19–27] In addition to this work other groups have observed collectively accelerated ions using vacuum diodes with dielectric anodes.[28–33] Peak proton energies $\lesssim 20$ MeV using the neutral gas method,[34] and ~ 45 MeV for the vacuum diode geometry have been reported.[35,36] Unfortunately, in both types of experiments the effective distance over which the acceleration is observed has been limited to a maximum of a few tens of centimeters. As a result there has been great interest in determining the acceleration mechanism, the appropriate scaling laws, and the development of possible methods for controlling the acceleration processes.

In addition to this work several novel collective acceleration schemes have been conceived in which ions are trapped and accelerated in large-amplitude cyclotron[37] and space charge waves[38,39] supported by a relativistic electron beam. In these techniques the phase velocity of the appropriate "negative-energy" mode is caused to increase in a controlled fashion, thus accelerating bunches of ions while keeping them trapped in the potential wells associated with the wave. Although there are several practical difficulties, "proof-of-principle" experiments are now in progress.

In subsequent sections of this chapter we will present only a brief summary of the neutral gas and vacuum diode experiments. Although these experiments are historically important, several excellent reviews already exist[40-42]; moreover, the significance of this work lies not so much in the experimental results themselves, but rather in the theoretical suggestions for extending the large collective fields over useful distances. Our primary intention is to provide a survey of the several new proposals for collective ion acceleration, including where possible the results of the most recent experiments and computer simulations.

7.2. Summary of Results for the Neutral Gas and Vacuum Diode Systems

7.2.1. Collective Acceleration in a Neutral Gas-Filled Drift Tube

In the typical experiment an electron beam of 1–10 MeV, 10–100 kA, and 10–100 nsec pulse duration is injected into a chamber filled with low-pressure neutral gas (~ 100 mTorr of H_2, D_2, N_2, Ar, air, etc.). A schematic description of the experimental arrangement, including representative diagnostics, is given in Fig. 7.1. As might be expected, the collective acceleration process is intimately associated with the physics of beam propagation in the neutral gas-filled cavity (Chapter 5). Experimentally, collective ion acceleration in neutral gas systems is apparently *never*

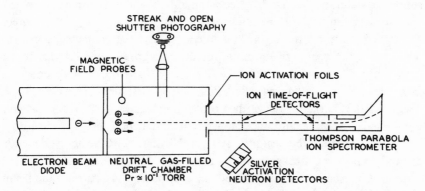

Figure 7.1. Schematic description of the neutral gas collective ion acceleration experiment, including representative diagnostics.

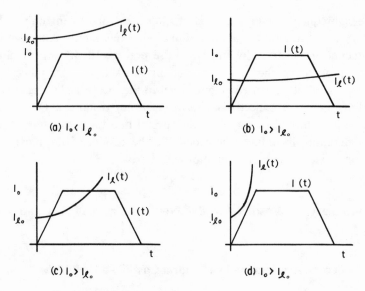

Figure 7.2. Idealized beam current pulse and space-charge-limiting current versus time for (a) the peak injected current I_0 less than the limiting current; (b) low-pressure gas ($\lesssim 10$ m Torr) —peak beam current exceeds the limiting current; (c) moderate gas pressure (100 m Torr– 500 m Torr)—peak beam current exceeds the limiting current; (d) high gas pressure ($\gtrsim 1$ Torr)—peak beam current never exceeds the space-charge-limiting current.

observed unless the peak beam current exceeds the so-called space-charge-limiting current,[23,24] a fact that was correctly predicted by Olson.[43,44]

To qualitatively analyze the initial stages of the process consider the simplified diagrams of Fig. 7.2.[24] Assuming the voltage rise time is zero, the space-charge-limiting current can be written as $I_l(t) = I_{l_0}[1 - f_e(t)]^{-1}$, where I_{l0} depends only on the beam kinetic energy and the geometry, and $f_e(t)$ depends on the degree of ionization of the background neutral gas. If the peak injected current does not exceed I_{l0} (Fig. 7.2a) the beam propagates from the anode region regardless of the gas pressure. (However, from Chapter 5 the characteristics of the propagating beam will obviously depend upon the degree of beam-induced ionization, and, hence, the gas pressure.)

If $I_0 > I_{l0}$, then the beam behavior in the vicinity of the anode depends critically upon the background fill pressure. During the time the rising portion of the beam current does not exceed the limiting current, the beam will propagate. Meanwhile, the background gas becomes ionized by electron impact ionization and ions are generated according to

$$\frac{\partial n_i(t)}{\partial t} = \frac{n_b(t)}{\tau_e} \tag{7.1}$$

where τ_e is the phenomenological electron ionization time, and n_b and n_i are the beam electron and ion densities. When the beam current exceeds the limiting current, a deep potential well (virtual cathode) forms in the vicinity of the anode and beam propagation ceases. Ions previously created by electron impact ionization are trapped in the well and can rapidly contribute to the ion density through ion impact ionization. The process can be thought of as ion avalanching with the ion density growing as

$$\frac{\partial n_i(t)}{\partial t} = \frac{n_b(t)}{\tau_e} + \frac{n_i(t)}{\tau_i} \tag{7.2}$$

where τ_i is the effective ion avalanche time. Useful estimates of τ_e and τ_i for hydrogen gas are given by[45]

$$\tau_e \approx 5[\, p(\text{Torr})]^{-1} \text{ nsec} \tag{7.3a}$$

$$\tau_i \approx 0.33[\, p(\text{Torr})]^{-1} \text{ nsec} \tag{7.3b}$$

In the low-pressure regime ($\lesssim 10$ mTorr), described by Fig. 7.2b, there is insufficient gas density for these processes to rapidly neutralize the beam space charge, and the major portion of the beam will not propagate. For somewhat higher pressures, as described by Fig. 7.2c, ionization of the gas (primarily by ion impact ionization) is rapid enough that the deep potential well collapses early in the pulse and a major portion of the beam can propagate. For still higher pressures ($\gtrsim 1$ Torr) the situation is described by Fig. 7.2d. Ionization of the gas due solely to electron impact ionization is rapid enough that the beam current never exceeds the space charge limiting current, i.e., $\tau_e \lesssim (I_{l0}/I_0)t_r$, where t_r is the current rise time. The deep potential well never forms, and the beam propagates immediately away from the anode.

From the previous discussion, if $I_0 < I_{l0}$ a deep well never forms in the vicinity of the anode, and beam electrons do not slow appreciably in the axial direction. If the transit time of the beam electrons through the drift tube is less than the characteristic charge neutralization time due to electron impact ionization alone (this is usually the case of experimental interest), then a potential minimum of magnitude $|e\phi| \ll (\gamma_0 - 1)mc^2$ forms near the geometric center of the drift space, and an axial electrostatic field of order

$$E_z \ll \frac{(\gamma_0 - 1)mc^2}{(L/2)} \tag{7.4}$$

exists to accelerate ions.

A reaction which competes with the ion impact ionization reaction discussed earlier is charge exchange. For example, the appropriate charge exchange reaction for the diatomic hydrogen gas molecule at room temperature is the nonresonant reaction

$$H^+ + H_2 \rightarrow H + H_2^+ \tag{7.5}$$

Olson[45] has noted that a crude, but useful, approximation is that for ion energies ε_i in excess of 50 keV, ion impact ionization is energetically possible and favored over charge exchange, while for $\varepsilon_i < 50$ keV, charge exchange dominates.

The important consequence of this result is that charge exchange can provide an effective barrier to ion acceleration if (roughly) the time required for a proton to be accelerated to 50 keV exceeds the characteristic time for charge exchange to occur. This criterion can be expressed as[45]

$$E(V)/p(\text{Torr}) \lesssim 10^6 \text{ V cm}^{-1} \text{ Torr}^{-1} \tag{7.6}$$

When Eq. (7.6) holds, charge exchange dominates and no accelerated ions should be seen. For 1-MeV electrons and a 1-m drift tube filled with H_2 at 100 mTorr, from Eq. (7.6), $E_z/p \ll 2 \times 10^5 \text{ V cm}^{-1} \text{ Torr}^{-1}$, and ion acceleration is effectively prohibited.

On the other hand, if the beam current exceeds the space charge limit, a virtual cathode forms in the vicinity of the anode and strong accelerating fields of order 10^6 V cm^{-1} exist. In this case charge exchange does not terminate the acceleration process and a large energetic ion population can result. Because of the oscillatory behavior of the virtual cathode a few ions can leak out and achieve high energy; however, most remain trapped in the vicinity of the anode and create a broad ion spectrum with a peak energy of $\lesssim 1.5$ times the electron kinetic energy ε_e and an average value $\sim \varepsilon_e$.

This condition exists until there is sufficient neutralization of the beam space charge to reduce the depth of the potential well below the injected electron beam kinetic energy. (Almost complete charge neutralization is required.) An estimate of this charge neutralization time τ_n is obtained by integrating Eq. (7.2) and setting $n_i = n_b$. For H_2 a useful approximate expression is[43]

$$\tau_n \simeq 3\tau_i \approx 1.0 \left[p(\text{Torr}) \right]^{-1} \text{nsec} \qquad (7.7)$$

When $t > \tau_n$ the deep potential well collapses, the beam propagates away from the anode region, and the trapped ions receive net energy. Since net acceleration occurs when the ions become untrapped the process is largely uncontrolled and does not appear capable of generating large numbers of high-energy ions.

Once beam propagation is established, there is a translating potential well of depth $\lesssim \varepsilon_e/e$ associated with the unneutralized space charge at the beam front. If some of the ions accelerated during the collapse of the virtual cathode were to become trapped in the translating well, the possibility exists for further ion acceleration and increased ion energies. In fact, in some of the early experiments fast ions ($\varepsilon_i > 3\varepsilon_e$) were observed to be moving at speeds equal to the beam front velocity, providing strong evidence for ion trapping in a traveling potential well.[19,20,22,46] The limits to which this process can be extended depend critically on the mechanisms which control the beam front velocity and the ion trapping efficiency.

During the virtual cathode stage a potential well of characteristic length $l \approx 2R$ forms and decays over the charge neutralization time τ_N; hence, the beam front could be considered to translate self-consistently with a front velocity of the order $\beta_f c \approx 2R/\tau_N$. However, if the background gas pressure is so low that $\beta_f c < \beta_{i0} c$ where $\beta_{i0} c$ corresponds to the velocity of the fastest ions created during the deep well stage, then it may be expected that the beam front velocity will be determined by these fast ions. In essence, the fast ions can extend the equilibrium well length such that the front propagates self-consistently with speed $\beta_{i0} c$. From these arguments the dependence of the characteristic beam front velocity $\beta_f c$ as a function of gas pressure is indicated schematically in Fig. 7.3.[43,44]

For electrons beams with $\nu/\gamma \lesssim 1$ this model gives good agreement with the experimental data; however, if $\nu/\gamma \gtrsim 1$ the beam front speed can be limited by a power balance argument.[43,44,47,48] For propagation of the

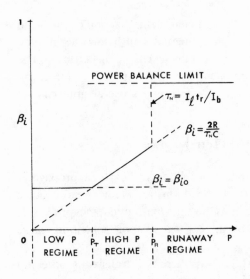

Figure 7.3. Variation of the characteristic beam front velocity as a function of gas pressure.

front the beam must supply power P_m to create the self-magnetic field, and power P_{se} to remove the secondary electrons for charge neutralization. In approximate terms

$$P_m \approx \beta_f c \int_0^R \frac{B_\theta^2}{8\pi} (2\pi r\, dr) = \beta_f \left(\frac{I_b^2}{4c} \right) \left(1 + 4\ln\frac{R}{r_b} \right) \qquad (7.8)$$

$$P_{se} \approx \kappa \varepsilon_e I_b / e \qquad (7.9)$$

where κ, a constant of order unity, denotes the average kinetic energy of a secondary electron in units of the primary electron energy. If the beam front moves at its self-consistent (constant) velocity, there is no beam power lost as the result of ion acceleration and the maximum beam front velocity is given by

$$\beta_{fm} \approx [\kappa + GX^{-1}]^{-1} \qquad (7.10)$$

where $G = [1/(4c)](1 + 4\ln R / r_b)$, and $X = (\varepsilon_e / eI_b)$. Although power balance must provide a limitation for maximum beam front velocities, attempts[47] to explain the results of neutral gas ion acceleration experiments solely in terms of a power balance model have resulted in unphysical best-fit parameters.[42]

The number of ions formed in the virtual cathode which can later become trapped in the translating well depends critically on the time-dependent ion velocity distribution, the final beam front velocity, and the details of the transition from the injection stage to the translating well stage. As might be expected, qualitative analyses of the trapping process indicate that the number of trapped ions should sharply decrease with increasing beam front velocity.[43]

On the basis of this discussion it is readily apparent that there are a number of serious limitations which restrict the practical application of collective acceleration achieved by simply injecting an intense electron beam into a low-pressure neutral gas. To summarize briefly: (i) ion acceleration does not occur unless the beam current exceeds the space charge limit; (ii) the number of ions trapped in the translating potential well associated with the beam front motion varies inversely with the front velocity; and (iii) there are several mechanisms which naturally limit the beam front velocity (fast ions produced during the deep well stage, characteristic charge neutralization time, and power balance), but none of these mechanisms offers a convenient means for controlling the motion of the beam front. Moreover, passive methods of beam front control, such as pressure gradients or variations in the guide tube radius, have been found to be ineffective.[49] It appears, therefore, that active means for controlling the beam front motion are necessary to extend the acceleration process.

7.2.2. Collective Acceleration in an Evacuated Drift Tube

In the typical experiment (Fig. 7.4), an intense electron beam is injected through a hole in the center of a dielectric anode into an evacuated drift tube. Since the anode is initially an insulator, the electrons emitted from the cathode at early times in the pulse experience strong radial electric fields as a result of both the accelerating structure and the beam self-fields. Hence, most of the electrons do not initially pass through the hole in the anode, but strike the insulator itself creating an anode plasma and charging the insulator until it eventually flashes over to ground. In essence, these processes lead to the condition of a grounded anode which shorts out the radial electric fields, and causes self-pinching in the diode and substantial beam transmission through the central hole. After the beam passes through the anode plasma the space charge of the beam creates an electrostatic field (of order ε_e/e if $I > I_l$) which accelerates ions from the anode plasma in the

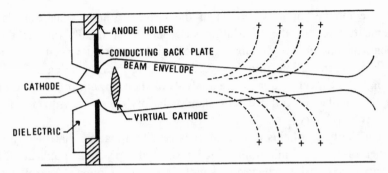

Figure 7.4. Schematic description of the evacuated drift tube (Luce diode) collective ion acceleration experiment.

direction of the electron beam. This represents an important distinction between these experiments and those of Plyutto in which the cathode plasma is the ion source. It is also important to note that since there is no background neutral gas, charge exchange processes cannot interfere with the accelerated ions. Hence, it is not necessary that the beam current exceed the space charge limit for ion acceleration to be observed; however, the stronger electric fields associated with virtual cathode formation should greatly enhance the acceleration mechanism.[31]

A wide variety of theories have been advanced to explain the ion acceleration results, and several are probably necessary to fully explain the detailed observations. For ions with kinetic energies on the order of the electron beam kinetic energy, an electrostatic mechanism appears to be the dominant mechanism,[31] although the important role of an inductive correction has been observed in numerical simulations of this geometry.[50] Careful measurements of the temporal dependence of the highest-energy ions indicate that peak kinetic energies are not observed until very late in the electron beam pulse.[33] This observation appears to clearly rule out a traveling potential well at the unneutralized beam front. The fact that strong rf emissions were also observed at times correlated with the high-energy ion acceleration is suggestive of a wave instability mechanism, with the electron–ion two-stream instability being a likely candidate.[33] Numerical simulations of this instability for a low ν/γ beam have yielded an exponential energy spectrum similar to that observed experimentally.[51]

Although rather high kinetic energy ions have been obtained using both the neutral gas and "Luce diode" configuration (up to 900-MeV Xe ions[52]), there does not appear to be a simple means of extending the large collective fields over distances of more than a few tens of centimeters in either case. As a result interest in these experiments has largely diminished, and collective acceleration efforts are now being focused on methods which appear to offer reasonable means for controlling the motion of a deep potential well or wells to permit ion acceleration over useful distances. These schemes can generally be classified as either wave methods or net electron space charge methods. Examples of the first class include the autoresonant accelerator (ARA) concept[37] (cyclotron mode) and the converging guide accelerator (CGA),[38] while the most significant of the second class is the ionization front accelerator (IFA) concept of Olson.[53] In subsequent sections these concepts are discussed in more detail.

7.3. The Ionization Front Accelerator (IFA)

The desired circumstance for collective acceleration is that in which ions become trapped in a deep potential well, and the motion of the well is then controlled to permit synchronized ion acceleration. In the ionization front accelerator concept the potential well associated with the beam front of an intense relativistic electron beam is controlled by the photoionization of a special working gas. The IFA was first proposed by C. L. Olson, and is based on his extensive analysis of the neutral gas experiments. The basic concept is illustrated in Fig. 7.5. A beam of constant voltage, γ_0, current, I_0, and radius, r_b, is injected into a drift tube of constant radius, R, filled with a low-pressure neutral gas. The condition for virtual cathode formation can be

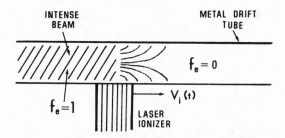

Figure 7.5. Schematic description of the ionization front accelerator (IFA).

written as[54]

$$I_0 > I_l(z, t) = \frac{\left(\gamma_0^{2/3} - 1\right)^{3/2}\left(mc^3/e\right)}{\left[1 + 2\ln(R/r_b)\right]\left(1 - f_e(z, t)\right)} \qquad (7.11)$$

which explicitly indicates the possibility for controlling the motion of a virtual cathode by controlling the axial and temporal variation of the charge neutralization fraction f_e. If the gas pressure is so low that beam-induced ionization is essentially negligible, then an external source, e.g., a laser, can be used to ionize the gas in a controlled fashion. Behind the ionization front $f_e \simeq 1$, while ahead of the front $f_e = 0$, and a virtual cathode exists. Ideally the beam will propagate through the ionized gas region in a charge neutral (but not current neutral) fashion, and then diverge ahead of the ionization front where there is no charge neutralization. A steep potential well of depth $(\gamma_0 - 1)mc^2/e$ and axial extent $\sim r_b$ is thus created and caused to move synchronously with the ionization front. Idealized numerical simulations in which f_e was specified as a function of position and time have, in fact, indicated the desired behavior.[55] Since the front essentially represents a moving ground plane (anode), the IFA is a direct extension of the naturally occurring acceleration process observed in the neutral gas experiments. With the IFA scheme it should be possible, in principle, to produce accelerating gradients of $\sim 10^6$ V/cm over interesting distances (1–10 m).

The current of the intense electron beam should be chosen such that $I < I_A$, where I_A is the Alfven limit, to permit propagation in the charge neutral state with no current neutralization. In addition, for reasonable well depths (1 MeV), the beam voltage would be greater than a few MeV.[53]

For appropriate control of the front motion it is necessary that the ionization source be powerful enough to neutralize the beam space charge in a time τ_N short compared to $r_b/\beta_{im}c$, i.e., the axial length scale of the potential well divided by the maximum desired accelerated ion velocity. The laser photon flux J (photons/cm^2 sec) required to photoionize the background gas on the time scale τ_N is given by

$$J \gtrsim \frac{n_b}{n_a \sigma \tau_N} = \frac{n_b}{n_a} \frac{\beta_{im} c}{\sigma r_b} \qquad (7.12)$$

where n_a is the atom density of the background gas and σ is the photoionization cross section. For example, choosing cesium vapor as the working gas and assuming single-step photoionization from the ground state ($\lambda \lesssim 3184$ Å, $\sigma \approx 0.2$ Mb) the required laser power for $n_b/n_a \sim 10^{-2}$ and 100-MeV protons is quite high (10^7–10^8 W), although the total laser energy is relatively small—a few joules.

The accelerated ion pulse produced by an IFA would be very short ($\Delta t_i \sim r_b/\beta_{im}c$) with a per particle ion kinetic energy of the order of $\varepsilon_i \approx Z\varepsilon_e(L/r_b)$, where Z is the ion charge and L is the effective acceleration length. An estimate for the number of ions in the pulse is $N_i = (4/3)\pi r_b^3 f_e n_b Z^{-1}$, where $f_e = n_i Z/n_b$ is the fractional neutralization of the electron beam space charge due to the collectively accelerated ions. The relative efficiency of the acceleration process $\varepsilon = $ (ion energy out)/(electron beam energy in) is

$$\varepsilon = \frac{N_i \varepsilon_i}{N_e \varepsilon_e} \sim f_e \frac{L}{\beta_e c \Delta t_b} \tag{7.13}$$

where Δt_b is the electron beam pulse duration required to accelerate the ions from an initial ion velocity $\beta_{i0}c$ to a final ion velocity $\beta_{im}c$; in the nonrelativistic limit

$$\Delta t_b = (\beta_{im} - \beta_{i0})c\left(\frac{M}{Ze}\right)E^{-1} \tag{7.14}$$

where $E \sim \varepsilon_0/er_b$ is the accelerating field strength. If $\beta_{im} \gg \beta_{i0}$, the $\Delta t_b \simeq 2L/\beta_{im}c$ and Eq. (7.13) becomes

$$\varepsilon \approx f_e \frac{\beta_{im}}{\beta_e} \tag{7.15}$$

Hence, the efficiency is directly proportional to both the ion loading and the final ion velocity.

These simple estimates of IFA performance depend on precise control of the beam front motion by the external ionization source. If there are other effects that limit the maximum beam front velocity, then the maximum ion kinetic energy and/or the efficiency will be less than optimum. Analyses of several limiting mechanisms (including power balance, axial momentum conservation, kinematic constraints, etc.) indicate that it may be possible to accelerate ions of atomic mass A to energies in excess of $100/A$ MeV at efficiencies of the order of 1%–10%.[42,56-58]

Table 7.1. A Summary of Design Parameters for the IFA I Experiment

Electron beam	Acceleration region	Proton pulse
0.6 MeV	50 MV/m	5 MeV
20 kA	1.2 cm diam	0.8 kA
0.5 cm radius	10 cm length	0.25 cm radius
10 nsec	6.5 nsec of	0.17 nsec
0.01 TW	e beam pulse	0.004 TW
0.1 kJ		0.7 J

An initial series of experiments has been performed by Olson to demonstrate the feasibility of the IFA mechanism.[59] A summary of the experimental design parameters is presented in Table 7.1. The photoionization scheme consisted of a two-step process in which a dye laser was first used to excite the cesium working gas. A frequency-doubled ruby laser was then used to photoionize the cesium from the excited state. The laser sweep was accomplished by injecting the dye laser into a programmed light pipe array. Although beset with system synchronization difficulties, rather precise laser control of the drift front velocity was demonstrated, and the data suggest that 5-MeV protons were achieved in an acceleration length of only 10 cm.

In addition to the previously mentioned synchronization difficulties, the experimental chamber had to be heated to $\sim 250°C$, to achieve the desired cesium operating pressure, which created a number of materials-related problems. Further, the use of the two-step photoionization process required that the frequency-doubled ruby laser be injected opposite the electron beam source. As a result the task of diagnosing the accelerated ion pulse was somewhat complicated. In an effort to alleviate these difficulties a search was made for an alternate working gas that would require only a single laser photoionization source. The present optimum choice appears to be dimethyl aniline (DMA), which can undergo multiphoton ionization using a Xe–Cl laser.[60] A second series of IFA experiments based on this scheme, plus the use of a more powerful, longer-pulse electron beam source is now underway.

7.4. Wave Collective Ion Acceleration Mechanisms

Collective acceleration mechanisms which rely on the normal modes of a relativistic electron-beam–magnetic-guide field-evacuated drift tube system have received considerable attention in recent years. To illustrate the underlying principles of these concepts we return to the dispersion relation derived in Section 3.7.1 for the rigid rotor equilibrium. In this case, and in general, the system eigenmodes can be identified as

$$\omega = k_z v_z \pm \left(\frac{k_z^2}{k^2} \frac{\omega_b^2}{\gamma_b^3} \right)^{1/2} \qquad \text{(space charge waves)} \qquad (7.16)$$

$$\omega = k_z v_z \pm \left(\Omega^2 / \gamma_b^2 \right)^{1/2} \qquad \text{(cyclotron waves)} \qquad (7.17)$$

Physically, the space charge modes correspond to longitudinal bunching of the beam space charge, while the cyclotron modes correspond to traveling constrictions. These waves and their associated electrostatic potential variations are shown in Fig. 7.6.

In order to accelerate ions using these system eigenmodes, ions must first be trapped in the potential troughs, and the velocity of the potential wells plus trapped ions must be systematically increased. Hence, the wave phase velocity of the eigenmodes should be variable from very small values up to near the electron beam velocity. Clearly, the fast waves do not satisfy this requirement; on the other hand, for the slow waves v_ϕ is given by

$$v_\phi \simeq v_z \left(\frac{\omega_0}{\omega_0 + \Omega / \gamma_b} \right) \qquad \text{(slow cyclotron wave)} \qquad (7.18)$$

$$v_\phi \simeq v_z \left(\frac{\omega_0}{\omega_0 + \omega_b / \gamma_b^{3/2}} \right) \qquad \text{(slow space charge wave)} \qquad (7.19)$$

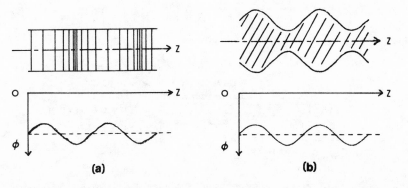

Figure 7.6. Physical representation of the space charge and cyclotron waves, and attendant potential variations. (a) Space charge wave; (b) cyclotron wave.

For the case of the slow cyclotron wave the wave phase velocity can be varied from nearly zero to essentially the beam velocity by simply decreasing the external magnetic field strength. This is the basis of the autoresonant accelerator concept (ARA) of Sloan and Drummond.[37] For the slow space charge waves v_ϕ can be similarly varied either by increasing the beam kinetic energy or by decreasing the beam plasma frequency.[38,39] Increasing the beam kinetic energy by increasing the ratio of the beam radius to the guide tube radius has given rise to the converging guide accelerator (CGA) concept of Sprangle, Drobot, and Manheimer.[38]

7.4.1. The Autoresonant Accelerator (ARA)

A schematic diagram of the ARA is shown in Fig. 7.7. An intense relativistic electron beam propagates in an evacuated drift tube; radial confinement is provided by a longitudinal magnetic field. The slow cyclotron wave is excited on the beam at frequency ω_0 and is grown to sufficient amplitude that ions can be trapped in the traveling potential wells. A qualitative ion trapping criterion is that[61]

$$\tfrac{1}{2}m_i v_0^2 < |e\phi_0| \tag{7.20}$$

where $|e\phi_0|$ is the magnitude of the wave potential at the position of ion injection and v_0 is the minimum wave phase velocity. If $|e\phi_0| \sim 100$ keV, then v_0 for protons must be less than approximately 4×10^8 cm/sec.

To maintain synchronism between the motion of the accelerated ions and the cyclotron wave, the wave phase velocity is required to vary according to

$$v_\phi = v_0 + at \tag{7.21}$$

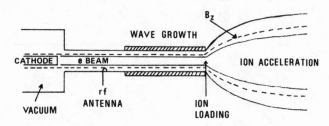

Figure 7.7. Schematic description of the autoresonant accelerator (ARA).

where

$$a = \frac{eE}{m_i} \approx \frac{e}{m_i} \frac{\phi_0}{(\lambda/4)}$$

and λ, the cyclotron wavelength, is given by

$$\lambda = \frac{2\pi}{k_z} = \frac{2\pi v_\phi}{\omega_0} \tag{7.22}$$

Equation (7.22) indicates an important result for all wave collective acceleration schemes. If the wave exciter is operated at constant frequency ω_0, then the accelerating electric field must decrease with length down the accelerator, i.e., as the wave phase velocity increases.

The location of a particular potential well will be given (nonrelativistically) by $z \simeq \frac{1}{2}at^2$, so that

$$v_\phi = v_0 + (2az)^{1/2} \simeq v_0 + 2\left(\frac{e\phi_0\omega_0 z}{\pi m_i v_\phi}\right)^{1/2} \tag{7.23}$$

or

$$v_\phi(v_\phi - v_0)^2 = \frac{4e\phi_0\omega_0 z}{\pi m_i} \tag{7.24}$$

From Eq. (7.18) the magnetic field decrease is given by

$$\Omega(z) = \omega_0\left(\frac{1 - v_\phi/c}{v_\phi/c}\right) \tag{7.25}$$

Early in the ion acceleration stage (z small),[62,63]

$$v_\phi \simeq v_0\left[1 + 2\left(\frac{e\phi_0\omega_0 z}{\pi m_i v_0^3}\right)^{1/2}\right] \tag{7.26}$$

and the required variation in magnetic field strength is approximately given by

$$\frac{\Omega(z)}{\Omega_0} \simeq \frac{v_0}{v_\phi} \simeq 1 + 2\left(\frac{e\phi_0\omega_0 z}{\pi m_i v_0^3}\right)^{1/2} \tag{7.27}$$

Late in the acceleration (z large, $v_\phi \gg v_0$)[62,63]

$$v_\phi \simeq \left(\frac{4e\phi_0\omega_0 z}{\pi m_i} \right)^{1/3} \tag{7.28}$$

and

$$\frac{\Omega(z)}{\Omega_0} = \left(\frac{\pi m_i v_0^3}{4e\phi_0\omega_0 z} \right)^{1/3} \left[1 - \left(\frac{4e\phi_0\omega_0 z}{\pi m_i c^3} \right)^{1/3} \right] \tag{7.29}$$

Graphs of the effective electric field, phase velocity, magnetic field variation, and ion kinetic energy ($T_i \simeq \frac{1}{2} m_i v_\phi^2$) are presented in Figs. 7.8a–7.8d.

An approximate bound on the maximum ion current can be obtained by requiring that the ion charge density be substantially less than the electron beam density, i.e.,[61]

$$\frac{n_i}{n_b} \sim \frac{I_i}{I_e} \cdot \frac{c}{v_i} \ll 1 \tag{7.30}$$

Although the above analysis is somewhat oversimplified (changes in the potential well depth and inductive corrections were ignored) it does indicate the general features of the ARA concept, and also illustrates the desirability of nonlinear operation. For example, a larger potential well depth at the entrance to the acceleration stages permits a higher minimum wave phase velocity, and increases the accelerating electric field. As a result a very detailed linear theory of cyclotron wave acceleration has been developed,[64,65] and numerous nonlinear saturation phenomena have been studied both analytically and computationally.[66–69]

At the present time the status of these complex issues has not been resolved. By way of illustration, a principal conclusion of the detailed linear theory was that radial nonuniformity in the beam kinetic energy due to beam space charge effects would change the character of the cyclotron eigenmode from a body mode to a surface mode for which the electron perturbations are localized near the beam surface.[64,65] Consequently, the potential well amplitude corresponding to the maximum radial perturbation of the beam electrons was predicted to decrease substantially.

However, the results of further investigations indicate that such surface behavior of the linear cyclotron eigenmode may be significantly modified by

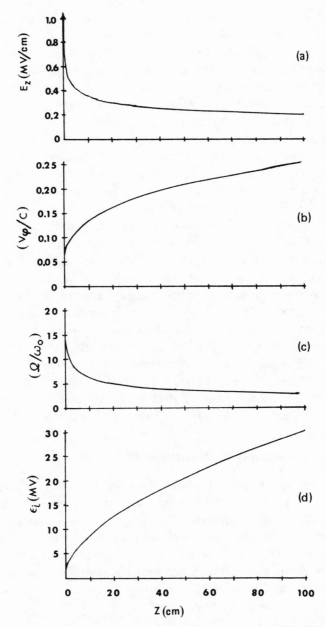

Figure 7.8. Approximate axial variation of (a) the effective electric field, (b) the wave phase velocity, (c) the external magnetic field strength, and (d) the ion kinetic energy in the autoresonant accelerator.

nonlinearities which alter the resonance response of the electrons.[66] In particular, the nonlinear decrease in the beam velocity due to finite perpendicular velocity perturbations can shift the electrons away from cyclotron resonance and reduce the surface character of the perturbations, thereby increasing the available potential well depth for a given maximum radial perturbation. Confirmation of such sensitive and important results must await the completion of further particle simulation studies performed in conjunction with careful laboratory experiments.

A "proof-of-principle" experimental study of the autoresonant accelerator concept is now in progress. To date the work has concentrated on producing a very cold, high-current (10–20-kA) relativistic (2.25 MV) electron beam and exciting a large-amplitude, low-phase-velocity, axisymmetric cyclotron wave.[70,71] The electron beam is produced in a foilless diode arrangement immersed in a uniform axial magnetic field. The cyclotron wave is launched by a high-Q half-wavelength resonant helical antenna structure driven in an axisymmetric mode by a 100-kW external oscillator. In order to suppress unstable high-frequency asymmetric modes the undriven end of the antenna is loaded with a lossy dielectric material. Using this configuration large-amplitude cyclotron waves have been excited, and wave potentials in excess of 120 kV (zero to peak) have been inferred, corresponding to on-axis accelerating fields of approximately 10 MV/m.[71] In addition, some control of the wave phase velocity has been demonstrated by spatially varying the magnetic field strength. However, the important tasks of ion injection and acceleration have yet to be performed.

7.4.2. The Converging Guide Accelerator (CGA)

A schematic diagram of the CGA is shown in Fig. 7.9. The slow space charge waves can be excited in an intense relativistic electron beam by a variety of slow wave amplifiers. Ion acceleration is achieved by first trapping ions in the electrostatic wells of the space charge wave, and then increasing the wave phase velocity by propagating the beam through a drift tube of decreasing radius along a uniform axial magnetic field. As a simple example consider the thin hollow beam of Section 3.3.1 for which the electrostatic potential is given by

$$\phi(r) = -\frac{2I}{v_z} \ln \frac{R}{r}, \qquad r_b \leq r \leq R \qquad (7.31)$$

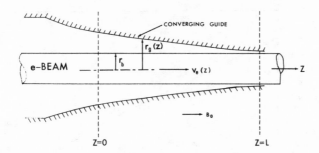

Figure 7.9. Schematic diagram of the converging guide accelerator (CGA).

where I is the total beam current carried in the annulus. Conservation of energy implies that $\gamma = \gamma_0 + e\phi/mc^2$, where $(\gamma_0 - 1)mc^2$ corresponds to the injected beam kinetic energy. If $R = R(z)$ is a decreasing function, then $\gamma(z)$ is increasing. Thus, the phase velocity of the slow space charge wave increases, and the trapped ions are accelerated.

The most serious practical problem with the CGA is to achieve low wave phase velocities for the low-energy stage of the acceleration process. As shown in Chapter 3 for the case of the one-dimensional drift space, the phase velocity of space charge waves does decrease to zero at the space charge limit, but with infinite slope (Fig. 7.10). This result has been derived for the case of a thin annular beam by Briggs (Ref. 72, Problem 7.5), and later generalized to arbitrary beam radial profiles by Godfrey.[73]

Since the phase velocity is so sensitive to beam current, slight diode voltage variations could cause the beam current to exceed the space charge limit, and attempts to grow a large-amplitude space charge wave could result in virtual cathode formation.[74] Because of these difficulties, ion trapping must occur at wave phase velocities of 0.15–$0.2c$, unless nonlinear large-amplitude wave effects provide further reduction of the phase velocity.[75] As a result, experiments to examine ion acceleration using the CGA concept have concentrated on the development of a high-energy ($\gtrsim 20$ MeV) proton injector and measurements of the wave phase velocity for beam currents near the limiting current.[76–80]

To summarize the CGA experimental results to date, a slow wave structure consisting of a disk-loaded waveguide has been successfully used to grow large-amplitude space charge waves on a relativistic electron beam.[80] In this type of structure wave growth occurs as a result of the interaction

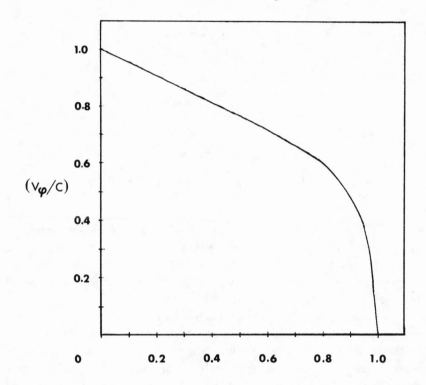

Figure 7.10. Variation of the phase velocity of the slow space charge wave versus the ratio of the beam current to space-charge-limiting current.

between the negative-energy slow space charge wave and the positive-energy TM electromagnetic modes of the system. Some measure of growth rate control was achieved by the use of resistive sheets in the disk-loaded cavities to dissipate electromagnetic energy. Further, variation of the wave phase velocity with I/I_l has been shown to compare well with linear theory estimates. Wave phase velocities as low as 0.2–$0.3c$ have been achieved, and the wave phase velocity was observed to vary with changes in the radius of the drift tube wall,[79] in an attempt to further decrease the phase velocity.

Although the most suitable ion injector system would probably be an induction linac, experiments to date have used a Luce diode as a source of protons for injection into a demonstration wave accelerator.[78] Using a double-diode configuration, high-energy ions from a Luce diode have been propagated through a second, lower-voltage electron beam diode, at the same time maintaining growth of a coherent space charge wave on the electron beam.

In summary, work on the CGA collective accelerator is at a relatively early stage. No ions have been trapped or accelerated, and speculations concerning achievable ion energies and currents must be regarded as premature. Substantial experimental and theoretical efforts will be necessary to understand such complex issues as ion loading effects, wave amplitude variations, the effects of beam temperature or voltage and current fluctuations, etc. in addition to evaluating new suggestions, such as the use of a frequency-dependent dielectric constant liner to effectively lower the phase velocity of the slow space charge wave.[74]

7.5. Summary

In conventional linear accelerators the accelerating electric fields are typically limited to values in the 1–10 MV/m range because of technological difficulties, e.g., rf power limits and dielectric breakdown strength. The promise of collective effect methods of acceleration is to substantially increase the effective accelerating gradient by using the very large self-fields associated with the intense space charge of a high-current, relativistic electron beam. In fact, collective accelerating fields of order 10^6 V/cm have already been demonstrated in neutral gas and vacuum diode beam injection experiments. Unfortunately, the effective accelerating length is limited to at most a few tens of centimeters; the processes are largely uncontrolled and do not appear capable of generating large numbers of high-energy ions.

More desirable is the situation in which ions become trapped in a deep potential well whose motion is then controlled to permit acceleration over substantial distances. In recent years several schemes have been proposed to achieve the desired effect. They can be generally classified as single potential well methods or as beam wave methods. The most important example of the first class is the ionization front accelerator (IFA), in which the motion of the potential well at the beam front is controlled by laser ionization of a background working gas. Examples of the second class include the autoresonant accelerator (ARA) and the converging guide accelerator (CGA). In the

Table 7.2. Comparison of the IFA, ARA, and CGA Collective Acceleration Methods

	Well depth	Average gradient	Ion current	Ion pulse waveform
IFA (beam front)	$(\gamma_0 - 1)mc^2$	$(\gamma_0 - 1)mc^2/r_b$	$I_i \sim 0.1 I_e \beta_{i\,max}$	Single pulse $\tau_p < 1$ nsec
ARA (cyclotron wave)	$\lesssim 0.3(\gamma_0 - 1)mc^2$	$\dfrac{\frac{1}{2}m_i(v_\phi)^2_{max}}{\int^{t_{max}}(v\phi)\,dt}$	$I_i < 0.1 I_e \beta_{i\,max}$	Continuous pulse train
CGA (space charge wave)	$\lesssim 0.3(\gamma_0 - 1)mc^2$	$\dfrac{\frac{1}{2}m_i(v_\phi)^2_{max}}{\int^{t_{max}}(v_\phi)\,dt}$	$I_i < 0.1 I_e \beta_{i\,max}$	Continuous pulse train

latter class large-amplitude waves are grown on an intense electron beam, and ions trapped in the potential troughs of the waves are accelerated by increasing the wave phase velocity.

Although a conclusive experimental demonstration of collectively accelerated ions has not been performed for any of these methods, "proof-of-principle" experiments have made considerable progress in achieving deep potential wells and controlling their motion. Recognizing this relatively early stage of development a tentative comparison of the three concepts is presented in Table 7.2. The primary advantage of the IFA is the high accelerating field associated with the deep potential well at the beam front. In contrast the maximum well depth available in the wave schemes is limited to some fraction of the electron beam kinetic energy. In addition, the effective accelerating gradient of the wave schemes decreases with accelerator length because the increase in wave phase velocity is achieved by increasing the wavelength. The output ion pulse of the IFA would consist of a single subnanosecond, high-power pulse. In contrast the wave methods would produce a continuous train of ion pulses throughout a major portion of the electron beam pulse.

Problems

7.1. Compute the time-dependent space-charge-limiting current, $I_l(t)/I_{l0} = [1 - f_e(t)]^{-1}$, for a background hydrogen pressure of (a) 10 mTorr; (b) 100 mTorr; (c) 1 Torr.

7.2. What are the consequences of a 10% beam voltage or current variation in the operation of the IFA?

7.3. Derive Eqs. (7.18) and (7.19).

7.4. The ARA (and CGA) has a high-frequency voltage ripple requirement for maintaining phase stability, i.e., a constant phase relation between the accelerated ions and their position in the potential wells of the cyclotron wave. Show that this ripple requirement scales inversely with the desired ion kinetic energy W_i approximately as

$$\left(\frac{\Delta\gamma}{\gamma}\right) \lesssim \frac{\pi}{2}\frac{\phi_0}{W_i}$$

where ϕ_0 is the potential well depth.

7.5. (a) Show that the electrostatic dispersion relation for space charge waves on a thin hollow beam of radius r_b propagating in a drift tube or radius R is given (in the $kR \ll 1$ limit) by

$$(\omega - kv)^2 = (k^2c^2 - \omega^2)I/\beta\gamma^3 I_0$$

where I is the beam current and $I_0 = (mc^3/e)/[2\ln(R/r_b)]$.

(b) Using this dispersion relation show that the phase velocity, $v_\phi = \omega/k$, tends to zero as the beam current approaches the space charge limit I_l according to

$$(v_\phi/c)^2 \simeq \tfrac{3}{2}(\gamma_0^{2/3} - 1)(I_l - I)/I_l$$

where $(\gamma_0 - 1)mc^2$ is the beam kinetic energy at injection.

8

Particle Beam Fusion Concepts

8.1. Introduction

In nuclear fusion energetic atomic nuclei of the light elements overcome the nuclear Coulomb barrier and merge to form heavier nuclei. In the process, one or more atomic particles are usually emitted; however, the mass of the resulting nuclei and the atomic particles is somewhat less than the mass of the original colliding nuclei, corresponding to a net energy release. Achieving a large number of fusion reactions to liberate substantial amounts of energy requires three conditions: (1) very high temperatures ($\sim 10^8$ °C) in order to create high ion thermal velocities, (2) a high density (n_i) of energetic ions to increase the collision probability, and (3) sufficiently long confinement times (τ_c) of the densely packed high-temperature nuclei to enable the fusion reaction to occur and sustain itself [n_i (No./cm^3) $\times \tau_c$ (sec) $\gtrsim 10^{14}$ sec/cm^3]. While such conditions exist in the stars owing to the tremendous gravitational forces and huge amounts of hydrogen, after three decades of effort earthbound fusion ignition has only been achieved in the hydrogen bomb by using an atomic bomb trigger.

Efforts to produce controlled thermonuclear fusion reactions have centered on magnetic confinement schemes and inertial confinement schemes. In magnetic confinement the charged particles of a hot deuterium–tritium (D–T) plasma are confined within a reactor vessel by strong magnetic fields. Because of practical limits to achieveable field strengths the plasma must be maintained at a fairly low density, and confinement times of the order of seconds to minutes are required for substantial fuel burnup. In the inertial confinement approach a small D–T pellet is rapidly heated and compressed to extreme densities by very high-power laser or particle beams. Although the compressed fuel reacts very rapidly, it is restrained by

293

its own inertia and substantial burning can occur before the pellet disassembles.

Application of intense particle beams to magnetic confinement systems include field reversal experiments using electron and ion rings, plasma heating by relativistic electron beams in solenoidal magnetic fields, and neutral beam heating of plasmas in toroidal devices, such as tokamaks.[1] Although such applications are important, in this chapter we will restrict our discussion to beam applications for inertial confinement systems.

Inertial fusion research directed toward scientific breakeven must solve two critical problems:[2]

(1) pellet design based on driver coupling

(2) high-energy driver technology

A clear understanding of the driver coupling to the fuel pellet, including energy absorption, compression, ignition, and burn is obviously necessary. In many ways lasers are ideal for irradiating such pellets; they generate very short pulses which can be easily focused onto the pellet surface by mirrors.[3] However, lasers tend to be expensive and relatively inefficient ($\sim 0.2\%$ for neodymium glass to $\lesssim 5\%$ for carbon dioxide lasers). Moreover, there is considerable debate concerning the ideal wavelength for a fusion laser, with much effort being directed toward the development of efficient short-wavelength lasers for better coupling to the pellet.

Interest in electron and light- and heavy-ion fusion drivers has arisen primarily because of their favorable cost and efficiency compared with lasers.[4] In this case the drivers can yield high total energy at relatively high efficiency ($\sim 30\%$). However, there are major problems to be solved regarding beam transport and focusing onto the target. After first considering some general pellet implosion criteria, in the remainder of this chapter we will examine in some detail the beam transport problems associated with the various particle beam approaches to inertial confinement fusion.

8.2. Pellet Implosion Criteria

The thermonuclear fuel which is most easily raised to fusion temperatures is a mixture of the hydrogen isotopes deuterium and tritium. The reaction is described by

$$D + T \rightarrow n + \alpha \tag{8.1}$$

A total energy of 17.6 MeV is liberated as kinetic energy of the neutron (14 MeV) and the alpha particle (3.6 MeV). The nuclear cross section has a resonance near zero energy, including the effect of the Coulomb barrier, and it has a peak of ~ 5 b at 125 keV (1 keV $= 1.16 \times 10^7$ °K). For a Maxwellian ion temperature of a few kilovolts, the reaction rate is appreciable.[5] In contrast the combined D–D reactions have a reaction rate about a factor of 30 lower at the same temperatures.

In inertial fusion it is conventional to replace the confinement time τ_c by the radial distance R traversed at sound speed v_s, and the ion number density n_i by the mass density ρ. Hence, the quantity (ρR) is often used as the inertial fusion figure-of-merit. At extremely high temperatures the sound speed is essentially given by the ion thermal speed $v_{\text{i.th.}} = (kT_i/m_i)^{1/2}$. In order to account for the expansion of a pellet of initial radius R during the time $R/v_{\text{i.th.}}$, and the fact that most of the mass is near the outside of the pellet in spherical geometry, it is customary to divide the pellet disassembly time by a factor of 4. Hence, the confinement time at an effective temperature of 10 keV is taken to be

$$\tau_c \simeq R/4v_{\text{i.th.}} \approx 0.3 \times 10^{-8} R \qquad (8.2)$$

In this case the Lawson condition that $n_i \tau_c \gtrsim 10^{14}$ sec/cm^3 becomes

$$\rho R \approx 4 m_i v_{\text{i.th.}}(n_i \tau_c) \simeq 0.15 \text{ g/cm}^2 \qquad (8.3)$$

for an average D–T ion mass of 4×10^{-24} g.

The efficiency of the thermonuclear burn can be estimated by calculating the burnup fraction, f_b, defined by

$$f_b(t) = 1 - n(t)/n_0 \qquad (8.4)$$

where n_0 is the fuel number density at the start of the burn. Differentiating Eq. (8.4), the rate of change of fuel density is given by

$$\frac{dn}{dt} = -n_0 \frac{df_b}{dt} \qquad (8.5)$$

The D–T reaction rate is defined by $R(t) = n_D n_T (\sigma v) = \frac{1}{4} n^2 \overline{(\sigma v)}$, where $\overline{(\sigma v)}$ denotes the product of the cross section σ and the relative velocity v of the reacting particles averaged over the Maxwellian ion temperature distributions. For D–T fuel, a numerical fit to the data is[6]

$$\left(\overline{\sigma v}\right)_{\mathrm{DT}} = 3.8 \times 10^{-12} T_i^{-2/3} \exp\left(-19.02 T_i^{-1/3}\right)(\mathrm{cm^3/sec}), \qquad T_i < 10 \text{ keV}$$

$$= 3.41 \times 10^{-14} T_i^{-2/3} \exp\left(-27.217 T_i^{-2/3} + 3.638 T_i^{-1/3}\right)(\mathrm{cm^3/sec}),$$

$$T_i > 10 \text{ keV} \qquad (8.6)$$

Since $R(t) = -(dn_{\mathrm{D}}/dt) = -(dn_{\mathrm{T}}/dt) = -\frac{1}{2}(dn/dt)$, we have

$$-\frac{dn}{dt} = \frac{1}{2} n^2 \left(\overline{\sigma v}\right)$$

or

$$\frac{df_b}{dt} = \frac{1}{2\tau_r}(1 - f_b)^2 \qquad (8.7)$$

where $\tau_r = [n_0(\overline{\sigma v})]^{-1}$ is the effective reaction time. Taking an average value of $(\overline{\sigma v}) \simeq 5 \times 10^{-16}$ cm^3/sec, $\tau_r \simeq 0.8 \times 10^{-8} \rho_0^{-1}$, where ρ_0 is in g/cm^3.

Assuming that τ_r is approximately constant, the solution of Eq. (8.7) is

$$f_b = (1 + 2\tau_r/t)^{-1} \qquad (8.8)$$

Evaluating f_b at the confinement time τ_c (equal to the fuel disassembly time)

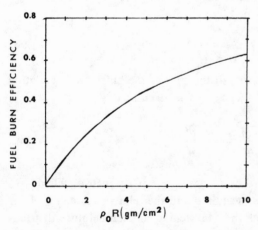

Figure 8.1. The variation of the fuel burn efficiency as a function of $(\rho_0 R)$.

yields[7]

$$f_b(\tau_c) \approx \frac{\rho_0 R}{6 + \rho_0 R} \qquad (8.9)$$

The fuel burn efficiency is graphed versus $\rho_0 R$ in Fig. 8.1.

For solid density D–T, $\rho_0 \simeq 0.213$ g/cm^3, the pellet radius required to achieve 10% burnup ($\rho_0 R \sim 0.67$ g/cm^2) is $R \sim 3.15$ cm, and the total amount of energy required to raise this fuel mass to an ignition temperature of 10 keV is approximately

$$E_{in}(J) = 3(\rho_0/m_i)(kT/2)(4\pi R^3/3)(1.6 \times 10^{-10} \text{ J/eV})$$

$$\approx 4 \times 10^{10} \text{ J} \qquad (8.10)$$

The controlled production of such large amounts of energy is currently impractical. However, since $\rho_0 R \sim R^{-2}$ and $E_{in} \sim \rho_0 R^3$, spherical compression of the pellet can dramatically lower the required input energy. For example, if the pellet is compressed such that the mass density $\rho_0 \simeq 1000$ g/cm^3, then the compressed pellet radius for 40% burn is 4×10^{-3} cm, the total full mass is reduced to about 250 μg, and the energy required to raise the fuel temperature to 10 keV is 3.2×10^5 J.

In addition to simple pellet compression two other effects can further decrease the driver energy requirement. First, for sufficiently large $\rho_0 R$ substantial self-heating can occur due to α-particle energy deposition. The α-particle range is determined by energy loss to electrons and to ions, although the loss to electrons is dominant for electron temperatures below approximately 40 keV. For electron temperatures between 4 and 20 keV, a good approximate expression for the α-particle range is[8]

$$\rho_0 R_\alpha = \frac{(2.6 \times 10^{-6}) T_e^{5/4}}{1 + (1.45 \times 10^{-6}) T_e^{5/4}} \left[1 + 0.31 \log(\rho_0/\rho_s) \right] \text{ g/cm}^2 \quad (8.11)$$

where ρ_s is the density of solid D–T. At $T_e = 10$ keV and $\rho_0 = 10^3$ g/cm^3, then $\rho_0 R_\alpha \sim 0.5$ g/cm^2.

If the driver energy is not deposited uniformly throughout the pellet a second important effect occurs. In this case spherically converging shock waves produce a sufficiently high temperature to ignite a central D–T core. The fusion α particles then efficiently heat the surrounding cold fuel leading to the formation of a spherically expanding "burn wave" which can result in almost compete fuel ignition.[3]

To minimize energy requirements only central ignition should occur while the remainder of the pellet is highly compressed at the lowest possible temperature. High compression ratios are desirable to maximize the reaction rate after ignition has occurred; however, excessive compression is inefficient because the fusion energy yield is eventually limited by fuel depletion. Moreover, very high compressions require that excessive work be done against the degeneracy pressure of the electrons. For example, at densities of 1000 g/cm³, the electrons are Fermi degenerate; i.e., the spatial volume available to the electrons is so small that the Pauli exclusion principle forces them to occupy high momentum states. In this case the average electron kinetic energy exceeds the electron temperature, and the electrons exert a pressure in excess of their thermal pressure given by[9,10]

$$P_F \approx \frac{2}{3} n_e \varepsilon_F \left[\frac{3}{5} + \frac{\pi^2}{4} \left(\frac{kT}{\varepsilon_F} \right)^2 \cdots \right] \qquad (8.12)$$

where

$$\varepsilon_F = \frac{h^2}{8m} \left(\frac{3n_e}{\pi} \right)^{2/3}$$

is the Fermi energy, and h is Planck's constant.

8.2.1. Charged Particle Energy Deposition

On the basis of the general remarks of the preceding section it is desirable to highly compress the pellet, but in such a manner that only the central core is raised to ignition temperature. The important problems which must be considered include generation of the driving pressure and the hydrodynamics of the resulting converging shocks. The detailed behavior of both effects requires a correct description of the charged particle energy deposition processes.

8.2.1.1. Electron Beam Energy Deposition. As energetic electrons traverse matter, they undergo many collisions with atomic electrons and nuclei, for each of which there are many possible energy losses and angular changes. For electrons with energies below ~10 MeV, the most significant interactions for predicting the resulting spatial, angular, and energy distributions are elastic nuclear (Coulomb) scattering, inelastic electron scattering in both individual collisions and collective (plasmon) interactions, and radiative (bremsstrahlung) interactions with both nuclei and atomic electrons. The nuclear Coulomb scattering cross section is very large and strongly peaked in the forward direction. Since the mass of the target nucleus is very much greater than the mass of the electron, the energy loss suffered by the electron is negligible, and these collisions may be considered as simple elastic scattering.

The essential features of the interaction between an energetic incident electron and an electron "free" of binding forces have been given in the classical treatments of Thomson and Bohr.[11] The calculation assumes that only a relatively small fraction of the incident electron energy is transferred in a single collision and that the incident electron direction does not change appreciably. Defining the cross section per free electron in terms of the impact parameter, b, as $d\sigma = 2\pi b\, db$, the calculation yields

$$\frac{d\sigma}{d(\Delta E)} = \frac{\pi e^4}{E}(\Delta E)^{-2} \tag{8.13}$$

When an incident electron travels a distance dx, the energy loss to all electrons with impact parameters between b and $b + db$ is given by $e^4/Eb^2(2\pi b\, db)\cdot(NZ)\cdot dx$, where N is the atomic density and Z is the atomic number. The total energy loss per unit length, termed the stopping power, is calculated from

$$-\frac{dE}{dx} = 2\pi NZ\frac{e^4}{E}\int\frac{db}{b} \tag{8.14}$$

In matter the atomic electrons are not free, but are bound in various atomic states. On the basis of a statistical model of the atom it is possible to define an average ionization potential I_{av} such that

$$-\frac{dE}{dx} \approx 2\pi NZ\frac{e^4}{4}\ln\frac{2E}{I_{av}} \tag{8.15}$$

I_{av} is approximately proportional to the atomic number of the material, with the constant of proportionality being of the order of the Rydberg energy, 13.5 eV.[12]

As the intense electron beam heats the pellet the material will ionize producing a number of plasma electrons which can participate in the slowing down process. In particular, collective mode effects can give rise to quantized energy losses which depend upon the free electron density and plasma dielectric properties.[13] Depending on the degree of ionization there are several possibilities. If the material is only weakly ionized then the inner atomic shell transition will remain unchanged, and the energy loss per interaction for the plasma oscillations will be much smaller in general than losses due to electron–electron interactions. However, if the material is almost completely ionized, then the quantity $(2E/I_{av})$, in the logarithm of Eq. (8.15) must be replaced by $(v/\lambda_D\omega_p)$, where v is the incident electron velocity, ω_p is the plasma electron frequency, and λ_D is the Debye length. When coupled with simple binary collision theory within the Debye sphere, the collective plasma wave excitation theory is normally considered to give a sufficiently accurate description of the plasma electron stopping power.

A complete treatment of electron energy losses due to bremsstrahlung has been given by Bethe and Heitler,[14] employing Dirac's equation for the electron and the Born approximation. The radiation probability depends on the effective distance from the electron to the nucleus because of the screening effect of the atomic electrons. For sufficiently high energies, E, of the incident particle, the screening can be considered complete for all photon frequencies, ν, whereas for low incident electron energies and high photon frequencies the screening may be neglected. For the latter case the cross section is given as

$$\sigma(E,\nu) = \frac{4Z^2}{137\nu}\left(\frac{e^2}{mc^2}\right)^2\left[1+\left(\frac{E}{E_0}\right)^2 - \frac{2}{3}\left(\frac{E}{E_0}\right)\right]$$

$$\times\left[\log\frac{2E_0E}{mc^2h\nu} - \frac{1}{2}\right] \tag{8.16}$$

where $E_0 = 0.511$ MeV.

In analogy with inelastic electron interactions, it is possible to develop a stopping power formula for the case of bremsstrahlung. Since the number of photons produced with frequency in the range ν to $\nu + d\nu$ when an electron of energy E traverses a thickness dx of material is $N\sigma(E,\nu)\,d\nu\,dx$, the energy loss by radiation per unit path is given by

$$-\left(\frac{dE}{dx}\right)_{\text{rad}} = N \int_0^{\nu_0} \hbar\nu\sigma(E,\nu)\,d\nu \qquad (8.17)$$

where $\hbar\nu_0 = E - mc^2$. Since σ is roughly proportional to ν^{-1}, the quantity $\hbar\nu\sigma(E,\nu)$ is approximately constant. Hence, the energy loss due to bremsstrahlung is proportional to Z^2 and increases nearly linearly with incident electron energy. In contrast, for the case of electron–electron collisions the energy loss is proportional to Z and increases only logarithmically with energy. From Ref. 14 the ratio of the radiative loss to the collisional loss is given approximately by

$$\frac{(dE/dx)_{\text{rad}}}{(dE/dx)_{\text{coll}}} = \frac{EZ}{1600mc^2} \qquad (8.18)$$

8.2.1.2. Ion Beam Energy Deposition. As for the case of electrons, the energy loss of ions in matter is primarily due to the processes of ionization and excitation, via the Coulomb force, of the electron cloud that surrounds the nucleus. In addition, at low ion energies nuclear scattering can also contribute appreciably to the energy loss. The stopping power formula of Bethe accounts for both ionization and excitation of the atomic electrons and has the form[15]

$$-\frac{dE}{dx} = -\frac{4\pi N_0 Z_{\text{eff}}^2 \rho e^4 Z_2}{m_e c^2 \beta_i^2 A_2}\left[\ln\frac{2m_e c^2 \beta_i^2 \gamma^2}{I_{\text{av}}} - \beta_i^2 - \frac{\Sigma_j C_j}{Z_2}\right] \qquad (8.19)$$

where N_0 is Avogadro's number, Z_{eff} is the effective charge of the projectile ion, and Z_2, A_2, and ρ are the atomic number, atomic weight, and mass density of the stopping medium. I_{av}, again, is the average ionization potential. A simple approximate expression for Z_{eff} that gives relatively good agreement with available experimental data is[16]

$$Z_{\text{eff}}/Z_1 \approx 1 - 1.034\exp\left(-\frac{137.04\beta_i}{Z_1^{0.69}}\right) \qquad (8.20)$$

The summation, $\Sigma C_j/Z_2$, denotes the effects of atomic shell corrections to the stopping power. Ignoring the shell corrections, the stopping power is observed to become singular when $2m_ec^2\beta_i^2\gamma^2 \leqslant I_{av}$. For gold, $I_{av} \sim 755$ eV. An upper bound on the applicability of the uncorrected Bethe formula is 0.34 MeV for protons, 4 MeV for carbon ions, and 82 MeV for ^{238}U ions. For light ion fusion 20-MeV carbon ions are of great interest. Hence, it is necessary to include the shell correction factors, especially in the 2-MeV/amu energy regime. A compatible set of shell correction coefficients have been tabulated by Andersen and Ziegler from proton slowing down data.[17] The shell corrections can also be applied to higher-mass ions with the same value of E/A_1, where A_1 is the atomic weight of the projectile ion.

For very low-energy ions even the shell-corrected Bethe formula is not very accurate, and it is necessary to use the so-called LSS model.[18] The electronic portion of this theory, assuming a Thomas–Fermi description of the electron clouds, predicts that the stopping power scales as the square root of the incident ion energy. The range of validity of this model is conservatively given by $Z_1^{1/3} > 137\beta$, which corresponds to ~ 1-MeV carbon ions. The smaller of the Bethe and LSS stopping powers is usually chosen as the more accurate bound electron stopping power estimate.[19]

Since the mass of the incident ion may not be negligible compared to the mass of a target nucleus it is necessary to include the effects of nuclear Coulomb stopping, especially for very small incident ion energies and for large Z_1 and Z_2. A least-squares fit[20] to the theory[21] gives

$$-\left(\frac{dE}{dR}\right)_{nuc} = C\varepsilon^{1/2}\exp\left[-45.2(C^1\varepsilon)^{0.277}\right] \left[\text{MeV}/(\text{g}/\text{cm}^2)\right] \quad (8.21)$$

where $R = \rho x$, $\varepsilon = E/A$, and C and C^1 are constants given by

$$C = \frac{A_1A_2}{(A_1+A_2)} \frac{\left(Z_1^{2/3} + Z_2^{2/3}\right)^{-1/2}}{Z_1Z_2} \quad (8.22)$$

$$C^1 = 4.14\times10^6 \left(\frac{A_1}{A_1+A_2}\right)^{3/2} \left(\frac{Z_1Z_2}{A_2}\right)^{1/2} \left(Z_1^{2/3} + Z_2^{2/3}\right)^{-3/4}$$

$$(8.23)$$

Taken together the Bethe, LSS, and nuclear stopping power theories provide an adequate description of the cold matter ion energy deposition, i.e.,[22]

$$\left(\frac{dE}{dx}\right)_{\text{cold}} = \min\left[\left(\frac{dE}{dx}\right)_{\text{Bethe}}, \left(\frac{dE}{dx}\right)_{\text{LSS}}\right] + \left(\frac{dE}{dx}\right)_{\text{nuc}} \quad (8.24)$$

Plasma effects are also important for the ion stopping power when the pellet material begins to ionize. An expression which includes the contributions of both binary collisions within a Debye radius and collective plasma oscillations outside the Debye radius is[23]

$$\left(\frac{dE}{dx}\right)_{\text{free}} = -\frac{\omega_\rho^2 Z_{\text{eff}}^2 e^2}{c^2 \beta_i^2} G(y_e)\ln \Lambda_{\text{free}} \quad (8.25)$$

where $\omega_\rho^2 = 4\pi\rho\bar{Z}_2 e^2 N_0/m_e A_z$, $\Lambda_{\text{free}} = 0.764\beta_i c/\omega_\rho b_{\text{min}}$. Here $b_{\text{min}} = \max[e^2 Z_1/m_{12}u^{-2}; h/2m_{12}\bar{u}]$, $m_{12} = m_1 m_2/(m_1 + m_2)$, and $\bar{u}$ is the relative speed between the ion and the target electrons. The function $G(y_e)$ is defined by

$$G(\xi) = \text{erf}(\xi^{1/2}) - 2(\xi/\pi)^{1/2} e^{-\xi} \quad (8.26)$$

with $y_e = (m_e c^2 \beta_i^2/2T_e)^{1/2}$. A similar expression exists for the plasma ion component of the stopping power, although it is usually negligible in comparison with Eq. (8.25) for most primary deposition problems. Preliminary calculations of the minimum ion range in heated material indicate range shortening by a factor of approximately 2, practically independent of the projectile ion species and the target material.[22]

8.2.2 Pellet Compression

Upon absorption, the driver energy subsequently appears as thermal energy in the absorption layer, and as kinetic energy of the expanding plasma. The pressure which normally drives the implosion arises from the reaction force due to ejection of material from the pellet surface (ablation), in the same manner that rocket engines generate thrust.

As an introduction to this problem consider the simple case of ablation into a vacuum.[10] Energy balance requires that the incident energy be equal to the energy in the ablated material. Assuming isothermal conditions the one-dimensional fluid equations of motion and continuity are

$$m_i n\left(\frac{\partial}{\partial t} + v\frac{\partial}{\partial z}\right)v = -kT\frac{\partial n}{\partial z} \qquad (8.27)$$

$$\frac{\partial n}{\partial t} + \frac{\partial(nv)}{\partial z} = 0 \qquad (8.28)$$

where z is the depth from the surface. The solution of Eqs. (8.27)–(8.28) is easily obtained as

$$n = n_0 e^{-z/v_{\text{i.th.}}t} \qquad (8.29)$$

$$v = v_{\text{i.th.}} + (z/t) \qquad (8.30)$$

where $v_{\text{i.th.}}$ is the (isothermal) sound velocity. The total energy in the ablation layer (expanding hot plasma) is

$$W = \int_0^\infty dz \tfrac{1}{2}m_i n(v_{\text{i.th.}} + z/t)^2 + \tfrac{3}{2}nkT$$

$$= 4n_0 v_{\text{i.th.}}(kT)t \qquad (8.31)$$

Hence, the driver power required for energy balance is simply $P = 4n_0 v_{\text{i.th.}}(kT)$. The mass flow into the vacuum region is $dm/dt = m_i v_{\text{i.th.}} n_0$, while the ablation pressure is $p_A = n_0 kT$.

In order to determine the acceleration of the surface, assume that the initial mass per unit area is m_0; then

$$m(t) = m_0 - t\frac{dm}{dt} \qquad (8.32)$$

and the equation of motion for the accelerating layer becomes

$$\left(m_0 - t\frac{dm}{dt}\right)\frac{dv}{dt} = p_A \qquad (8.33)$$

with the solution

$$v(t) = \frac{p_A}{dm/dt}\ln\left[\frac{m_0}{m(t)}\right] = v_{\text{i.th.}}\ln\left[\frac{m_0}{m(t)}\right] \qquad (8.34)$$

From Eq. (8.34) the kinetic energy of the accelerated layer is

$$\tfrac{1}{2}m(t)v^2(t) = \tfrac{1}{2}m(t)v_{i.th.}^2\{\ln[m_0/m(t)]\}^2 \tag{8.35}$$

and the ratio of the kinetic energy of the accelerated layer to the input driver energy is

$$\frac{\tfrac{1}{2}m(t)v^2(t)}{4nv_{i.th}kTt} = \frac{1}{8}\frac{m(t)}{m_0 - m(t)}\left[\ln\frac{m_0}{m(t)}\right]^2 \tag{8.36}$$

The maximum energy transfer is $\sim 8\%$, which occurs at $m_0/m(t)=5$. Although this estimate relies on several simplfying assumptions, more detailed numerical calculations for spherical geometry indicate that typical energy transfer efficiencies lie between 5% and 10%.

Since the velocity of the ablative material removal is determined by the local sound velocity, the time available for hydrodynamic motion of the accelerating layer is less than the sound transit time to the pellet center. The sound speed of the expanding layer must therefore be supersonic with respect to the cold material being accelerated, and shocks necessarily form. However, since shocks heat the inner material via irreversible processes such as viscosity and thermal conduction, the internal energy of the imploding material is increased, raising the minimum energy necessary for compression. Hence, the work required for a given compression ratio via a strong shock is greater than that required for an adiabatic (isentropic) compression. In fact, the maximum compression that can be achieved in spherical geometry by the passage of a single shock is a factor of 33.[24] In this case the beneficial effect of compression is essentially offset by an inefficient temperature distribution which reduces the fusion reaction rate in most of the fuel. The volume of the strongly heated and compressed center is too small and the compression time too short to produce an appreciable fusion yield.[10]

A conceptual solution to this problem is to shape the driver pulse[3] such that a succession of weak shock waves are arranged to coalesce at a radius approximately equal to the α-particle range in D–T. This generates the very large, nearly adiabatic compression required at the center of the pellet. The shocks which precede the compression preferentially heat only the center of the pellet because of spherical convergence. Hence, only mass in the central pellet region is ignited, and the cold, compressed outer fuel remains in the Fermi-degenerate state. The detailed theory of adiabatic (isentropic) implosion with selective shock heating of only the central pellet region has been developed by Kidder for the case of laser-driven ICF.[25]

Further improvements with regard to the compression problem are achieved by using larger-radius hollow spherical shells of D–T, rather than smaller solid targets.[26] The increased volume change on compression increases the hydrodynamic work for a given applied pressure. The resulting shock strengths are somewhat smaller, and it is easier to produce an optimized pressure and compression history.[10]

8.2.3. Rayleigh–Taylor Instability

The implosion arises from the acceleration of a dense layer by the pressure resulting from an ablation-created low-density plasma layer. Hence, this configuration is expected to be subject to the classic Rayleigh–Taylor hydrodynamic instability which occurs when a lower density fluid exerts force on a higher density fluid in a direction normal to the fluid interface.[27] If a perturbation of the form $\xi_0 \sin kx$ is applied to an interface at $z = z_0$, the amplitude ξ at time t is given by

$$\xi = \xi_0 e^{\gamma t} \tag{8.37}$$

The growth rate is given by $\gamma = (\alpha k a)^{1/2}$, where a denotes the acceleration of the interface, and α is the Atwood number determined by the densities ρ_1 and ρ_2 of the liquids according to

$$\alpha = \frac{\rho_2 - \rho_1}{\rho_2 + \rho_1} \tag{8.38}$$

Equation (8.37) is valid for instability wavelengths $\lambda = 2\pi/k \gtrsim \xi$; for $\xi > \lambda$, the growth becomes more nearly linear in time.[28, 29]

If the unstable interface is replaced by a region in which the density varies exponentially with characteristic length $(1/\beta)$ then the wave number k must be replaced by

$$k \to \frac{k\beta}{k + \beta} \tag{8.39}$$

Depending upon the details of the driver energy deposition, Rayleigh–Taylor instability considerations can play a critical role in pellet design. In addition to the ablation layer, the instability can also occur in target designs which have an initial density discontinuity, as well as near the end of an implosion when the pressure in a relatively light fuel region becomes sufficiently large to decelerate a dense pusher layer surrounding the fuel.[28]

8.2.4. Target Designs for Charged Particle Beam ICF

In preceding sections we have outlined various important processes that occur during the compression, ignition, and burn of ICF pellets. In this section we apply this background to the task of designing pellets for electron and ion beam drivers, with the intent of determining the driver system requirements (power, symmetry, etc.) necessary for the production of useful fusion energy.

Before turning to the detailed designs it is first useful to consider some general characteristics of particle beam targets. First, such targets will have absorption shells which are approximately one range thick.[30] As the range does not vary greatly with temperature or density, fixed voltage charged particle beams will deposit their energy in the same mass throughout the implosion. In addition, since the matter temperatures reached in the deposition layer are typically only a few hundred electron volts, electron conduction cannot efficiently remove the energy from the region in which it is absorbed.* Hence, particle beam implosions cannot rely on thermal conduction to symmetrize irregularities in the deposition, and the required uniformity of irradiation can be directly related to the allowable spread in implosion velocities according to

$$\frac{2\,\delta v}{v} \sim \frac{\delta E}{E} \tag{8.40}$$

where v is the implosion velocity and E is the beam energy flux on the target. The velocity spread will result in a difference in the radial compression of the fuel at the time τ_c of maximum compression of order

$$\delta r = \delta v \tau_c \tag{8.41}$$

For reasonable implosions this error must be less than the minimum compressed fuel radius. Combining Eqs. (8.40) and (8.41) gives the ap-

---

*Conduction smoothing in the corona is an important consideration in the design of laser-driven ICF pellets.

proximate symmetry requirement[31]

$$\frac{\delta E}{E} \lesssim \frac{R_c}{R_i} = \left(\frac{V_c}{V_i}\right)^{1/3} \qquad (8.42)$$

where R_i is the initial radius of the uncompressed fuel, and V_i and V_c are the initial and compressed fuel volumes, respectively. For example, a volume compression ratio of 1000 would require an irradiation variation of $\leqslant 10\%$.

For relatively long-range particles, such as 1-MeV electrons, high-Z ablator materials can be used because the resulting thermal radiation is largely self-trapped and radiation losses are limited. However, relatively short-range particles, such as 10-MeV α particles, should be absorbed in low-Z materials to limit thermal X-ray losses to tolerable levels.[30]

For a given mass of imploded material, the minimum energy required for implosion and ignition is approximately constant for the best low-entropy (adiabatic compression) designs. The required energy is of the order of 1 MJ/mg. Since energy is the time integral of the power

$$W \sim \langle P \rangle \tau \sim \frac{\langle P \rangle R_i}{\langle v \rangle} \qquad (8.43)$$

where $\langle P \rangle$ is the average power and $\langle v \rangle$ is the average implosion velocity required for ignition ($\gtrsim 2 \times 10^7$ cm/sec). For a constant implosion mass W is approximately constant so that the required power varies as R_i^{-1}; i.e., a larger pellet radius reduces the required driver power.

Since particle beam targets have deposition lengths which are roughly one range thick, the mass of the deposition region scales as $M \sim \rho \lambda_r R_i^2$. Hence, for a particle with a fixed range, the required power drops somewhat for decreasing radius, because of the mass decrease. Roughly $\langle P \rangle R_i \sim R_i^2$ for a constant deposition thickness so that $\langle P \rangle \sim R_i$. However, since $W \sim R_i^2$, and the pellet energy yield is proportional to the mass times the burnup fraction, which is proportional to R_i^4, the energy gain is roughly proportional to R_i^2. Hence, regardless of the specific design, achieving pellet implosions with long-range particles will require large targets, high powers, and large energies.[30]

For short-range particles the minimum deposition length is determined primarily by fluid stability and implosion symmetry criteria. For fixed aspect ratio (R_i/λ_r), the required power and energy scale as $\langle P \rangle \sim R_i^2$, $W \sim R_i^3$. Hence, decreasing the radius results in a dramatic decrease in the power and energy required for implosion, while the energy gain decreases only as R_i. To obtain this scaling requires short-range particles, however, since the particle range must be proportional to R_i.[30]

8.2.4.1. Electron Beam Fusion Pellet Designs. Because of the relatively long range of ~ 1-MeV electrons in solid D-T, pellet designs for electron beam fusion are structured, containing various layers of differing atomic number. The first targets considered (Fig. 8.2) consisted of three parts:[32] (1) the *ablator*, the outer part of a high-Z shell where most of the beam energy was deposited; (2) the *pusher*, the inner part of the high-Z shell which was driven inward by the ablator, and (3) the D-T *fuel*, which was compressed and heated by the imploding pusher. During the D-T burn phase, the pusher also served as a *tamper*, keeping the fuel inertially compressed for longer times.

For 1-MeV electrons and ~ 1-mm radius targets of this design, the power and energy required for breakeven were $\sim 8 \times 10^{14}$ W and ~ 4 MJ; it would be necessary to focus beam currents of $\sim 10^9$ A to a density of $\sim 10^{10}$ A/cm^2. This requirement is still well beyond the capability of present electron beam technology. A serious problem with these relatively simple targets is the bremsstrahlung produced by the interaction of the electron beam with the high-Z ablator. The bremsstrahlung is more penetrating than the electrons and it preheats the inner surface of the pusher, setting it on a high adiabat and decreasing its effectiveness.

From Eq. (8.15) the stopping power of relativistic electrons is relatively insensitive to beam voltage. Since the pressure in the ablator is proportional to $J(dE/dx)$, where J is the current density, it should also be relatively independent of beam voltage. However, since the effect of the bremsstrahlung becomes increasingly worse as the voltage is raised, use of a low-Z

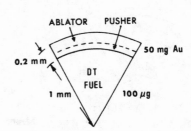

Figure 8.2. Initial pellet design for electron beam fusion (1-MV electrons).

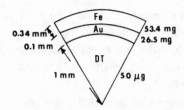

Figure 8.3. Low-Z ablator electron beam fusion target.

ablator layer (Fig. 8.3) should extend the voltage range and somewhat reduce the breakeven current. Computer simulations of electron beam fusion targets in which a lower-Z (iron) ablator surrounds a high-Z (gold), high-density pusher indicate that the breakeven power level for 1-MeV electrons drops to ~360 TW, primarily because of the reduced bremsstrahlung heating of the inner wall of the pusher.[33] The optimal iron thickness depends on the electron penetration depth, and is about 0.34 mm for 1-MeV electrons. At higher voltages, the current required for breakeven increases almost linearly with voltage.

Double-shell targets (Fig. 8.4) have also been examined for electron beam fusion.[30, 34] For such targets there is some velocity multiplication due to the collision between the two shells. For an elastic collision between two bodies of unequal mass, the velocity multiplication is $2/(1+\mu)$, where μ is the ratio of the Fe mass to the Au mass. The average implosion velocity can therefore be somewhat less than for the single-shell targets, and the initial target radius can be somewhat larger. Hence, the required power would be expected to decrease while the required energy would increase. Indeed, the double-shell design of Fig. 8.4 requires 250 TW and 45 MJ for breakeven.[30]

For 1-MeV electrons even the 250-MA current requirement for breakeven for the double-shell targets is quite high. If the deposition length of the electrons could be substantially shorter than the simple collisional deposition lengths used for most of the electron beam pellet calculations,[31]

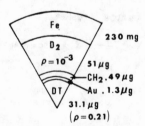

Figure 8.4. Double shell target for 1-MV electrons.

then the breakeven conditions would be considerably easier to attain. For example, if a 5-MeV electron beam were to deposit its energy in the same layer thickness as that assumed for 1-MeV electrons, then a target that would break even at 1 MeV, 250 kA should also break even at 5 MeV, 50 MA, which is more readily achievable, but still beyond present technology.

Several methods for obtaining such enhanced deposition have been proposed, including magnetic stopping in macroscopic fields,[35] scattering from micromagnetic turbulence,[36] electrostatic reflection,[37] and beam stagnation.[38] Numerical calculations of enhanced electron beam deposition in thin foils resulting from macroscopic electric and magnetic fields indicate an enhancement that is roughly proportional to I/I_A, provided that (1) the beam electrons are collisionless and (2) the magnetic field extends to a depth of the order of the electron Larmor radius on either side of the foil.[39]

8.2.4.2. Pellet Designs for Ion Beam Fusion. While it has been necessary to strive for enhanced deposition levels for electron beams, it is anticipated that ion beam energy coupling will only involve well-understood physical deposition processes at the power and energy levels required to drive efficient pellet implosions. In this regard ions appear to offer several distinct advantages over electrons from the standpoint of target design: (1) ions produce essentially no bremsstrahlung, thereby eliminating the pusher preheat problem; (2) there is very little ion scattering in the target so that reflection losses of ion beam energy are minimized; and (3) normally incident ions have a higher energy loss rate near the end of the range (the Bragg peak). This effect produces enhanced coupling of the ion beam energy into pusher kinetic energy. Moreover, nonrelativistic ions afford the possibility of tailoring not only the ionic species, but also the output voltage and current pulses to simultaneously optimize the beam generation and target coupling efficiencies. Finally, for a given deposition length ions can have higher kinetic energies (voltages) than electrons, thereby reducing the current requirement, and presumably simplifying the pulse power/accelerator task of high-power beam production for fusion.[31]

For ions it is relatively easy to convert the known range for a given particle to the range for another ion. The approximate scaling is

$$\lambda'_r(m'_i, Z', E') = \frac{m'_i Z^2}{m_i(Z')^2} \lambda_r\left(m, Z, \frac{m_i}{m'_i} E'\right) \tag{8.44}$$

where unprimed quantities correspond to the known quantities, and primed quantities correspond to the ion species for which the range is desired. λ_r denotes the range in g/cm^3, and m_i, Z, and E are the ion mass, charge state, and energy, respectively. From Eq. (8.19) it is easily observed that the stopping power varies approximately as Z_{eff}^2, and remains relatively insensitive to the stopping material when measured in units of g/cm^2. The normalization for Eq. (8.44) is provided from $\lambda_r \approx 0.005$ g/cm for hydrogen at 1 MeV. Thus, 10-MeV protons, 13-MeV deuterons, 40-MeV He^4 ions, and 10–20-GeV uranium ions all have about the same range and, moreover, the same energy deposition profile. Consequently, target calculations for one type of ion will be approximately valid for comparable ion species having the same beam power.

Because of multiple scattering and bremsstrahlung 1-MeV electrons have an extended deposition profile that produces considerable preheat and low thermonuclear gains, but also minimal sensitivity to the Rayleigh–Taylor instability.[40] In contrast, the Bragg peak, although smeared in reality, tends to produce more efficient, lower-power implosions, but with increased sensitivity to growth of fluid instabilities.[41] Hence, all ion beam target designs must be carefully checked for susceptibility to the Rayleigh–Taylor instability.

The first ion beam fusion targets to be considered were solid D–T spheres.[42] The breakeven implosion of such pellets with 0.5-MeV deuterons required about 4000 MA delivered with a rather precise pulse shape. However, the use of more structured targets substantially relaxes the beam power requirements. The first of such structured targets to be examined is shown in Fig. 8.5.[43] It consists of a single high-density shell surrounding the D–T fuel. Conceptually, the shell can be divided into three layers: (1) an outer *tamper* layer where the beam energy deposition is relatively low, (2) an intermediate layer encompassing the Bragg peak where the energy deposition is relatively high, and (3) an inner *pusher* layer which is imploded by the "explosion" of the intermediate layer, thereby compressing and heating

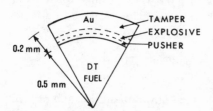

Figure 8.5. Initial pellet design for ion beam fusion. The Au shell thickness was optimized for 10-MV protons.

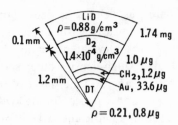

Figure 8.6. Low-power double-shell target designed
for breakeven with 9-MV α particles.

the fuel to ignition. The tamper effectively contains the explosion of the
intermediate layer, resulting in a more efficient conversion of the ion beam
energy into pusher kinetic energy than can be obtained with electron beam
(or laser) targets in which the energy deposition peaks at the pellet surface
(ablation) layer. For 10-MeV proton beams irradiating such targets,
breakeven is achieved with about 10 MA (100 TW), with a volume compression ratio of $\sim 3 \times 10^3$.[43] For higher-voltage beams the current required for
breakeven does not drop; as the voltage increases it is necessary to increase
the gold thickness to stop the protons. Below 10 MeV, the gold thickness
must be decreased and the tamping effect is reduced.

Since low-Z materials are more effective in stopping ions than high-Z
materials, various double-shell targets have been proposed.[30] These targets
have a separate low-Z, low-density ablator at a relatively large radius, which
reduces the focusing and power requirement while maintaining reasonable
aspect ratios. However, low-Z materials are poor pushers and tampers, and
a high-Z, high-density pusher shell is added at a much smaller radius. The
double-shell target of Fig. 8.6 gives a theoretical yield of 65 kJ with 45 kJ of
9-MeV α particles.[30] It requires a peak power of only 11 TW. For breakeven
at such low power and energy, it is necessary to carefully control the pulse

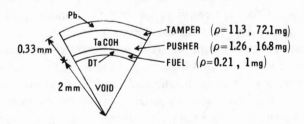

Figure 8.7. High-gain ion beam fusion target with low-density pusher.

shape, as well as the density of the material between the shells in order to obtain a nearly elastic collision.

A somewhat newer ion beam target is shown in Fig. 8.7. It has a high-Z, high-density outer shell, an intermediate layer of tantalum-doped plastic, and an inner layer of fuel. This target features an enhanced energy deposition explosion in the low-Z plastic that is tamped by the high-Z layer, resulting in efficient fuel compression. The pellet provides high gain (~ 100) for relatively modest beam power and energy requirements (~ 1.28 MJ and 240 TW), but also requires precise pulse shaping.[44]

8.3. Electron Beam Fusion Concepts

The first proposals for electron-beam fusion were due to D. Maxwell at Physics International and F. Winterberg.[45] The latter author suggested that 10 MeV electron beams could be used to drive a solid D–T target, invoking the two-stream instability to enhance the energy deposition. Although these early proposals suffered from several practical difficulties, they nevertheless stimulated research toward the development of more workable particle beam fusion schemes.[46, 47]

8.3.1. Electron Beam Fusion in Vacuum Diodes

From Section 8.2.4 it was seen that breakeven electron beam fusion targets will require several hundred megamperes of 1-MeV electrons with a total beam energy of a few megajoules. Early attempts to generate such beams involved self-pinched, large-aspect-ratio (low-impedance) vacuum diodes,[48] and further beam focusing within a space-charge-neutralizing plasma.[49] The concept is shown schematically in Fig. 8.8. Although such experiments have been extremely useful from the standpoint of providing experience in focusing very intense beams onto targets,[50-53] and even obtaining fusion neutrons,[54] the technique is apparently not able to meet the conditions necessary for fusion breakeven. As indicated in Chapter 2, such diodes are now known to be magnetically self-limiting; i.e., the pinched electron beam diode can be considered as a self-magnetically insulated diode in which the insulating field is produced by the diode current itself. An estimate of the maximum electron current that can be obtained from

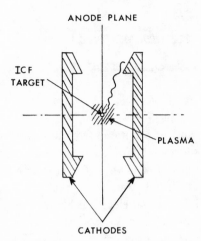

Figure 8.8. Spherically symmetric pellet irradiation in a double-electron-beam diode configuration.

such diodes (in the limit of infinite aspect ratio) is [see Eq. (2.129)]

$$I_e^{\text{max}} = 516 A^{1/2} (\gamma_0 + 1)^{1/2} \ln\left[\gamma_0 + (\gamma_0^2 + 1)^{1/2} \right] \text{ (kA)} \qquad (8.45)$$

where A is the atomic weight of the plasma in the diode. Taking $A^{1/2} \sim 14$, from Eq. (8.45) it requires $\gamma_0 \sim 34$ to produce 100 MA of electrons, which is in good agreement with the numerical estimates.[55] For such high-electron kinetic energies bremsstrahlung pellet preheat would make ICF fusion impossible.

8.3.2. Multiple Electron Beam Overlap Schemes

If single diodes cannot produce the requisite ~1-MeV electron beam power, a conceptual solution is provided by combining the electron beams from many diodes. This approach requires a means of transporting the beams without appreciable loss of energy or current density, and of efficiently overlapping the beams on the target. For near term proof-of-principle experiments, there is no requirement to protect the beam sources or target chamber from the blast of an ignited pellet, and the beam transport scheme must only meet the requirements for beam focusing onto the pellet. From Chapter 1 the electrically weakest element of the beam-producing pulse power system is the dielectric insulator which separates the energy transfer line from the diode vacuum. From simple area arguments this interface

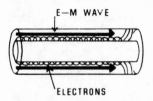

Figure 8.9. Schematic geometry of a self-magnetically insulated transmission line.

must be separated from the target region by approximately five meters.[56] A successful partial solution is provided by self-magnetically insulated vacuum lines (Fig. 8.9).[57, 58] As a very high-power electromagnetic pulse travels down the transmission line it sets up a transverse magnetic field that inhibits the electrical breakdown of the vacuum gap. Ahead of the pulse some electrons cross from cathode to anode; within the body of the pulse, however, the strong self-generated magnetic field confines the electron flow to the cathode surface. In magnetically insulated transmission line experiments at Sandia almost 100% power transport efficiency has been demonstrated at power levels of 800 MW.[57]

If the gap separation in the transmission line is much smaller than the transverse dimension of the line, then the line can be deformed in such a manner as to cause focusing of the electromagnetic energy and the production of intense electron beams. Because of anticipated electron beam transport difficulties it has been suggested that several magnetically insulated lines terminate in diodes placed near the fusion target itself.[56] This, in fact, is one primary approach being pursued by the Russian program at the Kurchatov Institute of Atomic Energy.

If the end of the vacuum line is removed from the immediate vicinity of the target in order to provide "stand off" from the effects of the pellet explosion, it is then necessary to devise some means of transporting several intense electron beams to the pellet. One method, first proposed by G. Yonas at Sandia, involves the creation of a dense magnetized plasma channel by means of a wire-initiated gas discharge.[59] In an idealized channel model, the plasma conductivity is essentially infinite at the time of beam injection and the channel is perfectly rigid in space. Hence, the beam will be completely charge and current neutralized and the beam transport properties can be analyzed on the basis of single-particle trajectories in the confining magnetic field of the discharge current.

The overlap of several intense electron beams transported in multiple current-carrying plasma channels has been studied by means of a spherical multi-"disk" model.[60] A relatively simple estimate of the increase, α, in the average current density incident on a target surface due to beam overlap can be obtained by assuming a cold uniform beam current distribution both in the channels and on the target surface. If all incident electrons reach the target, then the current density gain is simply proportional to the ratio of the total channel area to the target surface area. For N channels of radius r_c and a target radius r_t,

$$\alpha = \frac{N(\pi r_c^2)}{4\pi r_t^2} = \left(\frac{N}{4}\right)\left(\frac{r_c^2}{r_t^2}\right) \qquad (8.46)$$

An obvious upper limit to the current density gain is obtained for $r_c = r_t$, which yields $\alpha = N/4$.

In addition to the simple geometric effects indicated above there is a marked dependence of the gain on the beam velocity space distribution. In order to provide 50 MA of 2-MV electrons with even as many as 50 separate diodes, then the separate 1-MA beams still substantially exceed the Alfven electron current limit ($I_A \sim 80$ kA). Hence, the individual electron beams are "hot" as they emerge from the diodes in the sense that a large fraction of the electron kinetic energy resides in transverse motion. At the overlap radius (where the incoming channels touch each other—Fig. 8.10) the individual channel fields merge to produce a net field with a high degree of azimuthal symmetry.[61] Inside the overlap radius the beam particles are not confined by the channel fields, and the angular momentum of a beam particle is determined by its azimuthal velocity at the overlap radius. Since electrons which enter the overlap region with a large transverse velocity will spread rapidly, it is desirable to make the overlap radius as small as possible.[60]

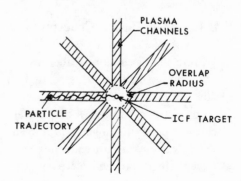

Figure 8.10. Schematic diagram of the overlap of several electron beams transported in magnetic plasma channels.

Numerical calculations for the case of a planar disk of 12 0.15-cm-radius channels, each carrying a plasma current of 60 kA with an injected electron beam energy of 1.5 MeV, have indicated that, in fact, the injected beam temperature does have a large effect on the transport efficiency.[60] When the velocity space angular distribution was specified as a modified cosine law with a truncation angle of 6° (cold beam) 90% of the electrons were able to reach a distance of closest approach to the origin of ~40% of the overlap radius. For a truncation angle of 90° (hot beam) the 90% figure of merit increased to 90% of the overlap radius.

The primary electron loss mechanism for the single planar disk was found to be the axial grad-B drift in the net azimuthal field dominated by the return current channels. By superposing a cusp magnetic field[56, 60] three particular improvements were noted: (1) the grad-B drift loss region was shifted to larger radius thereby decreasing electron loss from the central region; (2) axial electron reflexing inside the overlap region was observed; and (3) deeper penetration into the overlap region was observed.[60] For the single-disk geometry, the only one tested, agreement was obtained between experiment and theory.[62]

In addition to the 12-channel single disk, a 36-channel three-disk spherical calculation was also performed. If the beams from each disk simply overlapped a factor of 3 improvement would be obtained. Unfortunately, only about a 60% current density gain was observed, primarily because of the different magnetic field conditions in the two geometries.[60] While some improvement should result from the superposition of a mutiple cusp field, the basic problem is still the hot electron beams produced from the individual diodes. Unless a novel source of cold beams can be developed, only limited overlap gains will be achieved with electrons.

8.4. Ion Beam Fusion Concepts

From the discussion of the previous sections it is apparent that there are two major difficulties associated with the electron beam approach to inertial confinement fusion. First, the problem of electron target design is very complicated because of bremsstrahlung preheat and the relatively long electron deposition ranges. As a result several hundred megamperes of 1-MeV electron current is probably required to obtain pellet ignition. However, single pinched electron beam diodes cannot supply such large currents, and various analyses indicate severe problems with overlapping several hot electron beams.[60-63] These difficulties, coupled with the dramatic advance in the state of the art of generating very intense ion beams with existing pulse power hardware (Chapter 2), has led to a shift in emphasis in the US particle beam ICF community from electron beam to ion beam fusion concepts.[64]

There are now two major approaches to the production of the requisite ion beams: (1) heavy ion fusion (HIF) and (2) light ion fusion (LIF). In the heavy ion approach conventional high-energy accelerator technology and transport methods (bunchers and storage rings) combined with various complex beam manipulations would produce intense beams of heavy ions.[65] In contrast, the light ion approach is identified with pulsed power, high-current diode technology and novel methods of beam transport, including multistage acceleration, to achieve higher currents of relatively light ions of lower kinetic energy.[66]

8.4.1. Light Ion Fusion Approaches

In light ion fusion ion beams with atomic number 1–10 and kinetic energy 1–50 MeV will be required to deposit a few $\times 10^7$ J/g in the outer layers of a pellet in ~10 nsec. The necessary peak power is ~100 TW. Several techniques for meeting these criteria are believed possible, including a variety of single ion diode options,[67-70] as well as a multistage linear accelerator (Pulselac) approach.[71] With only one exception,[70] all of the options use magnetic fields to suppress electron flow while permitting ions to be accelerated freely.

A schematic of a magnetically insulated ion diode that produces a radially converging proton beam is shown in Fig. 8.11.[67] A 10–20-kG pulsed magnetic field completely penetrates the stainless steel cathode structures, but only slightly penetrates the anode. Electrons from the cathode plasma are constrained to drift in the axial and azimuthal direction forming a virtual cathode. An anode plasma is produced by electric field stress-induced surface breakdown along imbedded insulating hydrocarbon regions.[72] Protons from the anode plasma are accelerated radially inward by the virtual cathode. Using this diode on the Proto-II accelerator, Johnson et al. have produced 1.6-MA proton currents with 80% efficiency. The focused proton current density has approached 1 MA/cm^2.[67] Such beams have been used to perform "exploding-pusher" implosions[73] using aluminum foil cylin-

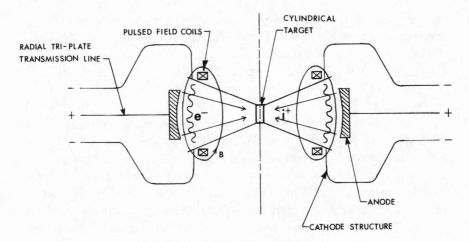

Figure 8.11. Schematic diagram of the radial ion diode.

drical and conical diagnostic targets which were less than one range thick for megavolt protons.[74]

Experiments at the Naval Research Laboratory have concentrated on the "pinch-reflex-diode" (Fig. 8.12), in which the magnetic insulation is provided by the self-magnetic field of the pinched electron flow (Chapter 2). Megampere currents of 2-MeV protons and deuterons have been extracted in 100-nsec pulses from 100-cm^2 diodes at ~ 70% efficiency. Using geometrical focusing deuteron current densities in excess of 300 kA/cm^2 have been produced in a configuration appropriate for injection into plasma transport channels.[68]

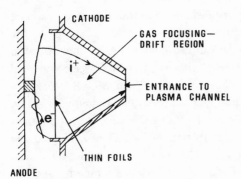

Figure 8.12. Schematic diagram of the pinch-reflex diode.

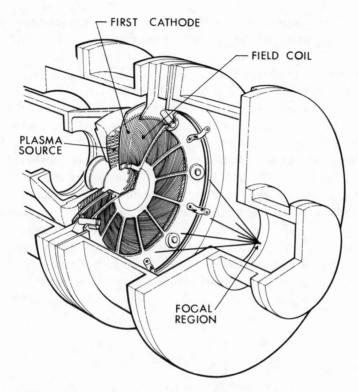

Figure 8.13. Schematic diagram of the Ampfion diode. (Courtesy C. Mendel.)

Another diode specifically designed for beam extraction is shown in Fig. 8.13.[69] This automagnetic plasma-filled field-insulated ion diode (Ampfion) approach utilizes a preformed plasma prior to application of high voltage to the accelerating structure. The insulating magnetic field arises from the diode current itself flowing through spiral turns. As ions are extracted from the anode plasma a sheath (double layer) develops, and the thickness of the sheath causes the diode impedance to rise, thereby bunching the extracted ion beam. Initial tests of this concept have verified several desirable features, including efficiency, locally small divergence, and impedance control; the major remaining problem is the generation of a uniform large area plasma of a single ion species and charge state.[69]

From the discussion of Section 8.3 it was observed that target overlap gains using magnetic plasma channel transport of electron beams were limited because of the large electron transverse energy; however, if the intense electron beams could be efficiently converted into ion beams at the overlap radius then much larger overlap gains might result. To illustrate this "light bulb" concept consider the geometry of Fig. 8.14 in which several electron beams are transported via plasma channels to a thin grounded foil located at the channel overlap radius.[70] The smaller target is separated from the foil by a vacuum. As the primary beam electrons penetrate the foil a virtual cathode is formed in the vacuum region and the electrons are reflected, in a manner similar to that which occurs in a reflex triode. It was initially conjectured that electron-to-focused-ion beam conversion efficiencies of the order of 50% might be possible.[75] More recent studies, however, assuming a uniform electron energy distribution at the channel overlap radius indicate rather severe difficulties: (1) the ion conversion efficiency is realtively low (~5%); and (2) the ion kinetic energy is substantially lower than the peak electron kinetic energy, remaining approximately constant for a wide range of electron diode voltages.[76] For such a broad distribution function a well-defined virtual cathode apparently cannot form.

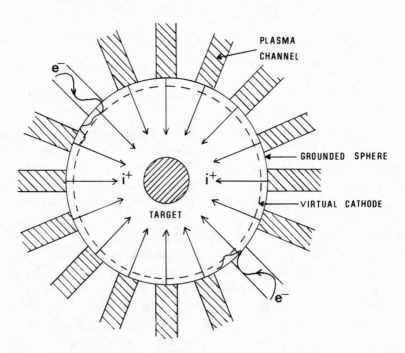

Figure 8.14. Schematic diagram of the "light-bulb" ion diode concept.

As for the case of electron beams the overlap of several intense ion beams transported in multiple plasma channels has also received much attention.[60-63] Ion diodes suitable for injecting beams into channels include a modified version of the radial ion diode,[77] the pinch-reflex diode, and the Ampfion diode. In contrast to the electron case the ion diode beam currents are typically small compared with the ion Alfven current, $I_{Ai} = 1.4 \times 10^6 (A_i V / Z_i)^{1/2}$ (A), where A_i and Z_i are the ion atomic weight and charge state, and V is the accelerating voltage in megavolts. Hence, the ion kinetic energy is primarily associated with the motion in the direction of propagation, and the ion diode beams are relatively cold. For such cold distributions, single-disk overlap calculations suggest that the $N/4$ geometrical limit for the ion current density overlap gain can be achieved.[60] Although experiments on multiple ion beam overlap have yet to be performed, single 100-kA proton beams extracted from pinch-reflex diodes have been transported over 1-m distances through 50-kA wall-stabilized Z discharges established in 1-Torr gas backgrounds.[78] For injection angles of $\sim 10°$ transport efficiencies of 50%–80% were obtained. Laser-initiated discharges have also been demonstrated,[79] and are the basis for some LIF reactor concepts.

For the case of ion beams the use of plasma channels can also provide a means for beam bunching and power compression.[80, 81] Suppose that the diode voltage waveform produces ions which exit the diode with a time-dependent axial velocity given by

$$v_i(t) = \frac{v_{i0}}{1 - t/T}, \qquad 0 < t \leqslant \tau < T \qquad (8.47)$$

where τ is the ion diode pulse duration at the diode and T determines the steepness of the voltage ramp. After propagating a distance Z, the beam pulse duration is reduced according to

$$\Delta t = \tau / \alpha \qquad (8.48)$$

where α, the power multiplication factor is given by[82]

$$\alpha = \left(1 - \frac{Z}{v_{i0} T}\right)^{-1} \qquad (8.49)$$

From Section 8.2, the total power absorbed by the pellet to produce breakeven should exceed $\sim 2 \times 10^{14}$ W, which can be satisfied by, e.g., an incident flux of 4×10^{13} W/cm^2 of 8-MeV He$^+$, assuming a target surface area $\lesssim 5$ cm^2 (corresponding to 5×10^6 A/cm^2). For a channel overlap gain of ~ 10 and a pulse compression ratio of 2–4, the channel injection current density must be of the order of 1–3×10^5 A/cm^2.

An accelerator approach that is consistent with the use of plasma channel overlap schemes would utilize separate, synchronized, modular accelerator units with each producing a light ion beam of a few terawatts.[83] The PBFA-I machine (Fig. 8.15) is the latest step in the evolution of such accelerators which began with Proto I in 1975 at the 1-TW level[84] and the Proto II in 1977 at the 8-TW level.[85] PBFA I provides the capability for 36 separate ion beams at almost 1 TW each.[86] Pulse power testing of PBFA I was completed in January 1981, and various ion diode and transport experiments will be performed. Construction of the projected 100 TW PBFA II is expected to commence in 1982.[74]

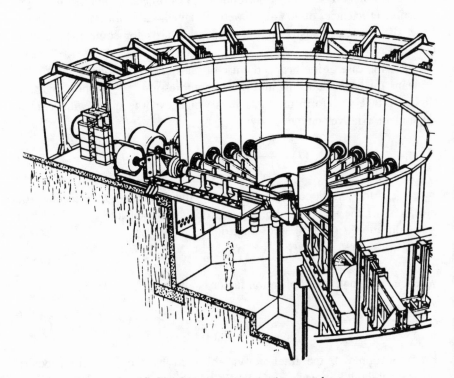

Figure 8.15. The PBFA-I modular ion beam accelerator.

A potentially serious problem with very light ion beams is that of magnetic deflections in the large self-fields of the high ion currents. In part for this reason present projections call for PBFA II to have $\lesssim$15-MV diodes capable of accelerating somewhat higher atomic number species. An alternate approach which extends this reasoning would use a linear accelerator to postaccelerate pure beams of intermediate-mass ions.[71] The major obstacle to the transport of high-current beams of such ions is the electrostatic self-repulsion of the ion space charge. In the Pulselac accelerator approach to light ion fusion, electrons are caused to flow into the ion beam volume to approximately balance the ion space charge. In regions without accelerating fields rapid neutralization has been observed in a number of experiments.[71, 87] In accelerating gap regions, however, the electrons must be prevented from crossing the applied potential and draining energy. An effective solution is the use of magnetically insulated gaps that provide transverse electron confinement, yet permit the ions to pass unimpeded.[71] By properly shaping the magnetic fields the trapped electron clouds can even act as virtual electrodes for ion beam focusing (Fig. 8.16).

Since the Pulselac concept is capable of generating relatively higher kinetic energies because of the modular accelerator approach, it is of interest to examine the various scaling laws in order to choose the optimum ion mass for inertial fusion.[66] If there is no emittance growth of the ion beam as it propagates through the accelerator, then the minimum value of the beam

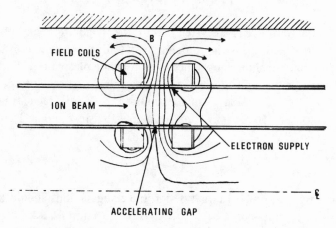

Figure 8.16. The Pulselac magnetically insulated accelerating gap.

divergence is determined by the ion source temperature, i.e.,

$$\Delta\theta = \frac{(v_\perp)_s}{v_i} \sim \left(\frac{T}{\varepsilon_i}\right)^{1/2} \tag{8.50}$$

where $(v_\perp)_s$ corresponds to the perpendicular ion velocity associated with the source, and v_i is the longitudinal ion velocity. Hence, at the final lens, which focuses the beam onto the target, the divergence angle scales as the reciprocal of the square root of the ion energy, $(\varepsilon_i)^{-1/2}$, for a fixed plasma source temperature.

For a given beam divergence angle the minimum distance D from the final focusing lens to the target of radius r_t is[88]

$$D = \frac{r_t}{\Delta\theta} \sim r_t(\varepsilon_i/T)^{1/2} \tag{8.51}$$

Assuming a current density j_i at the final lens, the total current that can be delivered to the target can be expressed as

$$I_t = j_i(4\pi D^2)F_a \tag{8.52}$$

where F_a is the area utilization factor, i.e., the total area of the beam at the position of the final lens divided by $4\pi D^2$. The maximum ion current that can be delivered to the target occurs for $F_a = 1$ and is given by

$$(I_t)_{\max} = j_i 4\pi r_t^2/(\Delta\theta)^2 \tag{8.53}$$

The total current is also related to the required beam power by $I_t = Z_a P/\varepsilon_i B$, where Z_a is the ion charge state in the accelerator and B is the longitudinal bunching factor. For typical anode plasma ion sources,[72, 89] the ion species are singly charged, and the available source current density scales as $A_i^{-1/2}$. Choosing normalizations of a 2° divergence angle for 2-MeV protons, a 500-A/cm^2 proton source density, and a 1-cm target diameter then

$$2.5\varepsilon_i F_a/A_i^{1/2} \approx Z_a P/\varepsilon_i B \tag{8.54}$$

For a given optimized target design the energetic ions should have the same range, regardless of the ion species. From Section 8.2.1.2, this implies that if the ions are fully stripped as they penetrate the target, the ion energy should scale approximately as $Z_i A_i^{1/2}$. For all light ions (except protons), $Z_i \sim A_i/2$, so that $E_i \sim \frac{1}{2} A_i^{3/2}$, and the ion velocity only varies weakly as $v_i \sim A_i^{1/4}$.

Assuming that $Z_a = 1$ and $B = 1$ (no bunching), then for a required beam power of 10^{14} W, the approximate area utilization factor scaling with atomic mass number is obtained as

$$F_a \approx 20 A_i^{-5/2} \tag{8.55}$$

For this set of assumptions it is apparent that the very lightest ions (p^+, d^+, t^+) cannot be used without longitudinal bunching since $F_a \geqslant 1$. On the other hand for 0^+, $F \approx 0.02$ without any longitudinal bunching. Hence, for intermediate-mass ions the present directed-flux plasma ion source technology may be adequate for fusion applications.[66]

It is also instructive to examine the ion deflections in their self-magnetic field, which will be of order $B_s \sim 2I_t/cD$. The difficulty imposed by self-deflection is indicated qualitatively by the ratio of the ion Larmor orbit to the distance D, i.e.,

$$\frac{r_L}{D} \sim \frac{\left(2m_i c^2 \varepsilon_i\right)^{1/2}}{eB_s D} \sim \frac{c^2}{eI_t}\left(\frac{m_i \varepsilon_i}{2}\right)^{1/2} \sim A_i^{11/4} \tag{8.56}$$

Hence, the self-deflection criterion is also much less stringent for intermediate-mass ions.[66] The success of Pulselac, however, will require improved understanding of the physics of multiple stage acceleration, particularly with regard to longitudinal phase instabilities.[90]

8.4.2. Heavy-Ion Fusion Approaches

Since 1976 there has been extensive study and experimentation directed toward the use of beams of heavy ions ($A_i \gtrsim 200$) as the driver for pellet implosions. Much of the progress in this field is summarized in the proceedings of annual workshops. [91-94] Although the technology of beam handling is generally rather well developed in the field of high-energy accelerators, the range-energy relations and target requirements cause the parameter range of interest to be somewhat low in terms of particle kinetic energy (10–20 GeV), but rather higher in beam current (10–20 kA) when compared with presently operating high-energy accelerators.

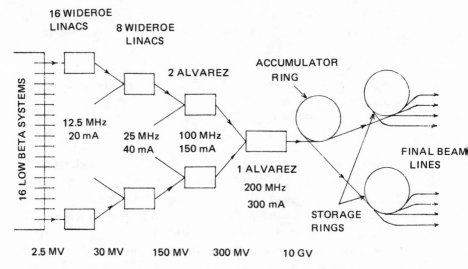

Figure 8.17. Schematic diagram of the rf accelerator/storage ring approach for heavy ion fusion.

The two primary accelerator choices are generally considered to be an rf linac operated in conjunction with a number of storage rings, and a single-pass induction linac. The overall rf accelerator approach is shown schematically in Fig. 8.17.[94] The basic concept involves funneling or "merging" of several relatively low-current beams produced by low-frequency linacs at the lowest energy into fewer linacs at higher energies with higher frequencies and currents, culminating in a single linac at the highest frequency and current.[92] Merging implies the filling of every rf "bucket," i.e., the region of stable accelerating phase of the rf waveform, at each higher frequency in order to realize the full space charge capability at every stage of the accelerator. With this approach the size and power consumption of each portion of the structure is minimized.

There are presently three candidate systems for the injector (low-beta) section: (1) conventional high-voltage injection into a low-frequency Wideroe linac;[93] (2) rf quadrupole (RFQ) accelerators;[95] and (3) arrays of small electrostatic quadrupoles (Meqalac).[96] The geometries of the RFQ and the Wideroe structures are particularly well suited to the merging concept. The most important problem to be investigated is that of emittance growth. The various experimental test programs at the several national laboratories should provide a broad experimental data base for comparison with what is believed to be an adequate rf linac simulation code capability. Similar remarks also apply to the higher-kinetic-energy linac sections. Since the beams are merged by frequency multiplying, it is possible, in principle, to conserve both the normalized transverse and longitudinal emittances. In practice it is anticipated that some emittance growth will occur at injection and at each merger, but detailed experimental measurements will be required.

As a result of the merging process 20-mA beams starting from 16 separate ion sources will become a single beam of $I_{lin} \sim 300$ mA. In order to increase the beam current to the level required for pellet fusion it is proposed to inject the linac beam into storage, or accumulation, rings. The most commonly proposed multiturn injection scheme has been to inject N_t turns into the horizontal plane of a ring used expressly for injection, transfer the accumulated beam into a storage ring after first interchanging the horizontal and vertical phase planes, and then repeat the process for N_s storage rings for a total multiplication of the linac current of $N_t N_s$. The beam inside each storage ring is then bunched into N_b bunches with a compression factor $C_{s.r.}$, and each bunch is ejected into a separate beam linac so that a total number of $N = N_b N_s$ beams are directed toward the target. With further compression by a factor C_b of the bunches in the beam lines the final current reaching the target is given by

$$I = (I_{lin}N_t)NC_{s.r.}C_b \qquad (8.57)$$

For a target design criterion of 15 kA, Eq. (8.57) implies that the product $(N_t NC_{s.r.}C_b) \sim 5 \times 10^4$. Current design goals require that $C_{s.r.} \sim C_b \sim 7$, and $N_t N \sim 1000$.

Although there do not appear to be any fundamental limitations for any of the beam manipulations outlined above, it is readily apparent that a fusion reactor based on the rf accelerator approach would be a very complex facility requiring rather large initial capital expenditures.

A comparatively simpler approach is to form a single high-current beam (12 A at 2 MV) from the injector and low β sections and accelerate it to the desired endpoint energy using a single induction linac accelerator. The beam would also be gradually compressed by shaping the accelerating pulses. Although this scheme is conceptually simpler and a high-current electron induction linac technology does exist, there is essentially no experience in accelerating high-current heavy ion beams in this type of accelerator, and several important questions remain to be answered. In particular, an analysis of the coupling of the longitudinal and transverse beam dynamics is expected to be very difficult.

Regardless of the particular accelerator approach, a high-current ion beam must be transported through an evacuated beam line to the reactor chamber, and then to the pellet. Stable transport in the evacuated lines is complicated by the rapid longitudinal beam compression and the resulting increase of space charge forces. Optimization of this transport system will require better estimates of allowable momentum spread. Until a chromatic correction system designed to include the effects of space charge has been demonstrated, a conservative estimate of allowable momentum spread is about $\pm 1\%$.

Present designs for final beam transport through the reactor chamber to the pellet call for ballistic focusing in a background gas at a pressure of 10^{-4}–10^{-3} Torr. In this pressure regime, the heavy ion beam can propagate in an essentially unneutralized state, and plasma effects (two-stream and filamentation instabilities) and gas effects (scattering) are still relatively unimportant. Moreover, such pressures are consistent with the lithium waterfall reactor concept.[97]

8.5. Summary

Achieving a self-sustaining thermonuclear reaction requires high plasma temperatures (10^8 °C), high ion densities, and sufficiently long confinement times. In magnetic confinement schemes a relatively low-density plasma is confined within a reactor vessel by strong magnetic fields for relatively long times. The figure of merit for scientific breakeven is the Lawson criterion, $n_i \tau_c \gtrsim 10^{14}$. By comparison, in inertial confinement fusion (ICF) a target is compressed to extreme densities by very powerful beams, and pellet disassembly is restrained by the inertia of the compressed matter. For ICF the Lawson criterion is usually restated in terms of the $\rho_0 R$ product, where ρ_0 is the mass density of the fuel and R is the radial distance traversed at sound speed. ($\rho_0 R = 1$ corresponds to a burn efficiency of $\sim 14\%$.) Interest in particle beam fusion drivers for ICF has arisen primarily because of the cost and inefficiency of lasers; however, there are major problems to be solved regarding particle beam transport and focusing onto the target.

From the physics of pellet compression it is desirable to highly compress the pellet in such a fashion that only the central core is raised to ignition temperature. The fusion α particles then efficiently heat the surrounding cold fuel forming a spherically expanding burn wave that can result in almost complete fuel ignition. The important problems which must be considered for pellet compression include generation of the driving pressure and the hydrodynamics of the converging shocks. A correct description of the charged particle energy deposition physics is necessary for predicting the detailed behavior of both effects.

From the standpoint of target design, ion drivers offer several distinct advantages over electron drivers. Since ions produce essentially no bremsstrahlung, there is no pellet preheat problem. Also, the Bragg specific ionization peak near the end of the ion range produces enhanced coupling of the ion beam energy into compression kinetic energy. Moreover, for a given energy deposition length ions can have higher kinetic energies than electrons, thereby reducing the beam current requirement, and presumably simplifying the tasks associated with beam production and acceleration. Finally, by tailoring the accelerator voltage pulses it is possible to cause bunching of the nonrelativistic ion pulse at the target, thereby reducing the instantaneous beam space charge and current handling problems in the accelerator.

In addition to the ion pellet implosion advantages, there are also severe problems with multiple electron beam transport and overlapping of several such beams onto a pellet. Because of these issues there has been a shift in emphasis in the US particle beam ICF community from electron to ion beam fusion concepts. Two major approaches to the generation of the requisite 100-TW ion beams are heavy ion fusion (HIF), and light ion fusion (LIF). The HIF approaches would use conventional high-energy accelerator technology and transport methods (storage rings and bunchers) combined with complex beam manipulations in order to produce intense beams of heavy ions ($A > 200$). Although there do not appear to be any fundamental limitations for this approach, relevant experiments to investigate key physics issues (emittance growth) will probably require a full voltage device.

In contrast to the heavy ion approaches, LIF concepts are generally associated with pulsed power, high-current diode technology, and novel methods of beam transport, including multistage acceleration, to achieve higher currents of light and medium weight nuclei ($A < 20$). Almost all of these LIF accelerator options use magnetic fields to suppress electron flow while permitting ions to be accelerated freely. For example, in the Pulselac linear accelerator approach the intense ion space charge is neutralized by permitting electrons to flow into the ion beam volume. The electrons are prevented from crossing accelerating gaps, however, by applied magnetic fields that provide transverse electron confinement.

The PBFA-I machine at Sandia National Laboratories is the latest step in the evolution of high-current ion accelerators which utilize several (36) separate, synchronized, modular accelerating units with each capable of delivering a power of 1 TW. This device will be used to test several independent ion diode options. Moreover, the power and energy may be sufficient to produce substantial fusion neutron and X-ray yields, as well as permitting important studies of pellet implosion dynamics.

Problems

8.1. Compare the average kinetic energy of the Fermi degenerate electrons as a function of temperature over the temperature range 100 eV $< T <$ 100 keV, assuming a compressed density of 1000 g/cm^3.

8.2. Plot Eq. (8.36) and show that the maximum energy transfer efficiency occurs for $\ln[m_0/m(t)] = 2[1 - m(t)/m_0]$.

8.3. For short-range particles (ions) the minimum range λ_r is determined primarily by fluid stability and implosion symmetry requirements. If the aspect ratio R_i/λ_r is held constant (R_i is the initial pellet radius), show that the power and energy required for implosion scale as $\langle P \rangle \sim R_i^2$, $W \sim R_i^3$, while the energy gain (pellet energy yield/driver energy input) scales as R_i.

8.4. Suppose that a 1-MV magnetically insulated ion diode produces ions which exit the diode with a time-dependent axial velocity given by

$$v_i(t) = v_{i0}/(1 - t/T)$$

(a) How does the ion beam power at the diode exit vary with time? (b) What is the time variation of the ion beam power at a distance Z from the diode?

8.5. The mean "scattering" angle for a charged particle beam undergoing betatron oscillations in a plasma discharge channel scales as $(I_d/I_A)^{1/2}$, where I_d is the channel discharge current and I_A is the particle Alfven current. For the same I_d compare the mean angle for 1-MV electrons and 20-MV singly charged carbon ions.

8.6. Plot the area utilization factor F_a as a function of the ion atomic weight for bunching factors of 1, 2, and 5, assuming the normalization factors in the text. What bunching factor is required to make protons viable for pellet fusion according to this analysis?

References

Chapter 1

1. J. A. Nation, *Part. Accel.* **10**, 1 (1979).
2. V. P. Smirnov, *Prib. Tekhnika Eksper.* **2**, 7 (1977).
3. T. H. Martin, K. R. Prestwich, and D. L. Johnson, Sandia Laboratories Report No. SC-RR-69-421 (1964).
4. T. H. Martin, J. P. Van Devender, D. L. Johnson, D. H. McDaniel, and M. Aker, Proc. Int'l. Top. Conf. E-Beam Res. and Technol., p. 450 (Albuquerque, New Mexico, 1975).
5. A. D. Blumlein, U.S. Pat. No. 2,465,840 (March 29, 1948).
6. C. Mendel, Jr., and S. A. Goldstein, *J. Appl. Phys.* **48**, 1004 (1977).
7. *High Voltage Technology*, L. L. Alston, ed., Oxford University Press, Oxford (1968).
8. J. C. Martin, Internal Report SSWA/JCM/703/27, AWRE, Aldermaston, England (1970).
9. J. D. Shipman, *IEEE Trans. Nucl. Sci.* **NS-18**, 294 (1971).
10. D. H. McDaniel, J. W. Poukey, K. D. Bergeron, J. P. Van Devender, and D. L. Johnson, Proc. 2nd Int'l. Top. Conf. E and I Beam Res. and Technol., Ithaca, New York, Vol. II, p. 819 (1977).
11. J. Creedon, *J. Appl. Phys.* **48**, 1070 (1977).
12. K. Bergeron, *J. Appl. Phys.* **48**, 3065 (1977).
13. J. Poukey and K. Bergeron, *Appl. Phys. Lett.* **32**, 8 (1978).
14. J. D. Lawson, *J. Elect. Control* **5**, 146 (1958).
15. F. C. Ford, W. T. Link, and J. Creedon, Physics Internations Rept. No. REP PIPB-9 (August, 1966).
16. S. E. Graybill and S. V. Nablo, *Appl. Phys. Lett.* **8**, 18 (1966).
17. G. Yonas and P. W. Spence, Physics International Rept. No. DASA 2175 (1968).
18. R. C. Davidson, *Theory of Nonneutral Plasmas*, W. A. Benjamin, Reading, Massachusetts (1974).
19. A. A. Vlasov, *J. Phys. (USSR)* **9**, 25 (1945).
20. N. A. Krall and A. W. Trivelpiece, *Principles of Plasma Physics*, McGraw-Hill, New York (1973).
21. L. B. Loeb, *Electrical Coronas*, Univ. of California Press, Berkeley (1965).

Chapter 2

1. S. Dushman, *Phys. Rev* **21**, 623 (1923).
2. A. L. Hughes and L. A. DuBridge, *Photoelectric Phenomena*, McGraw-Hill, New York (1932).

3. R. H. Good, Jr., and E. W. Müller, *Handbuch der Physik*, Vol. 21, Springer, Berlin (1956).
4. R. H. Fowler and L. W. Nordheim, *Proc. R. Soc. London*, *Ser. A* **119**, 173 (1929).
5. L. W. Nordheim, *Proc. R. Soc. London*, *Ser. A* **121**, 626 (1928).
6. L. W. Nordheim, *Z. Phys.* **30**, 177 (1929).
7. W. Schottky, *Z. Phys.* **14**, 63 (1923).
8. S. P. Bugaev, E. A. Litvinov, G. A. Mesyats, and D. J. Proskurovskii, *Sov. Phys. Usp.* **18**, 51 (1975).
9. H. E. Tomaschke and D. Alpert, *J. Vac. Sci. Technol.* **4**, 192 (1967).
10. A. P. Komar, V. P. Sanchenko, and V. N. Shrednik, *Sov. Phys. Dokl.* **4**, 1286 (1959).
11. G. A. Mesyats, G. P. Bazhenov, S. P. Bugaev, D. I. Proskurovskii, V. P. Rotshtein, and Ya. Ya. Yurike, *Sov. Phys. J.* **12**, 688 (1969).
12. D. Alpert, D. A. Lee, E. M. Lyman, and H. E. Tomaschke, *J. Vac. Sci. Technol.* **1**, 35 (1964).
13. F. M. Charbonnier, C. J. Bennette, and L. W. Swanson, *J. Appl. Phys.* **38**, 627 (1967).
14. W. B. Nottingham, *Phys. Rev.* **59**, 907 (1941).
15. W. W. Dolan, W. P. Dyke, and S. K. Trolan, *Phys. Rev.* **91**, 1054 (1953).
16. W. P. Dyke, J. K. Trolan, E. E. Martin, and J. P. Barbour, *Phys. Rev.* **91**, 1043 (1953).
17. H. R. Jory and A. W. Trivelpiece, *J. Appl. Phys.* **40**, 3924 (1969).
18. I. Langmuir, *Phys. Rev.* **3**, 238 (1931).
19. R. K. Parker, R. E. Anderson, and C. V. Duncan, *J. Appl. Phys.* **45**, 2463 (1974).
20. F. Friedlander, R. Hechtel, H. R. Jory, and C. Mosher, Varian Associates Report No. DASA-2173, 1965.
21. D. DePackh, Naval Research Laboratory Radiation Project Progress Report Nos. 5 and 17, 1968.
22. J. M. Creedon, *J. Appl. Phys.* **46**, 2946 (1975).
23. L. Brillouin, *Phys. Rev.* **67**, 260 (1945).
24. M. Friedman and M. Ury, *Rev. Sci. Instrum.* **41**, 1334 (1970).
25. A. A. Kolomenskii, E. G. Krastelev, A. M. Maine, V. A. Papadichev, and S. G. Rot, *Sov. Phys. Tech. Phys. Lett.* **2**, 265 (1976).
26. J. A. Nation and M. Read, *Appl. Phys. Lett.* **23**, (1973).
27. G. S. Kino and N. Taylor, *Trans. IRE* **ED-9**, 1 (1962).
28. V. S. Voronin and A. N. Lebedev, *Sov. Phys. Tech. Phys.* **18**, 1627 (1974).
29. M. Jones and L. Thode, Los Alamos Scientific Laboratory LA-UR-79-3107 (1979).
30. E. Ott, T. M. Antonsen, Jr., and R. V. Lovelace, *Phys. Fluids* **20**, 1180 (1977).
31. J. Chen and R. V. Lovelace, *Phys. Fluids* **21**, 1623 (1978).
32. R. B. Miller, K. R. Prestwich, J. W. Poukey, and S. L. Shope, *J. Appl. Phys.* **51**, 3506 (1980).
33. S. Humphries, Jr., *Nucl. Fusion* **20**, 1549 (1980).
34. S. Humphries, J. J. Lee, and R. N. Sudan, *Appl. Phys. Lett.* **25**, 20 (1974).
35. S. Humphries, J. J. Lee, and R. N. Sudan, *J. Appl. Phys.* **46**, 187 (1975).
36. T. M. Antonsen and E. Ott, *Appl. Phys. Lett.* **28**, 424 (1976).
37. J. M. Creedon, I. D. Smith, and D. S. Prono, *Phys. Rev. Lett.* **35**, 91 (1975).
38. D. S. Prono, J. M. Creedon, I. Smith, and N. Bergstrom, *J. Appl. Phys.* **46**, 3310 (1975).
39. S. Humphries, Jr., R. N. Sudan, and L. Wiley, *J. Appl. Phys.* **47**, 2382 (1976).
40. J. Golden, T. J. Orzechowski, and G. Bekefi, *J. Appl. Phys.* **45**, 3211 (1974).
41. R. N. Sudan and R. V. Lovelace, *Phys. Rev. Lett.* **31**, 1174 (1973).
42. K. D. Bergeron, *Appl. Phys. Lett.* **28**, 306 (1976).
43. T. M. Antonsen, Jr., and E. Ott, *Phys. Fluids* **19**, 52 (1976).
44. J. W. Poukey, *J. Vac. Sci. Technol.* **12**, 1214 (1975).
45. S. A. Goldstein, R. C. Davidson, R. Lee, and J. G. Siambis, 1st Topical Conf. Elect. Beam Res. and Technol., Sandia National Laboratory, SAND76-5122, p. 218 (1975).

46. S. A. Goldstein, R. C. Davidson, J. G. Siambis, and R. Lee, *Phys. Rev. Lett.* **33**, 1471 (1974).
47. A. E. Blaugrund, G. Cooperstein, and S. A. Goldstein, 1st Topical Conf. Elect. Beam Res. and Technol., Sandia National Laboratory, SAND76-5122, p. 233 (1976).
48. S. A. Goldstein and R. Lee, *Phys. Rev. Lett.* **35**, 1079 (1975).
49. J. W. Poukey, 1st Topical Conf. Elect. Beam Res. and Technol., Sandia National Laboratory, SAND76-5122, p. 247 (1976).
50. S. Humphries, *Plasma Phys.* **19**, 399 (1977).

Chapter 3

1. R. C. Davidson, *Theory of Nonneutral Plasmas*, Benjamin, New York (1974).
2. M. Reiser, *Phys. Fluids* **20**, 477 (1977).
3. B. Breizman and D. D. Ryutov, *Nucl. Fusion* **14**, 873 (1974).
4. V. S. Voronin, Yu. T. Zozulya, and A. N. Lebedev, *Sov. Phys. Tech. Phys.* **17**, 432 (1972).
5. C. E. Fay, A. L. Samuel, and W. Shockley, *Bell Syst. Tech. J.* **17**, 49 (1938).
6. T. C. Genoni and W. A. Proctor, IEEE Int'l. Conf. on Plasma Science, Monterey, California (1978), p. 94.
7. M. L. Sloan and J. R. Thompson, I-ARA-77-U-54 (August, 1977).
8. T. C. Genoni and W. A. Proctor, *J. Plasma Phys.* **23**, 129 (1980).
9. R. B. Miller and D. C. Straw, *J. Appl. Phys.* **48**, 1061 (1977).
10. F. B. Llewellyn, *Electron Inertia Effects*, Cambridge Univ. Press, London and New York (1941).
11. R. J. Lomax, *Proc. IEEE Part C* **108**, 119 (1961).
12. A. V. Pashchenko and B. N. Rutkevich, *Sov. J. Plasma Phys.* **3**, 437 (1977).
13. B. B. Godfrey, *IEEE Trans. Plasma Sci.* **PS-6**, 380 (1978).
14. R. J. Briggs, *Phys. Fluids* **19**, 1257 (1976).
15. W. B. Bridges and C. K. Birdsall, *J. Appl. Phys.* **34**, 2946 (1963).
16. C. K. Birdsall and W. B. Bridges, *Electron Dynamics of Diode Regions*, Academic Press, New York (1966).
17. P. J. Hart and D. L. Wolford, *J. Appl. Phys.* **43**, 2698 (1972).
18. D. A. Dunn and I. T. Ho, *AIAA J.* **1**, 2770 (1963).
19. P. C. deJagher, H. J. Hopman, and B. Jurgens, Proc. 12th Int'l. Conf. Phenomena in Ionized Gases, Eindhoven, the Netherlands (Aug. 18–22, 1975).
20. B. B. Godfrey, *J. Comp. Phys.* **19**, 58 (1975).
21. J. W. Poukey and J. R. Freeman, *Phys. Fluids* **17**, 1917 (1974).
22. A. Drobot, private communication.
23. B. B. Godfrey, *IEEE Trans. Plasma Sci.* **PS-7**, 53 (1979).
24. B. L. Bogema, Univ. of Maryland Tech. Report No. 72-037, Univ. of Maryland (1971).
25. L. E. Thode, B. B. Godfrey, and W. R. Shanahan, *Phys. Fluids* **22**, 747 (1979).
26. A. V. Agafonov, V. S. Voronin, A. N. Lebedev, and K. N. Pazin, *Sov. Phys. Tech. Phys.* **19**, 1188 (1975).
27. P. Diament, *Phys. Rev. Lett.* **37**, 168 (1976).
28. T. C. Genoni and R. B. Miller, *Phys. Fluids* **24**, 1397 (1981).
29. D. Mosher, G. Cooperstein, S. J. Stephanakis, S. A. Goldstein, D. G. Colombant, and R. Lee, NRL Report No. 3658 (1977).
30. S. Humphries, Jr., *Appl. Phys. Lett.* **32**, 792 (1978).
31. S. Humphries, Jr., *Nucl. Fusion* **20**, 1549 (1980).
32. See, for instance, E. Stuhlinger, *Ion Propulsion for Space Flight*, McGraw-Hill, New York, p. 232 (1964).

33. J. W. Poukey and S. Humphries, Jr., *Appl. Phys. Lett.* **33**, 122 (1978).
34. H. V. Wong, M. L. Sloan, J. R. Thompson, and A. T. Drobot, *Phys. Fluids* **16**, 902 (1973).
35. W. E. Drummond, G. I. Bourianoff, E. P. Cornet, D. E. Hasti, W. W. Rienstra, M. L. Sloan, H. V. Wong, J. R. Thompson, and J. R. Uglum, AFWL-TR-75-296, Air Force Weapons Lab., Albuquerque, New Mexico (1976).
36. O. Buneman, R. H. Levy, and L. M. Linson, *J. Appl. Phys.* **37**, 3203 (1966).
37. R. H. Levy, *Phys. Fluids* **8**, 1288 (1965).
38. W. Knauer, *J. Appl. Phys.* **37**, 602 (1966).
39. I. B. Bernstein and S. K. Trehan, *Nucl. Fusion* **1**, 3 (1960).
40. M. Friedman and D. A. Hammer, *Appl. Phys. Lett.* **21**, 174 (1972).
41. R. B. Miller and D. C. Straw, *J. Appl. Phys.* **47**, 1897 (1976).
42. L. S. Bogdankevich and A. A. Rukhadze, *Sov. Phys. Usp.* **14**, 163 (1971).
43. J. W. Poukey and N. Rostoker, *Plasma Phys.* **13**, 897 (1971).

Chapter 4

1. H. Alfven, *Phys. Rev.* **55**, 425 (1939).
2. D. A. Hammer and N. Rostoker, *Phys. Fluids* **13**, 1831 (1970).
3. A. A. Rukhadze and V. G. Rukhlin, *Sov. Phys. JETP* **34**, 93 (1972).
4. J. L. Cox and W. H. Bennett, *Phys. Fluids* **13**, 182 (1970).
5. S. Putnam, Physics International Report No. DNA 2849F (PIFR-72-105) (July, 1972).
6. D. S. Prono and E. P. Lee, *Plasma Phys.* **15**, 691 (1973).
7. S. E. Rosinskii, A. A. Rukhadze, V. G. Rukhlin, and Ya. G. Epel'baum, *Sov. Phys. Tech. Phys.* **17**, 737 (1972).
8. R. Lee and R. N. Sudan, *Phys. Fluids* **14**, 1213 (1971).
9. S. E. Rosinskii and V. G. Rukhlin, *Sov. Phys. JETP* **37**, 436 (1973).
10. W. H. Bennett, *Phys. Rev.* **45**, 890 (1934).
11. W. H. Bennett, *Phys. Rev.* **98**, 1584 (1955).
12. G. Kuppers, A. Salat, and H. K. Wimmel, *Plasma Phys.* **15**, 441 (1973).
13. J. L. Cox, Jr., and W. H. Bennett, *Phys. Fluids* **13**, 182 (1970).
14. J. Benford and B. Ecker, *Phys. Rev. Lett.* **26**, 1160 (1971).
15. J. Benford and B. Ecker, *Phys. Rev. Lett.* **28**, 10 (1972).
16. P. A. Miller, R. I. Butler, M. Cowan, J. R. Freeman, J. W. Poukey, T. P. Wright, and G. Yonas, *Phys. Rev. Lett.* **39**, 92 (1977).
17. V. P. Grigorev, A. N. Didenko, and N. S. Shulaev, *Sov. Phys. Tech. Phys.* **23**, 747 (1978).
18. S. V. Yadavalli, *Z. Phys.* **196**, 255 (1966).
19. M. N. Rosenbluth, *Phys. Fluids* **3**, 932 (1960).
20. S. Weinberg, *J. Math Phys.* **5**, 1371 (1964).
21. S. Weinberg, *J. Math Phys.* **8**, 614 (1967).
22. E. P. Lee, *Phys. Fluids* **16**, 1072 (1973).
23. E. P. Lee, Lawrence Livermore Lab. Report No. UCID-16268 (May, 1973).
24. R. C. Mjolsness, J. Enoch, and C. L. Longmire, *Phys. Fluids* **6**, 1741 (1963).
25. K. G. Moses, R. W. Bauer, and S. R. Winter, *Phys. Fluids* **16**, 436 (1973).
26. E. J. Lauer, R. J. Briggs, T. J. Fessenden, R. E. Hester, and E. P. Lee, *Phys. Fluids* **21**, 1344 (1978).
27. E. P. Lee, *Phys. Fluids* **21**, 1327 (1978).
28. H. S. Uhm and M. Lampe, NRL Memo Report No. 4111 (November 1979).

29. L. S. Bogdankevich and A. A. Rukhadze, *Sov. Phys. Usp.* **14**, 163 (1971).
30. O. Buneman, *Phys. Rev.* **115**, 503 (1959).
31. L. S. Bogdankevich and A. A. Rukhadze, *Sov. Phys. JETP* **35**, 126 (1972).
32. B. B. Godfrey, W. R. Shanahan, and L. E. Thode, *Phys. Fluids* **18**, 346 (1975).
33. E. S. Weibel, *Phys. Rev. Lett.* **31**, 1390 (1973).
34. R. Lee and M. Lampe, *Phys. Rev. Lett.* **31**, 1390 (1973).
35. C. A. Kapetanakos, *Appl. Phys. Lett.* **25**, 484 (1974).
36. K. F. Lee and J. C. Armstrong, *Phys. Rev. A* **4**, 2087 (1971).
37. G. Benford, *Plasma Phys.* **15**, 483 (1973).
38. K. Molvig, *Phys. Rev. Lett.* **35**, 1504 (1975).
39. L. E. Thode and R. N. Sudan, *Phys. Fluids* **18**, 1564 (1975).
40. E. K. Zavoisky, B. A. Demidov, Yu. G. Kalinin, A. G. Plakhov, L. I. Rudakov, V. E. Rusanov, V. A. Skoryupin, G. Ye. Smolkin, A. V. Titov, S. D. Franchenko, V. V. Shapkin, and G. V. Sholin, *Plasma Physics and Controlled Fusion Research 1971*, Vol. 2, p. 3 (1971).
41. R. V. Lovelace and R. N. Sudan, *Phys. Rev. Lett.* **27**, 1256 (1971).
42. W. E. Drummond, J. R. Thompson, G. I. Bourianoff, D. E. Hasti, M. L. Sloan, and H. V. Wong, Austin Research Assoc, Report No. I-ARA-75-U-139 (August, 1975).
43. L. E. Thode, *Phys. Fluids* **19**, 305 (1976).
44. L. E. Thode and R. N. Sudan, *Phys. Fluids* **18**, 1552 (1975).
45. L. E. Thode and B. B. Godfrey, *Phys. Fluids* **19**, 316 (1976).
46. A. T. Altynstev, A. G. Es'kov, O. A. Zolotovskii, V. I. Koroteev, R. Kh. Kurtmullaev, V. D. Masalov, and V. N. Semenov, *Sov. Phys.-JETP Lett.* **13**, 139 (1971).
47. A. T. Altyntsev, B. N. Breizman, A. G. Es'kov, A. O. Zolotovsky, V. I. Koroteev, R. Kh. Kurtmullaev, V. L. Maslov, D. D. Ryutov, and V. N. Semenov, *Nucl. Fusion Suppl.*, p. 161 (1972).
48. P. A. Miller and G. W. Kuswa, *Phys. Rev. Lett.* **30**, 958 (1973).
49. C. A. Kapetankos and D. A. Hammer, *Appl. Phys. Lett.* **23**, 17 (1973).
50. Yu. I. Abrashitov, V. S. Koidan, V. V. Konyukhov, V. M. Lagunov, V. N. Luk'yanov, and K. I. Mekler, *Sov. Phys.-JETP Lett.* **18**, 395 (1973).
51. G. C. Goldenbaum, W. F. Dove, K. A. Gerber, and B. G. Logan, *Phys. Rev. Lett.* **32**, 830 (1974); B. G. Logan, W. F. Dove, K. A. Gerber, and G. C. Goldenbaum, *IEEE Trans. Plasma Sci.* **2**, 182 (1974).
52. C. Ekdahl, M. Greenspan, R. E. Kribel, J. Sethian, and C. B. Wharton, *Phys. Rev. Lett.* **33**, 346 (1974).

Chapter 5

1. R. B. Miller and D. C. Straw, *J. Appl. Phys.* **47**, 1987 (1976).
2. S. Putnam, Physics International Report PIFR-105 (1970).
3. C. L. Olson, *Phys. Rev. A* **11**, 288 (1975).
4. P. Felsenthal and J. M. Proud, *Phys. Rev.* **139**, A1796 (1965).
5. D. W. Swain, *J. Appl. Phys.* **43**, 396 (1972).
6. R. Briggs, J. Clark, T. Fessenden, E. Lee, and E. Lauer, Lawrence Livermore Laboratory Report No. UCID-17516 (1977).
7. E. P. Lee, UCID-16268 (1973).
8. E. P. Lee, *Phys. Fluids* **19**, 60 (1976).
9. C. L. Olson, *Phys. Fluids* **18**, 585 (1975).

10. G. Yonas, P. Spence, D. Pellinen, B. Ecker, and S. Heurlin, Physics International Report No. PITR-106-1 (1969).
11. J. Creedon, Physics International Report No. PITR-16-67 (1967).
12. D. A. McArthur and J. W. Poukey, *Phys. Rev. Lett.* **27**, 1765 (1971).
13. R. J. Briggs, J. C. Clark, T. J. Fessenden, R. E. Hester, and E. J. Lauer, Proc, 2nd Int'l. Top. Conf. High Power Electron and Ion Beam Res. and Technol., Cornell Univ., Vol. 1, p. 319 (1977).
14. E. P. Lee, F. W. Chambers, L. L. Lodestro, and S. S. Yu, 2nd Int'l Top. Conf. High Power Electron and Ion Beam Res. and Technol., Cornell Univ., Vol. 1, p. 381 (1977).
15. R. J. Briggs, in *Advances in Plasma Physics*, Vol. 4, A. Simon and W. B. Thompson eds., Interscience Publishers, New York, p. 43 (1971).
16. P. A. Miller and J. B. Gerardo, *J. Appl. Phys.* **43**, 3008 (1972).

Chapter 6

1. *Microwave Magnetrons*, G. B. Collins ed., McGraw-Hill, New York (1948).
2. *Crossed-Field Microwave Devices*, F. Okress ed., Academic Press, New York, Vols. 1 and 2, (1961).
3. T. J. Orzechowski and G. Bekefi, *Phys. Fluids* **22**, 978 (1979).
4. A. N. Didenko, A. S. Sulaskshin, C. P. Fomenko, Yu. G. Shtein, and Yu. G. Yushkov, *Sov. Tech. Phys. Lett.* **4**, 3 (1978).
5. A. N. Didenko, A. S. Sulaskshin, C. P. Fomenko, V. I. Tsvetkov, Yu. G. Shtein, and Yu. G. Yushkov, *Sov. Tech. Phys. Lett.* **3**, 43 (1978).
6. A. Palevsky and G. Bekefi, *Phys. Fluids* **19**, 986 (1979).
7. G. Craig, J. Pettibone, and D. Ensley, Proc. IEEE Int'l. Conf. Plasma Science No. 79CH1410-ONP 5, p. 44 (1979).
8. A. W. Hull, *Phys. Rev.* **18**, 31 (1922).
9. L. Brillouin, *Phys. Rev.* **60**, 385 (1941); **65**, 166 (1942).
10. O. Buneman, *J. Elect. Control* **3**, 1 (1957); **3**, 507 (1957).
11. E. Ott and R. V. Lovelace, *Appl. Phys. Lett.* **27**, 378 (1975).
12. J. M. Creedon, *J. Appl. Phys.* **46**, 2946 (1975); **48**, 1070 (1977).
13. D. R. Hartree, *CVD Rept. Mag.* **1** (1941).
14. J. Slater, *Microwave Electronics*, Van Nostrand, New York (1950).
15. D. Gabor and G. D. Sims, *J. Electron.* **1**, 25 (1955).
16. R. Q. Twiss, *J. Electron.* **1**, 1 (1955).
17. R. V. Lovelace and E. Ott, *Phys. Fluids* **17**, 1263 (1974).
18. A. Palevsky, Ph.D. thesis, MIT (1980).
19. O. Buneman, in *Crossed Field Microwave Devices*, F. Okress ed., Academic Press, New York, Vol. 1, p. 209 (1961).
20. K. D. Bergeron, private communication (1978).
21. N. Kroll and W. Lamb, *J. Appl. Phys.* **19**, 183 (1948), Appendix I.
22. T. Shimizu, in *Crossed Field Microwave Devices*, F. Okress ed., Academic Press, New York, Vol. 1, p. 589 (1961).
23. R. Q. Twiss, *Australian J. Phys.* **11**, 564 (1958).
24. A. V. Gapanov, *Izv. VUZ Radiofizika* **2**, 450 (1959).
25. J. Schneider, *Phys. Rev. Lett.* **2**, 504 (1959).
26. R. H. Pantell, *Proc. IRE* **47**, 1146 (1959).

References

341

27. J. L. Hirshfield and J. M. Wachtel, *Phys. Rev. Lett.* **12**, 533 (1964).
28. D. A. Hammer, M. Friedman, V. L. Granatstein, M. Herndon, W. M. Mannheimer, R. K. Parker, and P. Sprangle, *Ann. N.Y. Acad. Sci.* **251**, 441 (1975).
29. V. L. Granatstein, R. K. Parker, and P. Sprangle, Proc. Intl. Top. Conf. Electron Beam Res. and Tech., SAND 76-5122, p. 401 (1975).
30. V. A. Flyagin, A. V. Gaponov, M. I. Petelin, and V. K. Yulpatov, *IEEE Trans. Microwave Theory Technol.* **MTT-25**, 514 (1977).
31. E. Ott and W. Manheimer, *IEEE Trans. Plasma Sci.* **PS-3**, 1 (1975).
32. P. Sprangle and A. T. Drobot, *IEEE Trans. Microwave Theory Tech.* **MTT-25**, 528 (1977).
33. A. Mondelli, private communication (1980).
34. M. Friedman, D. A. Hammer, W. M. Manheimer, and P. Sprangle, *Phys. Rev. Lett.* **31**, 752 (1973).
35. V. L. Granatstein, M. Herndon, P. Sprangle, Y. Carmel, and J. A. Nation, *Plasma Phys.* **17**, 23 (1975).
36. V. L. Granatstein, P. Sprangle, M. Herndon, R. K. Parker, and S. P. Schlesinger, *J. Appl. Phys.* **46**, 3800 (1975).
37. V. L. Granatstein, P. Sprangle, R. K. Parker, and M. Herndon, *J. Appl. Phys.* **46**, 2021 (1975).
38. P. Sprangle, R. A. Smith, and V. L. Granatstein, *Infrared and Submillimeter Waves*, K. J. Button, ed., Vol. I, p. 279, Academic Press, New York (1979).
39. A. Gover and A. Yariv, *Appl. Phys.* **16**, 121 (1978).
40. T. C. Marshall, S. P. Schlesinger, and D. B. McDermott, *Advances in Electronics and Electron Physics*, Vol. 8, Academic Press, New York (1980).
41. P. L. Kapitza and P. A. M. Dirac, *Proc. Cambridge Phil. Soc.* **29**, 297 (1933).
42. H. Motz, *J. Appl. Phys.* **22**, 527 (1951).
43. R. N. Phillips, *IRE Trans. Electron Devices* **ED-7**, 231 (1960).
44. R. H. Pantell, G. Soncini, and H. E. Puthoff, *IEEE J. Quantum. Electron.* **4**, 905 (1968).
45. J. F. Drake, P. K. Kaw, Y. C. Lee, G. Schmidt, C. S. Liu, and M. N. Rosenbluth, *Phys. Fluids* **17**, 778 (1974).
46. D. W. Forslund, J. M. Kindel, and E. L. Lindman, *Phys. Fluids* **78**, 1002 (1975).
47. W. M. Manheimer and E. Ott, *Phys. Fluids* **17**, 1413 (1974).
48. V. L. Granatstein and P. Sprangle, *IEEE Trans. Microwave Theory Tech.* **MTT-25**, 545 (1977).
49. L. R. Elias, W. M. Fairbank, J. M. J. Madey, H. A. Schwettman, and T. I. Smith, *Phys. Rev. Lett.* **36**, 717 (1976).
50. D. A. G. Deacon, L. R. Elias, J. M. J. Madey, G. J. Ramian, H. A. Schwettman, and T. I. Smith, *Phys. Rev. Lett.* **38**, 892 (1977).
51. V. L. Granatstein, M. Herndon, R. K. Parker, and S. P. Schlesinger, *IEEE Trans. Microwave Theory Tech.* **MTT-22** 1000 (1974).
52. V. L. Granatstein, S. P. Schlesinger, M. Herndon, R. K. Parker, and J. A. Pasour, *Appl. Phys. Lett.* **30**, 384 (1977).
53. T. C. Marshall, S. Talmadge, P. Efthimion, *Appl. Phys. Lett.* **31**, 320 (1977).
54. D. B. McDermott, T. C. Marshall, S. P. Schlesinger, R. K. Parker, and V. L. Granatstein, *Phys. Rev. Lett.* **41**, 1368 (1978).
55. J. D. Jackson, *Classical Electrodynamics*, Wiley, New York (1962). Ibid.
56. P. Sprangle, V. L. Granatstein, and L. Baker, *Phys. Rev. A* **12**, 1697 (1975).
57. N. A. Kroll and W. A. McMullin, *Phys. Rev. A* **17**, 300 (1978).
58. A. Hasegawa, *Bell System Tech. J.* **57**, 3069 (1978).
59. I. A. Bernstein and J. L. Hirshfield, *Phys. Rev. A* **20**, 1661 (1979).
60. P. Sprangle and A. T. Drobot, *J. Appl. Phys.* **50**, 2652 (1979).

61. T. Kwan, J. M. Dawson, and A. T. Lin, *Phys. Fluids* **20**, 581 (1977).
62. A. T. Lin and J. M. Dawson, *Phys. Rev. Lett.* **42**, 1670 (1979).
63. T. Kwan and J. M. Dawson, *Phys. Fluids* **22**, 1089 (1979).
64. P. C. Efthimion and S. P. Schlesinger, *Phys. Rev. A* **16**, 633 (1977).
65. P. Sprangle, C. M. Tang, and W. M. Manheimer, *Phys. Rev. Lett.* **43**, 1932 (1979).

Chapter 7

1. V. I. Veksler, "Coherent Principle of Acceleration of Charged Particles," Proceedings of the CERN Symposium on High Energy Acceleration and Ion Physics, Vol. I, Geneva (1956).
2. V. I. Veksler, Proceedings of the Sixth International Conference on High Energy Accelerators (1967).
3. L. J. Laslett, *IEEE Trans. Nucl. Sci.* **NS-20**, 271 (1973).
4. D. Keefe, Proc. IVth All-Union Conf. on Particle Accelerators (Nov. 1974), Nauka, Moscow, Vol. I, p. 109 (1975).
5. M. Reiser, *IEEE Trans. Nucl. Sci.* **NS-20**, 310 (1973).
6. U. Schumacher, C. Andelfinger, M. Ulrich, *IEEE Trans. Nucl. Sci.* **NS-22**, 989 (1975).
7. I. N. Ivanov and V. P. Sarantsev, *Proceedings of the First International School of Young Physicists on the Problems of Charged Particle Accelerators*, Uzhgohod (Sept. 4–15, 1975), JINR, Dubna (1976).
8. D. Keefe, "Particle Acceleration by Collective Effects," 1976 Proton Linear Accelerator Conference, Chalk River, Ontario, Canada, LBL-5536 (September 14–17, 1976).
9. A. A. Plyutto, *Sov. Phys. JETP* **12**, 1106 (1961).
10. A. A. Plyutto, P. E. Belensov, E. D. Korop, G. P. Mkheidze, V. N. Ryzhnov, K. V. Suladze, and S. M. Temchin, *JETP Lett.* **6**, 61 (1967).
11. A. A. Plyutto, K. V. Suladze, S. M. Temchin, and E. D. Korop, *Sov. J. At. Energy* **27**, 1197 (1969).
12. J. P. Mkheidze, A. A. Plyutto, and E. D. Korop, *Sov. Phys. Tech. Phys.* **16**, 749 (1971).
13. A. A. Plyutto, K. V. Suladze, S. M. Temchin, G. P. Mkheidze, E. D. Korop, B. A. Tskhadaya, and I. V. Golovia, *Sov. Phys. Tech. Phys.* **18**, 1026 (1974).
14. A. A. Plyutto, Sukhumi Institute of Physics and Technology, SFTI-1 (1977).
15. M. S. Rabinovich, "Collective Methods of Acceleration," Reprint No. 36, Lebedev Physical Institute (March, 1969).
16. S. Graybill and J. Uglum, *J. Appl. Phys.* **41**, 236 (1970).
17. S. Graybill, *IEEE Trans. Nucl. Sci.* **NS-18**, 438 (1971).
18. S. Graybill, *IEEE Trans. Nucl. Sci.* **NS-19**, 292 (1972).
19. J. Rander, B. Ecker, G. Yonas, and D. Drickey, *Phys. Rev. Lett.* **24**, 283 (1970).
20. J. Rander, *Phys. Rev. Lett.* **25**, 893 (1970).
21. G. W. Kuswa, L. P. Bradley, and G. Yonas, *IEEE Trans. Nucl. Sci.* **NS-20**, 305 (1973).
22. B. Ecker and S. Putnam, *IEEE Trans. Nucl. Sci.* **NS-20**, 301 (1973).
23. D. C. Straw and R. B. Miller, *Appl. Phys. Lett.* **25**, 379 (1974).
24. R. B. Miller and D. C. Straw, *J. Appl. Phys.* **47**, 1897 (1976).
25. A. A. Kolomensky, Proc. Ninth Int'l. Accel. Conf. High Energy Accel., SLAC, p. 254 (1974).
26. A. A. Kolomensky et al., *Sov. Phys. JETP* **41**, 26 (1975).
27. S. Graybill and F. Young, APS Div. Plasma Physics Mtg., San Francisco (November 15–19, 1976).

28. J. S. Luce and H. L. Sahlin, *IEEE Trans. Nucl. Sci.* **NS-20**, 336 (1973).
29. L. P. Bradley and G. W. Kuswa, *Phys. Rev. Lett.* **29**, 1441 (1972).
30. G. T. Zorn, H. Kim, and C. N. Boyer, *IEEE Trans. Nucl. Sci.* **NS-22**, 1006 (1975).
31. R. F. Hoeberling and D. N. Payton, III, *J. Appl. Phys.* **48**, 2079 (1977).
32. J. L. Adamski, P. S. P. Wei, J. R. Beymer, R. L. Gray, and R. L. Copeland, Proc. 2nd Intl. Top. Conf. High Power Electron and Ion Beam Res. and Technol., Ithaca, New York, 497 (1977).
33. R. J. Adler, J. A. Nation, and V. Serlin, to be published.
34. R. B. Miller and D. C. Straw, *IEEE Trans. Nucl. Sci.* **NS-22**, 1022 (1975).
35. J. S. Luce, W. H. Bostick, and V. Nardi, Proc. Conf. on Plasma Heating, Verona, Italy (1976).
36. W. W. Destler, R. F. Hoeberling, H. Kim, and W. H. Bostick, *Appl. Phys. Lett.* **35**, 296 (1979).
37. M. L. Sloan and W. E. Drummond, *Phys. Rev. Lett.* **31**, 1234 (1974).
38. P. Sprangle, A. T. Drobot, and W. M. Manheimer, *Phys. Rev. Lett.* **36**, 1180 (1976).
39. S. V. Yadavalli, *Appl. Phys. Lett.* **29**, 272 (1976).
40. G. Yonas, *Part. Accel.* **5**, 81 (1973).
41. C. L. Olson, *Part. Accel.* **6**, 107 (1975).
42. C. L. Olson and U. Schumacher, *Springer Tracts in Modern Physics: Collective Ion Acceleration*, Vol. 84, G. Hohler, ed., Springer, New York (1979).
43. C. L. Olson, *Phys. Fluids* **18**, 585 (1975).
44. C. L. Olson, *Phys. Fluids* **18**, 598 (1975).
45. C. L. Olson, *Phys. Rev. A* **11**, 288 (1975).
46. R. B. Miller and D. C. Straw, Proc. Intl. Top. Conf. E-Beam Res. and Technol., Albuquerque, New Mexico, Vol. 2, p. 368 (1976).
47. B. Ecker and S. Putnam, Symp. Coll. Methods of Accel., Dubna, (October, 1976).
48. S. E. Graybill, J. R. Uglum, W. H. McNeill, J. E. Rizzo, R. Lowell, and G. Ames, Ion Physics Corporation Report No. DASA 2477 (1970).
49. G. W. Kuswa, *Ann. N.Y. Acad. Sci.* **251**, 514 (1975).
50. R. J. Faehl, in Report No. LA-7734-PR, Los Alamos Scientific Laboratory, p. 100 (1979).
51. B. B. Godfrey and L. E. Thode, *IEEE Trans. Plasma Sci.* **PS-3**, 201 (1975).
52. L. E. Floyd, W. W. Destler, M. Reiser, and H. M. Shiu, *J. Appl. Phys.* **52**, 693 (1981).
53. C. L. Olson, Proc. IX Int'l. Conf. High Energy Accelerators, Stanford Linear Accelerator Center p. 272 (1974).
54. R. B. Miller, IEEE Proc. Int'l Conf. Plasma Science, p. 130 (March 24–26, 1976).
55. C. L. Olson, J. W. Poukey, J. P. Van Devender, and J. S. Perlman, *IEEE Trans. Nucl. Sci.* **NS-24**, 1659 (1977).
56. V. N. Tsytovich and K. V. Khodataev, *Comments Plasma Phys. Controlled Fusion* **3**, 71 (1977).
57. S. Putman, Symp. Collective Methods Accel., Dubna, (October, 1976).
58. R. J. Adler, *J. Appl. Phys.* **52**, 3099 (1981).
59. C. L. Olson, *IEEE Trans. Nucl. Sci.* **NS-26**, 4231 (1979).
60. C. L. Olson, IEEE Particle Accel. Conf., Wash. D.C. (March 1981).
61. W. E. Drummond, G. I. Bourianoff, D. E. Hasti, W. W. Rienstra, M. L. Sloan, and J. R. Thomson, Air Force Weapons Laboratory Report No. AFWL-TR-74-343 (1974).
62. W. W. Rienstra, Air Force Weapons Laboratory Report No. AFWL-TR-74-343 (1974), Appendix A.
63. W. A. Proctor and T. C. Genoni, *J. Appl. Phys.* **49**, 910 (1978).
64. B. B. Godfrey and B. S. Newberger, *J. Appl. Phys.* **50**, 2470 (1979).

65. B. B. Godfrey, *IEEE Trans. Plasma Sci.* **PS-7**, 53 (1979).
66. W. E. Drummond et al., Austin Research Associates Report No. I-ARA-79-U-72 (1979).
67. M. L. Sloan, E. P. Cornet, W. W. Rienstra, J. R. Thomson, and H. V. Wong, 3rd Intl. Conf. Collective Methods of Acceleration, N. Rostoker and M. Reiser, eds., p. 145 (1978).
68. G. I. Bourianoff, B. N. Moore, and B. R. Penumalli, 3rd Intl. Conf. Collective Methods of Acceleration, N. Rostoker and M. Reiser, eds., p. 191 (1978).
69. R. J. Faehl, W. R. Shanahan, and B. B. Godfrey, 3rd Intl. Conf. Collective Methods of Acceleration, N. Rostoker and M. Reiser, eds., p. 211 (1978).
70. H. A. Davis and E. Cornet, *Rev. Sci. Instrum.* **51**, 1176 (1980).
71. E. Cornet, H. A. Davis, W. W. Rienstra, M. L. Sloan, T. P. Starke, and J. R. Uglum, to be published in *Physical Review Letters.*
72. R. J. Briggs, *Phys. Fluids* **19**, 1257 (1976).
73. B. B. Godfrey, *IEEE Plasma Sci.* **PS-6**, 380 (1976).
74. P. Sprangle and A. T. Drobot, NRL Memo Report No. 3660 (1980).
75. D. Sullivan, private communication.
76. G. Gamel, J. A. Nation, and M. E. Read, *Rev. Sci. Instrum.* **49**, 507 (1978).
77. R. Adler, G. Gamel, J. A. Nation, M. E. Read, R. Williams, P. Sprangle, and A. Drobot, Proc. 2nd Intl. Conf. High Power Electron and Ion Beam Res. and Technol., Vol. II, p. 509 (1977).
78. R. Adler, G. Gamel, J. A. Nation, G. Providakes, and R. Williams, Proc. 3rd Int. Conf. Collective Methods of Acceleration (1978).
79. R. Adler, G. Gamel, J. A. Nation, J. Ivers, G. Providakes, and V. Serlin, *IEEE Trans. Nucl. Sci.* **NS-26**, 4223 (1979).
80. G. Gamel, J. A. Nation, and M. Read, *J. Appl. Phys.* **50**, 5603 (1979).

Chapter 8

1. D. D. Ryutov, *Nucl. Fusion* **19**, 1685 (1979).
2. C. Yamanaka, *Nucl. Fusion* **20**, 1084 (1980).
3. J. Nuckolls, L. Wood, A. Thiessen, and G. Zimmerman, *Nature* **239**, 139 (1972).
4. G. Yonas, *Sci. Am.* **239**, 50 (1978).
5. C. F. Wandel, T. Hesselberg, and O. Kofoed-Hansen, *Nucl. Instrum. Methods* **4**, 239 (1959).
6. W. R. Arnold, J. A. Phillips, G. A. Sawyer, E. J. Stoval, and J. L. Tuck, *Phys. Rev.* **93**, 483 (1954).
7. J. Nuckolls, Laser Interaction and Related Phenomena, Vol. 3B, H. J. Schwarz and H. Hora, eds., Plenum Press, New York, p. 399 (1974).
8. G. S. Fraley, E. J. Linnebur, R. J. Mason, and R. L. Morse, Los Alamos Sci. Lab. Report LA-5403-MS (1973).
9. J. Mayer and M. Mayer, *Statistical Mechanics*, Wiley, New York, p. 385 (1940).
10. K. A. Brueckner and S. Jorna, *Rev. Mod. Phys.* **46**, 325 (1974).
11. N. Bohr, *Philos. Mag.* **25**, 10 (1913).
12. F. Bloch, *Ann. Phys.* (Leipzig) **16**, 285 (1933).
13. D. Pines and D. Bohm, *Phys. Rev.* **85**, 338 (1952).
14. W. Heitler, *Quantum Theory of Radiation*, 3rd ed., Oxford University Press (1954).
15. J. F. Janni, Air Force Weapons Laboratory Report No. AFWL-TR-65-150 (1966).
16. M. D. Brown and C. D. Moak, *Phys. Rev. B* **6**, 90 (1972).

17. H. H. Andersen and J. F. Ziegler, *Hydrogen-Stopping Powers and Ranges in All Elements*, Pergamon Press, New York (1977).

18. J. Linhard, M. Scharff, and H. E. Schiott, *Kgl. Danske Videnskab. Selskab Mat. Fys. Medd.* **33**, No. 14 (1963).

19. C. Varelas and J. P. Biersack, *Nucl. Instrum. Methods* **79**, 213 (1970).

20. P. G. Steward and R. W. Wallace, University of California Radiation Laboratory UCRL-19128 (1970).

21. J. Linhard and M. Scharff, *Phys. Rev.* **124**, 128 (1964).

22. T. A. Mehlhorn, Sandia National Laboratory SAND 80-0038 (1980).

23. J. D. Jackson, *Classical Electrodynamics*, John Wiley, New York, p. 643 (1975).

24. Von G. Guderley, *Luftfahrt-Forschung* **9**, 302 (1942).

25. R. E. Kidder, *Nucl. Fusion* **19**, 223 (1979).

26. J. H. Nuckolls, R. O. Bangerter, J. D. Lindl, W. C. Mead, and Y. L. Pau, Eur. Conf. on Laser Interaction with Matter, Oxford, England (1977).

27. G. I. Taylor, *Proc. R. Soc. London Ser. A* **201**, 192 (1950).

28. R. O. Bangerter, J. D. Lindl, C. E. Max, and W. C. Mead, Int'l Conf. Electron Beam Res. and Technol., SAND 76-5122, Vol. I, p. 15 (1976).

29. R. Le Levier, G. Lasher, and F. Bjorklung, Lawrence Radiation Laboratory Report No. UCRL-4459 (1955).

30. J. D. Lindl and R. O. Bangerter, Int'l. Conf. Electron Beam Res. Technol., SAND-76-5122, Vol. I, p. 37 (1976).

31. M. J. Clauser and M. A. Sweeney, Int'l. Conf. Electron Beam Res. Technol., SAND-76-5122, Vol. I, p. 135 (1976).

32. M. J. Clauser, *Phys. Rev. Lett.* **34**, 570 (1975).

33. M. A. Sweeney and M. J. Clauser, *Appl. Phys. Lett.* **27**, 483 (1975).

34. R. C. Kirkpatrick, C. C. Cremer, L. C. Madsen, H. S. Rogers, and R. S. Cooper, *Nucl. Fusion* **15**, 333 (1975).

35. G. Yonas, J. W. Poukey, J. R. Freeman, K. R. Prestwich, A. J. Toepfer, M. J. Clauser, and E. H. Beckner, *Proc. VI Europ. Conf. on Plasma Phys.* **1**, Moscow, p. 383 (1973).

36. D. A. Tidman, *Phys. Rev. Lett.* **35**, 1228 (1975).

37. S. Humphries, Jr., R. N. Sudan, and W. C. Condit, Jr., *Appl. Phys. Lett.* **26**, 667 (1975).

38. M. J. Clauser, M. A. Sweeney, and A. V. Farnsworth, Jr., IEEE Conf. Plasma Sci., Paper 6A3 (May, 1976).

39. M. M. Widner, J. W. Poukey, and J. A. Halblieb, Sr., *Phys. Rev. Lett.* **38**, 548 (1977).

40. L. Baker, M. J. Clauser, J. R. Freeman, L. P. Mix, J. N. Olsen, F. C. Perry, A. J. Toepfer, and M. M. Widner, 2nd Int'l. Top. Conf. Electron. Ion Beam Res. and Technol., Vol. I, p. 169 (1977).

41. R. O. Bangerter, J. D. Lindl, C. E. Max, and W. C. Mead, Int'l. Conf. Electron Beam Res. and Technol., SAND-76-5122, Vol. I, p. 15 (1976).

42. J. W. Shearer, Lawrence Livermore Laboratory Report No. UCRL-76519 (1976).

43. M. J. Clauser, *Phys. Rev. Lett.* **35**, 848 (1975).

44. R. O. Bangerter and D. J. Meeker, 2nd Int'l. Top. Conf. Electron Ion Beam Res. and Technol. Vol. I, p. 183 (1977).

45. F. Winterberg, *Phys. Rev.* **174**, 212 (1968).

46. M. V. Babykin, E. K. Zavoiskii, A. A. Ivanov, and L. I. Rudakov, Plasma Phys. Contr. Thermonuclear Fusion Res. (Proc. 4th Int'l. Conf.), Vol. 1, IAEA, Vienna, p. 635 (1971).

47. G. Yonas, J. W. Poukey, K. R. Prestwich, J. R. Freeman, A. J. Toepfer, M. J. Clauser, *Nucl. Fusion* **14**, 731 (1974).

48. G. Yonas and P. Spence, Physics International Report No. PIFR 106 (1968).

49. G. Yonas, K. R. Prestwich, J. W. Poukey, and J. R. Freeman, *Phys. Rev. Lett.* **30**, 164 (1973).

50. J. Chang, M. J. Clauser, J. R. Freeman, G. R. Hadley, J. A. Halblieb, D. L. Johnson, J. G. Kelly, G. W. Kuswa, T. H. Martin, P. A. Miller, L. P. Mix, F. C. Perry, J. W. Poukey, K. R. Prestwich, S. L. Shope, D. W. Swain, A. J. Toepfer, W. H. Vandevender, M. M. Widner, T. P. Wright, and G. Yonas, Proc. 5th Conf. Plasma Phys. Cont. Thermonuclear Fusion Res., Vol. 2, p. 347 (1974).

51. F. C. Perry and M. M. Widner, *J. Appl. Phys.* **47**, 127 (1976).

52. M. M. Widner, F. C. Perry, L. P. Mix, J. Chang, and A. J. Toepfer, *J. Appl. Phys.* **48**, 1047 (1977).

53. J. Chang, M. M. Widner, G. W. Kuswa, and G. Yonas, *Phys. Rev. Lett.* **34**, 1266 (1975).

54. J. Chang, M. M. Widner, A. V. Farnsworth, Jr., R. J. Leeper, T. S. Prevender, L. Baker, and J. N. Olsen, Proc. 2nd Int'l. Conf. Electron and Ion Beam Res. and Technol. (1977).

55. J. W. Poukey, Int'l. Conf. Electron Beam Res. Technol., SAND-76-5122 (1976).

56. L. I. Rudakov, *Sov. J. Plasma Phys.* **4**, 40 (1978).

57. D. H. McDaniel, J. W. Poukey, K. D. Bergeron, J. P. Vandevender, and D. L. Johnson, Proc. 2nd Int'l. Conf. Electron and Ion Beam Res. and Technol., Vol. II, p. 819 (1977).

58. E. I. Baranchikov, A. V. Gordeev, V. D. Korolev, V. P. Smirnov, and A. S. Chernenko, *Sov. Phys. Tech. Phys.* **21**, 628 (1976).

59. P. A. Miller, R. I. Butler, M. Cowan, J. R. Freeman, J. W. Poukey, T. P. Wright, and G. Yonas, *Phys. Rev. Lett.* **39**, 92 (1977).

60. T. P. Wright and J. A. Halblieb, Sr., *Phys. Fluids* **23**, 1603 (1980).

61. T. P. Wright, *J. Appl. Phys.* **49**, 3842 (1978).

62. J. A. Halblieb, Sr., P. A. Miller, L. P. Mix, and T. P. Wright, *Nature* **286**, 366 (1980).

63. J. A. Halblieb, Sr., and T. P. Wright, *Phys. Fluids* **23**, 1612 (1980).

64. G. Yonas, *IEEE Trans. Nucl. Sci.* **NS-26**, 4061 (1979).

65. R. C. Arnold, *Nature* **276**, 19 (1978).

66. S. Humphries, *Comments Plasma Phys. Cont. Fusion* **6**, 45 (1980).

67. D. J. Johnson, G. W. Kuswa, A. V. Farnsworth, J. P. Quintenz, R. J. Leeper, E. J. T. Burns, and S. Humphries, *Phys. Rev. Lett.* **42**, 610 (1979).

68. G. Cooperstein, S. A. Goldstein, D. Mosher, R. J. Barker, J. R. Boller, D. G. Colombant, A. Drobot, R. A. Meger, W. F. Oliphant, P. F. Ottinger, F. L. Sandel, S. J. Stephanakis, and F. C. Young, Proc. 5th Workshop Laser Inter. Plasma Phen., Rochester, New York (November 5–9, 1979).

69. C. W. Mendel, Proc. IEEE Conf. Plasma Sci., Madison, Wisconsin (May 19–21, 1980).

70. P. A. Miller, Sandia National Laboratories Report No. SAND80-0974 (1980).

71. S. Humphries, Jr., J. R. Freeman, J. Greenly, G. W. Kuswa, C. W. Mendel, J. W. Poukey, and D. M. Woodall, *J. Appl. Phys.* **51**, 1876 (1980).

72. P. Dreike, C. Eichenberger, S. Humphries, and R. N. Sudan, *J. Appl. Phys.* **47**, 85 (1976).

73. D. J. Meeker, J. H. Nuckolls, and R. O. Bangerter, *Bull. Am. Phys. Soc.* **20**, 1352 (1975).

74. G. W. Kuswa, 8th Int'l. Conf. Plasma Phys. Cont. Nucl. Fusion Res., Brussels (July 1–10, 1980).

75. S. A. Goldstein and J. Guillory, *Phys. Rev. Lett.* **35**, 1160 (1975).

76. T. S. T. Young and R. D. Genuario, *J. Appl. Phys.* **51**, 4595 (1980).

77. G. W. Kuswa, E. J. T. Burns, A. V. Farnsworth, Jr., D. L. Fehl, S. Humphries, D. J. Johnson, R. J. Leeper, C. W. Mendel, P. A. Miller, L. P. Mix, J. W. Poukey, J. P. Quintenz, and T. P. Wright, Proc. 3rd Int'l. Conf. High Power Elect. and Ion Beams, Novosibirsk, USSR (1979).

78. F. L. Sandel, F. C. Young, S. J. Stephanakis, W. F. Oliphant, G. Cooperstein, S. A. Goldstein, and D. Mosher, *Bull. Am. Phys. Soc.* **24**, 1031 (1979).

79. J. N. Olsen and L. Baker, *Bull. Am. Phys. Soc.* **25**, 899 (1980).

80. A. J. Gale, US Patent Office (1963).
81. P. A. Miller, D. J. Johnson, T. P. Wright, and G. W. Kuswa, *Comments Plasma Phys.* **5**, 95 (1979).
82. D. Mosher, D. G. Colombant, and S. A. Goldstein, *Comments Plasma Phys.* **6**, 101 (1981).
83. G. Yonas, Proc. ANS 4th Top. Mtg. Tech. Controlled Nucl. Fusion, King of Prussia, Pennsylvania (October, 1980).
84. K. R. Prestwich, *IEEE Trans. Nucl. Sci.* **NS-22**, 3 (1975).
85. T. H. Martin, J. P. Vandevender, D. L. Johnson, D. H. McDaniel, and M. Aker, Proc. 1st Int'l. Top. Conf. Electron Beam Res. and Tech., Vol. 1, p. 450, Albuquerque, New Mexico (February, 1976).
86. T. H. Martin, G. W. Barr, J. P. Vandevender, R. A. White, and D. L. Johnson, Proc. IEEE 14th Pulse Power Modulator Symp., Orlando, Florida (June, 1970).
87. D. J. Johnson, A. V. Farnsworth, Jr., D. L. Fehl, R. J. Leeper, and G. W. Kuswa, *J. Appl. Phys.* **50**, 4524 (1979).
88. S. Humphries, Jr., R. N. Sudan, and L. Wiley, *J. Appl. Phys.* **47**, 2382 (1976).
89. S. Humphries, Jr., C. W. Mendel, G. W. Kuswa, and S. A. Goldstein, *Rev. Sci. Instrum.* **50**, 993 (1979).
90. S. Humphries, Jr., *Nucl. Fusion* **20**, 1549 (1980).
91. ERDA Summer Study of Heavy Ions for Inertial Fusion, LBL-5543 (July, 1976).
92. Proc. Heavy Ion Fusion Workshop, Brookhaven National Laboratory, BNL-50769, (October, 1977).
93. Proc. Heavy Ion Fusion Workshop, Argonne National Laboratory, ANL Report No. 79-41 (September 1978).
94. Proc. Heavy Ion Fusion Workshop, Lawrence Berkeley Laboratory, LBL-10301 (November, 1979).
95. R. H. Stokes, K. R. Crandall, J. E. Stovall, and D. A. Swenson, *IEEE Trans. Nucl. Sci.* **NS-26**, 3469 (1979).
96. A. W. Maschke, Brookhaven National Laboratory, BNL-51029 (June, 1979).
97. M. Monsler, J. Blink, J. Hovingh, W. Meier, and P. Walker, Proc. Heavy Ion Fusion Workshop, ANL Report No. 79-41, p. 225 (September, 1978).

Index